RICHARD BOWDLER SHARPE

CATALOGUE OF THE STRIGES

OR

NOCTURNAL BIRDS OF PREY

IN THE

COLLECTION OF THE BRITISH MUSEUM

Elibron Classics
www.elibron.com

Elibron Classics series.

ISBN 1-4021-5878-5 (paperback)
ISBN 1-4021-2207-1 (hardcover)

This Elibron Classics Replica Edition is an unabridged facsimile of the edition published in 1875, London.

CATALOGUE

OF THE

BIRDS

IN THE

BRITISH MUSEUM.

VOLUME II.

LONDON:
PRINTED BY ORDER OF THE TRUSTEES.
1875.

CATALOGUE

OF THE

STRIGES,

OR

NOCTURNAL BIRDS OF PREY,

IN THE

COLLECTION

OF THE

BRITISH MUSEUM.

BY

R. BOWDLER SHARPE.

LONDON:
PRINTED BY ORDER OF THE TRUSTEES.
1875.

PRINTED BY TAYLOR AND FRANCIS,
RED LION COURT, FLEET STREET.

PREFACE.

The Second Volume of the 'Catalogue of Birds in the British Museum' has been prepared by Mr. R. B. Sharpe, one of the Senior Assistants in the Zoological Department; it contains an account of all the species of Nocturnal Birds of Prey known at present, with a complete list of references to the literature. Every effort has been made to render the Collection in the British Museum as complete as possible, and to enable the author to overcome the unusual difficulties that present themselves in this family of Birds as regards the discrimination of the species, as well as their description.

ALBERT GÜNTHER,
Keeper of the Department of Zoology.

British Museum,
Dec. 1875.

INTRODUCTION.

The total number of the species of Owls is now about 190. The British Museum contains 153, so that 37 are still wanting to complete the collection; the specific distinction of perhaps 10 of these is dubious.

The exact number of specimens of *Striges* enumerated in this volume is 1090. Besides the series acquired by the purchase of Mr. A. R. Wallace's collection, which contained many species before unrepresented, in addition to numerous types, no great collections of Owls have been obtained by the British Museum, as there is always great difficulty in procuring specimens of these birds. The list of desiderata, however, has been considerably decreased by purchases, as well as presents from the following gentlemen:—Viscount Walden, Lord Lilford, Baron A. von Hügel, Major Hayes Lloyd, Captain G. E. Shelley, Professor Schlegel, Messrs. John Gould, W. Wilson Saunders, Osbert Salvin, Howard Saunders, Dr. B. Hinde, Messrs. T. E. Buckley, H. F. Blissett, and A. Bouvier.

The special thanks of the author for valuable assistance and loan of specimens are due to Viscount Walden, Captain Shelley, Dr. Otto Finsch, Messrs. Salvin and Godman, Howard Saunders, R. Swinhoe, Herr C. Godeffroy, of Hamburgh, Count Salvadori, of Turin; but, above all, to Mr. Allan Hume, who was so kind as to send to England a complete series of the smaller Indian Owls, including the types of many species not yet represented in European museums. The authorities of the Norwich Museum have also forwarded to London all the specimens which it was necessary to examine, including some of the rarest species.

As in the former volume, the affixes to the names of the persons from whom the specimens were obtained may be explained as follows:—"[P.]" means "Presented by;" "[C.]"="Collected by;" "[E.]"="Obtained by exchange." Where none of the above signs are used, the specimens were purchased.

R. B. SHARPE.

British Museum,
Nov. 22, 1875.

SYSTEMATIC INDEX.

CATALOGUE

OF

BIRDS.

Suborder STRIGES.

Outer toe reversible; eyes directed forwards and encircled by a facial disk; nostrils generally hidden by stiff bristles; plumage soft and fluffy; tibia more than double the length of tarsus.

Cf. Ridgway, B. N. Am. iii. pp. 5, 6; Sundev. Av. Meth. Tent. p. 103.

Synopsis of Families.

a. Hinder margin of sternum with two or more distinct fissures or clefts; furcula free, not attached to keel of sternum; inner margin of claw on middle toe *not* serrated; middle toe always longer than the inner one .... **Bubonidæ**, p. 1.

b. Hinder margin of sternum entire, with an emargination but no distinct cleft; furcula joined to keel of sternum; inner margin of claw on middle toe serrated; inner and middle toes about equal in length ...... **Strigidæ**, p. 289.

Family BUBONIDÆ.

Hinder margin of sternum always deeply cleft, two or more notches being present; furcula free; inner margin of middle claw not serrated; middle toe always longer than the inner one.

a. Ear-conch not larger than the eye, without an operculum; facial disk unequal, the portion below the eye being always much greater than the area above the latter .... *Subfam. BUBONINÆ*, p. 2.
b. Ear-conch much larger than the eye, with very large operculum shutting in the ear; facial disk always distinct, and extending as far above the eye as it does below it........ *Subfam. SYRNIINÆ*, p. 225.

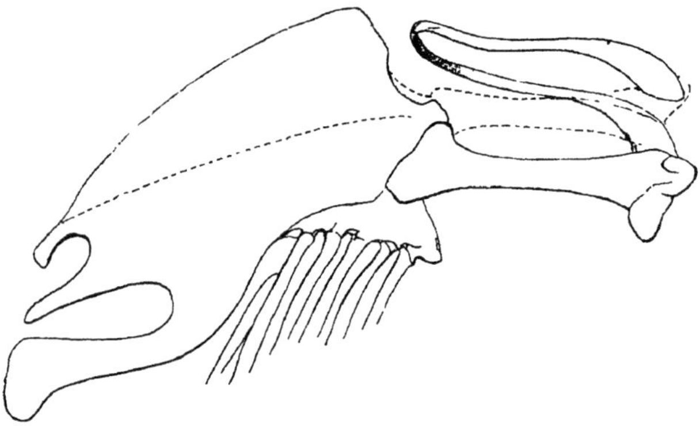

Sternum of *Bubo maximus*, to show hinder margin and furcula.

Subfam. I. BUBONINÆ.

Key to the Genera.

a. Soles of feet with tiny spicules, as in the Ospreys; tarsi generally naked.
  a'. With large ear-tufts .... 1. KETUPA, p. 4.
  b'. With no ear-tufts .... 2. SCOTOPELIA, p. 9.
b. Soles of feet ordinary, without any strongly developed spicules.
  c'. Cere not inflated; nostrils oval, and situated in the anterior margin of the cere.
    a''. With distinct ear-tufts.
      a'''. Wings short, falling short of the tail by more than the length of the middle toe; toes sometimes bare or scantily feathered; tarsi never naked .... 3. BUBO, p. 12.
      b'''. Wings very long, reaching nearly or quite to the tip of the tail, or falling short by less than the length of the middle toe; toes generally feathered; tarsi sometimes entirely naked .... 4. SCOPS, p. 43.
    b''. Ear-tufts rudimentary or absent.

c'''. Under tail-coverts produced nearly to the end of the tail; the latter short, only about half the length of the wing .... 5. NYCTEA, p. 125.
d'''. Tail long and graduated, approaching the wing in length, the difference between them scarcely greater than the length of tarsus........ 6. SURNIA, p. 129.
d'. Cere swollen and inflated, generally pea-shaped, the nostril pierced therein.
c''. With lengthened first primary, the distance between its tip and that of the third primary not so great as the length of tarsus.
e'''. Fifth primary escalioped on outer web like the fourth; toes at base always thickly enveloped in the tarsal feathers; hind part of tarsus always hidden by plumes.
a⁴. Nostril pierced near the anterior margin of the cere .... 7. CARINE, p. 132.
b⁴. Nostril situated in the centre of the cere .... 8. HETEROGLAUX, p. 141.
f'''. Fifth primary not escalloped, the fourth scarcely so; junction of toes as well as hinder part of tarsus always bare .... 9. SPEOTYTO, p. 142.
d''. With shortened first primary, the distance between its tip and that of the third primary equal to or greater than length of tarsus.
g'''. Wing long and pointed, the distance between it and the tip of the tail equal to or less than tarsus.
c⁴. Bare part of tarsus longer than middle toe and claw.......... 10. GYMNASIO, p. 149.
d⁴. Bare part of tarsus not so long as middle toe and claw.
a⁵. Tarsus not twice length of middle toe................ 11. NINOX, p. 151.
b⁵. Tarsus twice length of middle toe .... 12. SCELOGLAUX, p. 187.
h'''. Wing short and rounded, the distance between it and the tip of the tail much greater than tarsus.
e⁴. Tarsus about equal to middle toe, densely feathered; tail much more than half the wing, rounded.................... 13. GLAUCIDIUM, p. 188.
f⁴. Tarsus a little longer than middle toe, scantily haired; tail less than half the wing, even...... 14. MICRATHENE*, p. 224.

* The comparative characters of this last genus are drawn from Mr. Ridgway's work, as I have never seen *Micrathene*.

1. KETUPA.

	Type.
Ketupa, *Less. Traité d'Orn.* p. 114 (1831)............	K. javanensis.
Cultrunguis, *Hodgs. J. A. S. B.* v. p. 364 (1836)	K. flavipes.
Smilonyx, *Sund. Av. Meth. Tent.* p. 105	K. javanensis.

Range. Palestine; Indian peninsula and Himalayas; Indo-Malayan subregion; China.

Key to the Species.

a. Feet dark; under surface fulvous fawn-colour, mesially streaked with broad black centres to the feathers, all of which have dull rufescent cross bars, giving a vermiculated appearance to the parts .... *ceylonensis*, p. 4.

b. Feet yellowish (in skin); under surface bright tawny, streaked with black, with no cross vermiculations.

  a'. Much larger: wing over 18·5 inches; bands on tail (counting the apical one) 6 .... *flavipes*, p. 5.

  b'. Much smaller: wing not exceeding 13·5 inches; tail with only 4 or 5 bands .... *javanensis*, p. 8.

1. Ketupa ceylonensis.

Great Ceylonese Owl, *Brown, Illustr. Zool.* pl. 4.
Ceylonese Eared Owl, *Lath. Gen. Syn.* i. p. 120.
Strix ceylonensis, *Gm. S. N.* i. p. 287.
Strix leschenaulti, *Temm. Pl. Col.* ii. pl. 20.
Scops leschenaulti, *Steph. Gen. Zool.* xiii. pt. 2, p. 571; *Less. Man. d'Orn.* i. p. 118.
Scops ceylonensis, *Steph. l. c.* p. 54.
Ketupa leschenaulti, *Less. Traité*, p. 114.
Strix hardwickii, *J. E. Gray, Ill. Ind. Zool.* ii. pl. 31.
Strix dumeticola, *Tick. J. A. S. B.* ii. p. 571.
Cultrunguis nigripes, *Hodgs. J. A. S. B.* v. p. 364.
Cultrunguis leschenaulti, *Jerd. Madr. Journ.* x. p. 90.
Ketupa ceylonensis, *Gray, Gen. B.* i. p. 38; *Cass. Cat. Strigidæ Philad. Mus.* p. 9; *Blyth, Cat. B. Mus. A. S. B.* p. 37; *Bp. Consp.* i. p. 44; *Horsf. & Moore, Cat. B. Mus. E. I. Co.* i. p. 77; *Strickl. Orn. Syn.* p. 196; *Jerd. B. Ind.* i. p. 133; *Blyth, Ibis*, 1863, p. 28; *Tristr. Ibis*, 1865, p. 261; *Blyth, Ibis*, 1866, p. 255; *Gray, Hand-l. B.* i. p. 45; *Hume, Rough Notes*, ii. p. 379; *Swinh. P. Z. S.* 1871, p. 343; *Holdsw. P. Z. S.* 1872, p. 417; *Hume, Nests and Eggs Ind. B.* i. p. 64; *id. Str. F.* i. p. 431, ii. p. 469; *Ball, t. c.* p. 382.
Bubo ceylonensis, *Kaup, Tr. Z. S.* iv. p. 241; *id. Contr. Orn.* 1852, p. 116; *Schl. Mus. P.-B.* Oti, p. 19; *id. Revue Accipitr.* p. 7.

Adult. Above fawn-brown, the feathers broadly streaked with black down their centres, and largely mottled on both webs with fulvous, sometimes taking the form of spots; feathers of hind neck with faint vermiculated bars of brown; lower back and rump more uniform than rest of upper surface, with narrow black shaft-streaks, and faintly spotted on either web with light fulvous; outer scapulars mottled with paler fulvous than rest of back, contrasting somewhat with the latter; least wing-coverts coloured like back, the greater ones with broad creamy buff margins externally, the outermost of the latter series and the primary coverts dark brown, with distinct fawn-coloured spots; quills dark brown, the primaries tipped and distinctly spotted on the outer web with creamy buff, the corresponding bars on the inner webs being dull fawn-brown, the outer secondaries brown like the primaries, but the spots and bars less distinct and more mottled with brown, the innermost secondaries coloured like the back; tail dark brown, tipped and crossed with four bars of light fulvous fawn; loral plumes whitish; cheeks and ear-coverts dull rufous, with narrow black shaft-lines; auricular tufts coloured like head; chin and throat white, streaked with dark brown, the lower throat purer white, and forming a gular patch; rest of under surface of body fulvous fawn-colour, deeper and more rufescent on the chest, all the feathers with broad longitudinal streaks of black down their centres, and crossed with more or less distinct barrings of light fulvous, both the bars and mesial streaks being less distinct on the under tail-coverts; thighs uniform light fawn with obsolete rufous bars; under wing-coverts light fawn, barred and streaked like the breast, the greater series whitish, with narrow black shaft-lines, the outermost coverts largely spotted with black, and tipped and mottled with fulvous; greater series blackish brown at tip, yellowish white at base, like the inner lining of the wing, which is barred with these colours; cere greenish grey; bill dingy greenish, point of upper mandible blackish horny, of the lower mandible yellowish; feet dingy greenish grey, light greenish, or plumbeous; iris bright yellow. Total length 24·5 inches, culmen 2·5, wing 16, tail 8, tarsus 2·65.

Adult male. For full dimensions of the two sexes of this species see Hume's 'Rough Notes,' ii. p. 379, where every comparative measurement is given. The wing of the male is there said to measure from 16·5 to 18 inches.

Hab. India generally and Ceylon, extending eastwards into Assam, Arakan, and Tenasserim. Found also near Hong-Kong in China.

a. Ad. sk.	Madras.	Rev. H. H. Baber [P.].
b, c. Ad. sk.	Nepal.	B. H. Hodgson, Esq. [P.].
d. Ad. st.	Nepal.	B. H. Hodgson, Esq. [P.].
e. Ad. sk.	Behar.	B. H. Hodgson, Esq. [P.].
f. Ad. st.	India.	Major-Gen. Hardwicke [P.].
g. Ad. st.	Hong-Kong.	Purchased.

2. Ketupa flavipes.

Cultrunguis flavipes, *Hodgs. J. A. S. B.* v. p. 364.
Ketupa flavipes, *Gray, Gen. B.* i. p. 38; *Cass. Cat. Strigidæ Philad. Mus.* p. 10; *Blyth, Cat. B. Mus. A. S. B.* p. 37; *Bp. Consp.* i. p. 45; *Horsf. & Moore, Cat. B. Mus. E. I. Co.* i. p. 76; *Strickl. Orn. Syn.* p. 196; *Jerd. B. Ind.* i. p. 135; *Gray, Hand-l. B.* i. p. 45; *Hume, Rough Notes,* i. p. 355.

Bubo ketupa, *Kaup, Tr. Z. S.* iv. p. 242, et *Contr. Orn.* 1852, p. 117; *Schl. Mus. P.-B.* Oti, p. 18.
Ketupa magnifica, *Swinh. Ibis,* 1873, p. 127.

Adult female (type of species). Above tawny, broadly streaked with black down the centre of the feathers, the tawny colour being restricted to spots or broad lateral margins; scapulars more fulvous than the back, with less black in the centres of the plumes, the entire outer web tawny fulvous, contrasting with the dorsal plumes; wing-coverts rather blacker than the back, the tawny colour restricted to the external web and spots and bars on the inner, all the feathers broadly tipped with tawny; quills brown, chequered with tawny spots on the outer web, represented by bars on the inner, the tawny bars paler on the secondaries, but more completely traversing the feathers, which are likewise tipped with tawny buff; innermost secondaries pale brown in centre, the brown cross bars not complete, and often replaced with pale tawny brown, scarcely darker than the fulvous interspace, thus producing a mottled appearance; lower back, rump, and upper tail-coverts deep tawny, the centres of the feathers black, forming irregular bars towards the tips of the tail-coverts; tail blackish brown, tipped with fulvous and crossed with bands of pale tawny rufous; head and neck, including the auricular tufts, deep tawny, the feathers black down the centres; loral plumes whitish with white shafts, and tipped with blackish; ear-coverts tawny rufous, slightly varied with fulvescent centres to the feathers; throat fulvescent; fore neck conspicuously white, with narrow black central streaks to the feathers in the shape of a reversed spear-head; remainder of under surface of body golden tawny, with broad black streaks down the centre of the breast-feathers, becoming less pronounced on the flanks and much narrower on the under tail-coverts; under wing-coverts tawny, with brown centres to the feathers, the lower series buffy white at base, ashy brown at tips, thus resembling somewhat the inner lining of the wing; cere bluish; bill also bluish, blacker at tip; feet greyish, the reticulations whiter, soles of feet pinkish; claws yellow, pinkish on margins; iris bright yellow*. Total length 26·5 inches, culmen 2·5, wing 18·5, tail 9·5, tarsus 3·8, auricular tufts 3·2.

Young. Much paler than the adult and more fulvous, the streaks down the feathers of the upper surface not nearly so black or strongly pronounced, the under surface also pale fulvous, with light brown central streaks to the feathers.

Hab. Himalayas; Ningpo in China.

a. ♀ ad. st.	Nepal.	B. H. Hodgson, Esq. [P.].	Type of species.
b. ♂ juv. st.	Nepal.	B. H. Hodgson, Esq. [P.].	
c. Bones (various).	Nepal.	B. H. Hodgson, Esq. [P.].	

Obs. I cannot consider Mr. Swinhoe's *Ketupa magnifica* to be distinct from *K. flavipes*; it seems to me to be identical in every respect. I subjoin the description which I made of the type of the former bird, most obligingly lent to me by Mr. Swinhoe.

* Soft parts from Hodgson's original painting of type.

Adult female (type of species). General colour tawny variegated with black, all the feathers of the upper surface centred with black, the head streaked with this colour; the dorsal feathers black, barred near the base and spotted near the tip with tawny; all these tawny markings very irregular in shape, and giving a mottled appearance to the whole of the upper surface; scapulars externally uniform tawny, some of them inclining to white on the outer web, and giving slight indications of a shoulder-patch; least wing-coverts black, slightly spotted with tawny, and hence appearing darker than the back, the greater and median series black, notched and barred with tawny, just like the back, the greater series whitish at tip; primary coverts blackish, tipped with tawny and crossed with two rows of tawny spots; quills blackish, with a broad terminal band of light brown, shading off into whitish, especially on the primaries, which are tipped with whitish only; all the quills barred or deeply notched with tawny on both webs, these not forming continuous bands across the feathers, but the spots or notches paler and more fulvous on the outer web, especially on the primaries, the notches on the secondaries often obscured with dusky brown shading; tail blackish, with pale tawny-coloured tip, and crossed with five other bands of deeper tawny colour; ear-tufts very long, measuring 3·25 inches, tawny in colour centred with black, the latter colour predominating on the outer feathers; lores pure white, with hair-like black shaft-lines; cheeks, ear-coverts, and sides of neck with black shaft-lines, broader on the latter; feathers just behind the eye black; chin white; the throat tawny, with very narrow black shaft-streaks, broader on the fore neck, which is again pure white; remainder of under surface clear tawny, all the feathers broadly centred with black streaks, the flanks and abdominal plumes indistinctly waved with lighter bars, the remains of faintly indicated lateral spots of buffy white; under tail-coverts narrowly centred with black; under wing-coverts deep tawny, the inner plumes and axillars with narrow blackish shaft-lines, the outer ones marked with black near the edge of the wing, which is white, the greater series dusky brown, tawny buff at base, thus resembling the inner lining of the quills, which are blackish brown below, barred with pale tawny buff, more plainly on the inner web, which is entirely tawny buff at base; bill deep bluish black, lighter at the base, which is concealed by protruding feathers; inside of mouth pale yellowish flesh-colour; rim of eyelid deep purplish brown; iris fine orange-yellow; exposed portion of tarsus and toes dingy, with a tinge of clay-colour, yellowish on terminal scutes of toes; claws bluish black, fading to light leaden at their bases. Total length 25 inches, wing 18·8, tail 10·5, tarsus 3·1, feathered to within 1·1 inch of the base of toes. (*Mus. R. Swinhoe.*)

Hab. Ningpo, China.

Obs. Mr. Swinhoe has given an excellent description of this species, drawn from the birds as they lay in the flesh. My measurements are different from Mr. Swinhoe's, doubtless because they are taken from the skin. It will be better, therefore, to reproduce his detailed measurements, as they represent the natural size of the bird. "*Measurements of the female.* Length 25 inches; wing very much arched, across arc 16·75, pressed flat 17½, in rounded state 3 inches short of tail-tip; 3rd to 6th primaries equal and longest; 2nd 1·2 shorter than the same. Tail rounded, of twelve shortish hog-backed feathers 9½ inches long; outer 0·75 shorter than centrals. Tarse 3·3 inches long, feathered with down for 2 inches in front, a little lower behind, and ¼ inch less on sides. Middle toe 1·8, its claw 1·1; outer toe 1·5, its claw 1·05; inner toe 1·7, its claw 1·35; hind toe 1·2, its claw 1·1. The male has the tarse feathered in front to within 0·75 inch of where it meets the toes, at the back to 0·25.

"The female looks rather larger than its mate, and is perhaps an inch longer; but I find no appreciable difference in the other measurements. The male has more white behind the ears, and a larger white breast-patch; and its tarse is plumed for a greater length; but the fine down with which it is clothed seems to wear off."

3. **Ketupa javanensis.**

Strix ketupa, *Horsf. Tr. Linn. Soc.* xiii. p. 141.
Strix ceylonensis (nec *Gm.*), *Temm. Pl. Col.* ii. pl. 74; *Vigors, App. Raffl. Life*, p. 650.
Scops ceylonensis, *Less. Man. d'Orn.* i. p. 118.
Scops ketupa, *Cuv. Règne An.* i. p. 347 (1829).
Ketupa javanensis, *Less. Traité*, p. 114; *Gray, Gen. B.* i. p. 38; *Cass. Cat. Vult. Philad. Mus.* p. 9; *Blyth, Cat. B. Mus. A. S. B.* p. 37; *Bp. Consp.* i. p. 45; *Horsf. & Moore, Cat. B. Mus. E. I. Co.* i. p. 76; *Strickl. Orn. Syn.* p. 195; *Blyth, Ibis*, 1863, p. 28; *Pelz. Reis. Novara, Vög.* p. 26; *Wall. Ibis*, 1868, p. 25; *Gray, Hand-l. B.* i. p. 45; *Hume, Rough Notes*, ii. p. 384; *Salvad. Ucc. Born.* p. 20.
Bubo ketupa, *Kaup, Tr. Z. S.* iv. p. 242; *id. Contr. Orn.* 1852, p. 117.
Ketupa javensis, *Bp. Rev. et Mag. de Zool.* 1854, p. 542; *Hume, Str. F.* ii. p. 469.
Bubo javaensis, *Schl. Mus. P.-B.* Oti, p. 18; *id. Dierent.* p. 10, fig. 4; *id. Revue Accipitr.* p. 6.
Ketupa javanica, *Scl. P. Z. S.* 1863, p. 206.
Smilonyx javensis, *Sundev. Av. Meth. Tent.* p. 105.

Adult. Above tawny with very broad brown centres to the feathers, the upper surface somewhat varied with fulvous spots, which appear on either web of the dorsal and scapular feathers; the head and hind neck as well as the auriculars slightly paler than the back, but streaked with black in the same manner; wing-coverts black, margined with tawny and spotted on both webs with the same, the spots and margins on the median and greater series paler and more fulvescent; quills dark brown, tipped with fulvous, the primaries with large fulvous spots on outer web, continued into a paler brown band on the inner one, the secondaries similarly but more numerously barred than the primaries, the outer bars, however, more strongly tinged with tawny; lower back, rump, and upper tail-coverts deep tawny, varied with broad brown centres to the feathers, and with rounded spots of pale fulvous on both webs; tail fulvous brown, tipped with fulvous and crossed with two complete bands of light tawny and an indication of a third basal one; sides of face pale tawny, the cheeks varied with whitish shaft-lines; sides of neck and under surface of body tawny, with long and rather narrow streaks of black, the lower throat mottled with white bases to the feathers, causing an inconspicuous patch; under wing-coverts tawny, the least ones narrowly streaked with black, the outer ones broadly and longitudinally spotted; lower series uniform ashy brown; wing-lining ashy brown, barred with yellowish white; iris yellow. Total length 20 inches, culmen 2·1, wing 13·1, tail 6·5, tarsus 2·45, auricular tufts 2.

Young. Plumage more fluffy than in adult, and dark brownish tawny above and below, the streaks dusky brown and rather dull; tail with three complete bands of tawny fulvous.

Hab. Java, Sumatra, Borneo, and the Malayan peninsula northwards to Aracan.

a. Ad. st.	Java.	Hon. East-India Co. [P.].
b. ♂ ad. sk.	E. Java.	A. R. Wallace, Esq. [C.].
c. ♂ ad. sk.	Malacca.	A. R. Wallace, Esq. [C.].
d. Juv. sk.	Malacca.	Purchased.
e. ♀ juv. sk.	Sarawak.	A. R. Wallace, Esq. [C.].
f. Ad. sk.	Borneo.	J. Gould, Esq. [P.].
g. Skeleton.		Purchased.
h. Skeleton (without skull).		Purchased.

2. **SCOTOPELIA.**

	Type.
Scotopelia, *Bp. Consp.* i. p. 44 (1850)	S. peli.
Megapelia, *Kaup* (*teste G. R. Gray*)	S. peli.
Scotoglaux, *Heine, J. f. O.* 1860, p. 196	S. peli.

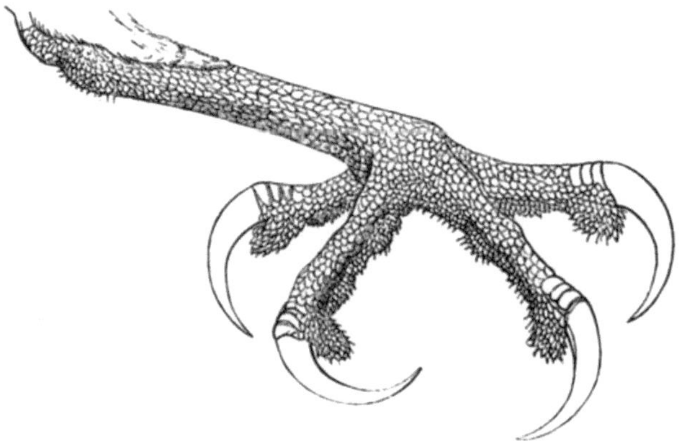

Leg of *Scotopelia ussheri*, to show spicules on the foot.

Range. Africa: from Senegambia to Gaboon on the west coast; Zambesi on the east coast.

Key to the Species.

a. Larger; wing 16·5 inches; above rufous, banded across with black .... *peli*, p. 10.
b. Smaller; wing 13·6 inches; above uniform rufous .... *ussheri*, p. 11.
c. Smaller; wing 13·0 inches; above dusky brown, vermiculated with dull tawny rufous all over the upper surface .... *bouvieri*, p. 11.

1. Scotopelia peli.

Scotopelia peli, *Bp. Consp.* i. p. 44 (*ex Temm. MS. in Mus. Lugd.*); *Strickl. Orn. Syn.* p. 197; *Hartl. J. f. O.* 1855, pp. 358, 360; *id. Orn. W.-Afr.* p. 18; *Cass. Pr. Phil. Acad.* 1859, p. 32; *Gurney, Ibis*, 1859, p. 445, pl. 15; *Hartl. J. f. O.* 1861, p. 101; *Kirk, Ibis*, 1864, p. 317; *Finsch u. Hartl. Vög. Ost-Afr.* p. 100; *Murie, J. Anat. Phys.* vi. p. 70; *Sharpe, ed. Layard, B. S. Afr.* p. 69.
Bubo peli, *Kaup, Contr. Orn.* 1852, p. 117.
Scotopelia typica, *Bp. Rev. et Mag. de Zool.* 1854, p. 543.
Strix pelii, *Schl. Hand-l. Dierk.* i. p. 176, pl. 1. fig. 10.
Ulula pelii, *Schl. Mus. P.-B.* Striges, p. 23 (1862); *id. Dierent.* p. 14, fig. 6.
Scotoglaux peli, *Heine, J. f. O.* 1860, p. 196.
Ketupa peli, *Gray, Hand-l. B.* i. p. 45 (1869).

Adult. Above deep rufous bay, crossed with numerous irregular bars of black, fainter on the head, which is more fulvescent; wing-coverts and entire wing bay barred with black, exactly like the back, the under surface of the wing rufous and barred exactly like the upper surface; tail rufous fawn-colour, rather lighter than back, banded across with black; under surface of body light bay, with distinct cordiform bars of black, somewhat irregular in shape; under wing-coverts rufous bay with a few black spots and bars, more distinct on the lower series, which are lighter, and resemble the inner lining of the wing; cere bluish lead-colour; bill similar to cere but darker, excepting towards tip; tarsi dirty white, tinged with bluish pink; talons light horn-colour with a tinge of dull blue; iris extremely dark brown. Total length 23·5 inches, culmen 2·7, wing 16·5, tail 10, tarsus 2·7.

Obs. The birds from the Zambesi are a little larger than the Gaboon example described above. They are, moreover, differently coloured underneath, being fulvous with longitudinal black shaft-stripes and an occasional cordiform spot on the flanks; some of the breast-feathers with subterminal black spots. As, however, this appearance is seen in Mr. Gurney's plate of a West-African bird (*l. c.*), it is probably not a specific character, but the sign of nonage.

Hab. W. Africa, from Senegambia to Gaboon. Zambesi region in S.E. Africa.

Ad. sk.	River Ogówe, Gaboon (*Marche et De Compiègne*).	M. A. Bouvier.

b Ad. st.	Zambesi.	Dr. Kirk [C.].
c. Juv. st.	Zambesi.	Dr. Kirk [C.].

2. Scotopelia ussheri.

Scotopelia ussheri, *Sharpe, Ibis*, 1871, pp. 101, 417, pl. xii.; *Ussher, Ibis*, 1874, p. 46.

Adult. Above bright bay, a few of the feathers dusky brown towards the tips, the head and hind neck browner than the back, as also the sides of the face; wing-coverts bay, like the back, the least ones very bright, the median and greater series paler and more fulvescent, somewhat shaded with dusky brown towards the tips, and faintly barred with darker brown; quills bay, rather more fulvous than the back, very broadly barred with dark brown on inner web, extending on the outer one, but more narrowly and faintly, under surface much paler than the upper, the broad bars very distinct; tail dusky fulvous, crossed with five dark brown bars, not of equal breadth nor strictly conterminal; under surface of body pale fawn-colour, the throat paler and more fulvous, the entire breast-feathers streaked down the centre with dark brown, inclining to rufous on either side of the shaft-streak; thighs and under tail-coverts not streaked; under wing-coverts also uniform fawn-colour with no streaks, the lower series brownish at tip, resembling the wing-lining; cere yellow; bill horn-brown, yellowish green at junction with cere; feet yellow, claws yellowish horn-colour. Total length 20·5 inches, culmen 1·65, wing 13·6, tail 7·8, tarsus 2·35.

Hab. W. Africa: forests of Denkera, in the interior of Fantee.

a. Ad. sk.	Denkera (*Aubinn*).	R. B. Sharpe, Esq. [P.

3. Scotopelia bouvieri. (Plate I.)

Scotopelia bouvieri, *Sharpe, Ibis*, 1875, p. 261.

Adult. Above dusky brown, with blackish shaft-lines, plainly indicated on the head and hind neck: all the feathers of the upper surface mottled with dull tawny vermiculations, these predominating on the head, which has consequently a paler appearance than the back; scapulars externally fulvous, forming a tolerably distinct shoulder-patch; least wing-coverts blacker than the back, the tawny vermiculations not so distinct on this series, but very plain on the median and greater series, which are consequently much more tawny on the outer web and at the tip, the latter having also an indication of a whitish spot; primary coverts blackish, only slightly vermiculated with tawny; quills blackish, tipped narrowly with fulvous, and barred with tawny in about equal proportions to the black, the tawny bars not conterminous on the primaries, being much more dingy and shaded with blackish on the inner webs, the outer webs paler and clearer, the intermediate blackish markings very indistinct, and dissolving into obscure vermiculations, the secondaries much more blackish in tint and resembling the back, the tawny bands tolerably well indicated on the outer feathers, but everywhere obscured with blackish vermiculations; upper tail-coverts resembling the back, and vermiculated in the same manner; tail blackish, tipped with pale tawny, shading off into fulvous on the apical margin, the two centre feathers blackish, only irregularly mottled with tawny near the shaft and on the margins, clearer at tip, the other feathers crossed with tawny bars not strictly conterminous, of which about four can be counted without including the pale terminal bar; lores whitish, as also the fore part of the cheeks, all the feathers with narrow blackish shaft-lines; ear-coverts and sides of face tawny like the head, with the same blackish shaft-stripes, the bases to the feathers silvery white; under surface tawny buff, broadly streaked with black, these streaks linear near the base and dilating towards the tips, all the feathers more or less white at base and on their margins, giving a very whitish appearance to the chest, which disappears on the abdomen, only a few of the feathers of which are laterally whitish; vent and under tail-coverts tawny buff, only a few plumes mesially streaked with black; under wing-coverts tawny buff, most of the feathers whitish at base, some of the outer ones with a few tiny marks of brown, the lower series blackish, fulvous at base, and thus resembling the inner lining of the wing, which is blackish, barred with fulvous on the inner web; bill horn-colour, yellowish at base of both mandibles; legs yellowish, claws dark horn. Total length 17 inches, culmen 1·8, wing 13, tail 8, tarsus 2·2.

Hab. Interior of Gaboon.

a.	Ad. sk.	Lopé, Oyowé River, Gaboon.	MM. Marche and De Compiègne [C.]. Type of species.

3. BUBO*.

	Type.
Bubo, *Cuv. Règne An.* 1817, p. 331	B. maximus.
Bubotus, *Rafinesque (teste G. R. Gray)*	B. maximus.
Ascalaphia, *Geofr. St.-Hilaire,* 1830 (*teste G. R. Gray*)	B. ascalaphus.
Huhua, *Hodgs. As. Research.* xix. p. 173 (1836)	B. nipalensis.
Heliaptex, *Swains. Classif. B.* ii. p. 217 (1837)	B. arcticus.
Urrua, *Hodgs. J. A. S. B.* vi. p. 372 (1837)	B. bengalensis.
Etoglaux (*pro* Huhua), *Hodgs. J. A. S. B.* x. p. 28 (1841)	B. nipalensis.
Mesomorpha (*pro* Urrua), *Hodgs. l. c.* p. 28	B. bengalensis.
Aibryas, *Gloger, Handb. Naturf.* p. 228 (1842)	B. arcticus.
Nyctaetus, *Gloger, l. c.* p. 228 (1842)	B. lacteus.
Pseudoptynx, *Kaup, Isis,* 1848, p. 770	B. philippensis.
Megaptynx, *Bp. Rev. et Mag. de Zool.* 1854, p. 542	B. magellanicus.
Pachyptynx, *Bp. l. c.* p. 542	B. lacteus.
Nisuella, *Bp. l. c.* p. 542	B. maculosus.
Ptiloskelos, *Tickell, J. A. S. B.* xxviii. p. 448 (1859)	B. nipalensis.

Range. The entire New World; the whole of the Old World, excepting the Moluccas, Australia, and Oceania.

* *Cf.* Sharpe, 'Ibis,' 1875, p. 327.

Key to the Species.

a. Toes so thickly enveloped in plumes, that even their last joint is hidden.
 a'. Larger: wing 18·5–19·5 inches; hind neck yellowish, broadly streaked with black, the head and neck decidedly lighter than the back .... *ignavus*, p. 14; *turcomanus*, p. 17.
 b'. Smaller: wing 14·5–16 inches; head and neck resembling the back, which is blackish, freckled with white mottlings, the bases of the feathers tawny buff .... *virginianus*, p. 19.
b. Toes more or less scantily clothed with feathers, but the final joint always bare; sometimes the entire toe bare.
 c'. Bill blackish.
  a''. Hind neck yellowish, broadly streaked with black, forming as it were a wig.
   a'''. Underneath ocellated, the light markings very broad, and in form of spots .. *ascalaphus*, p. 24.
   b'''. Underneath narrowly barred .... *bengalensis*, p. 25.
  b''. Hind neck buff or grey, but always barred, bearing a great resemblance to the rest of the upper surface.
   c'''. General tone of plumage tawny buff; size large, wing 14·5–15·5; ear-coverts tawny buff, only slightly spotted with brown.
    *a*⁴. Broadly barred below with white or buff and black, the light bars arranged in duplicate bands on the breast; quills dark brown, the light spots on the outer webs of the primaries much narrower than the dark ones ...... *capensis*, p. 27.
    *b*⁴. Very narrowly barred below with buff and dark brown; primaries tawny buff, barred with brown, the buff interspaces much the broadest ...... *magellanicus*, p. 29.
   d'''. General tone of plumage brown, or greyish brown; size small; wing 12·8–13·5; ear-coverts greyish white, barred with brown.
    *c*⁴. Iris yellow, below largely spotted and mottled with white; toes thickly feathered .... *maculosus*, p. 30.
    *d*⁴. Iris brown; below mottled with small bars and spots; toes scantily feathered .... *cinerascens*, p. 32.
 d'. Bill yellow or yellowish white.
  c''. Breast greyish, finely vermiculated with brown; head greyish, with black ear-tufts.
   e'''. Larger: wing 17·5–19·3 inches; no mesial dark brown streaks to the feathers of the head and breast ...... *lacteus*, p. 33.
   f'''. Smaller: wing 16·4, with distinct cen-

tral markings of dark brown on the feathers of the head and breast ...... *coromandus*, p. 35.

d″. Under surface barred across with dark brown, or black and white.

g‴. Ear-coverts dusky fulvous, barred with black or dull brown; head blackish, sometimes with bars of dull rufous or buffy white.

e^1. Toes plumed; size large, wing 18–19·4 inches.

a^5. Primary coverts uniform; head and ear-coverts uniform blackish .... *shelleyi*, p. 37.

b^5. Primary coverts with distinct lighter brown bars; head with wavy bars of buff........................ *nipalensis*, p. 37.

f^1. Size small: wing 13–13·5 inches.

c^5. Toes bare; under wing-coverts white, spotted with brown; leg-feathers white, spotted or barred with brown; primary coverts uniform dark brown; bars on tail about 6.. *orientalis*, p. 39.

d^5. Toes scantily haired; under wing-coverts dark brown, very slightly varied with spots or bars of white; leg-feathers dark brown, barred with fulvous; primary coverts dark brown, with rufescent irregular bars; bars on tail about 8 ...... *leucostictus*, p. 41.

h‴. Ear-coverts rufous; general colour rufous, barred with black; tail rufous brown, with about 9 black cross bands ...... *poensis*, p. 42.

e″. Breast white, longitudinally streaked with black .... *philippinensis*, p. 43.

1. Bubo ignavus.

The Great Horn-Owl or Eagle Owl, *Albin, Birds*, ii. pl. 9.
Great Horned Owl, from Athens, *Edwards, Glean.* pl. 227.
Le Grand Duc, *Briss. Orn.* i. p. 477; *Buff. Pl. Enl.* i. pl. 435.
Le Grand Duc d'Italie, *Briss. Orn.* i. p. 482.
Strix bubo, *Linn. S. N.* i. p. 131; *Shaw, Mus. Lever.* p. 119, pl. 29; *Temm. Man. d'Orn.* i. p. 100; *Roux, Orn. Prov.* p. 69, pl. 46; *Naum. Vög. Deutschl.* i. Taf. 44; *Werner, Atlas*, Rapaces, pl. 41; *Schl. & Susem. Vög. Eur.* Taf. 50; *Yarr. Brit. B.* i. p. 107; *Kjærb. Orn. Dan.* pl. vii. fig. 3; *Midd. Sibir. Reis.* p. 131; *Sundev. Sv. Fogl.* p. 188, pl. xxiv. fig. 4.
Bubo microcephalus, *Leach, Syst. Cat. Mamm. &c. Brit. Mus.* p. 11 (1816); *Steph. Gen. Zool.* xiii. pt. 2, p. 55.
Bubo ignavus, *Forster, Syn. Cat. Brit. B.* p. 3; *Newt. ed. Yarr. Brit. B.* i. p. 168; *Shelley, B. Egypt*, p. 180; *Dresser, B. Eur.* part xxii.
Bubo maximus, *Fleming, Brit. An.* p. 57; *Gould, B. Eur.* i. pl. 37; *Bp. Comp. List B. Eur. & N. Am.* p. 6; *Macgill. Brit. B.* iii. p. 428; *Gray, Gen. B.* i. p. 37; *Kaup, Tr. Z. S.* iv. p. 239, et *Contr. Orn.* 1852, p. 116; *Horsf. & Moore, Cat. B. Mus. E. I. Co.* i. p. 72; *Strickl. Orn. Syn.* p. 212; *Hewits. Eggs Br. B.* i. p. 50, pl. xvii. fig. 4; *Fritsch, Vög. Eur.* tab. 12. fig. 5; *Jaub. & Barth.*

Lapom. Rich. Orn. p. 80; *Schl. Mus.P.-B.* Oti, p. 7; *Newt. Ooth. Wolley.* p. 160; *Degl. & Gerbe, Orn. Eur.* i. p. 141; *Loche, Expl. Sci. d'Algér. Ois.* p. 100; *Gray, Hand-l. B.* i. p. 43; *Heugl. Orn. N.O.-Afr.* i. p. 110; *Gould, B. Gt. Br.* pl. xxx.; *Hume, Rough Notes,* ii. p. 374; *Bettoni, Ucc. Lomb.* i. tav. 107; *Salvad. Faun. Ital. Ucc.* p. 32.

Bubo germanicus, *Brehm, Vög. Deutschl.* p. 119.
Bubo septentrionalis, *Brehm, Vög. Deutschl.* p. 120.
Bubo europæus, *Less. Traité,* p. 115, pl. 17. fig. 1.
Asio bubo, *Swains. Classif. B.* ii. p. 217.
Otus bubo, *Schl. Rev. Crit.* p. xiii.
Bubo atheniensis, *Bp. Consp.* i. p. 48; *Taczan. J. f. O.* 1870, p. 39.
Bubo grandis, *Brehm, J. f. O.* 1853, p. 346.
Bubo bubo, *Licht. Nomencl. Av.* p. 7.
Bubo melanotus, *Brehm, Naum.* 1855, p. 270.

Nestling. Covered with down of a dull white colour, the incipient feathers being of a dull tawny buff-colour, barred across with blackish.

Adult male. General colour above blackish, mottled and varied with yellowish tawny colour; head blackish, the tawny vermiculations being confined to the terminal margins and a few lateral bars on each side of the feather, so that the centre of the latter remains blackish; ear-tufts black, 3·2 inches long, excepting near the basal half of the inner web of the interior plumes, which are tawny buff, with narrow blackish vermiculations; nape and hind neck much paler than the crown, the feathers being for the most part tawny buff with broad black centres, from which spring on each side narrow black transvermiculating lines; back again darker, the feathers being mostly black, the half-concealed bases tawny buff, vermiculated and irrorated coarsely with black; scapulars whitish externally or fulvous, sparingly transvermiculated with black, and forming an indistinct shoulder-patch; wing-coverts blackish, the least series very slightly varied with tawny buff, the median and greater series more frequently mottled with this character, the latter whitish at tip, with narrow irregular cross lines of black, forming an indistinct bar across the wing; primary coverts blackish, only slightly mottled with fulvous near the base of the outer web; quills dark brown, regularly barred with tawny buff, all the bars very minutely dotted with black on the outer web, the inner web for the most part tawny, most irregularly mottled with wavy lines, dots, and markings of black; the secondaries not so distinctly barred with tawny buff, all the lighter spots on the outer web obscured by minute spots and markings of dark brown, the inner webs much more plainly barred with tawny, none of these bars, however, being without brown vermiculations; the innermost secondaries blackish, mottled with fulvous all over, the tip whitish, as also of all the secondaries; lower back, rump, and upper tail-coverts more tawny than the back, the former coarsely barred, and the latter vermiculated with blackish, especially on the tail-coverts, which are whitish at tip; tail brown, the centre feathers vermiculated all round the margins with fulvous, and having three or four irregularly indicated bands, consisting of a few whitish spots of irregular shape: all the other feathers more or less distinctly barred with deep tawny buff, the inner webs for the most part bright tawny, with irregular lines of blackish, more thickly distributed towards the tips of the feathers, which are here thickly though minutely spotted with brown, as are all the tawny bands on the outer webs; lores and feathers in front of and below the eye whitish, with narrow blackish shaft-lines; above the eye a patch of black feathers; sides of face dull tawny, irrorated with narrow circular bars of brown; sides of neck coloured like the hind neck, but less strongly marked; chin pure white, as also the fore neck, separated from each other by a narrow cordon of tawny feathers centred with black, and having small lateral bars of the same; crop covered with tawny buff down, succeeded by a patch of white feathers in the centre of the breast; rest of under surface of body light tawny buff, most of the feathers with a slight glistening of silvery white, all of the chest-feathers very broadly streaked down the centre with black, and likewise laterally spotted or barred with irregular lines of black, these central black streaks much narrower on the breast and abdomen, which are likewise very regularly, though narrowly, barred across with blackish; under tail-coverts deep tawny, barred across with narrow lines of black; leg-feathers deep tawny, with irregular transverse washings of blackish; under wing-coverts tawny, the inner ones whitish, with narrow zigzag cross lines of black, the lower series brown, tawny at base; inner lining of wing dull brown, barred with tawny buff on inner web, most of the quills entirely tawny buff at base of the latter, more or less minutely speckled with brown; bill and claws blackish horn-colour; iris rich orange. Total length 26 inches, wing 18·6, tail 11·2, tarsus 3·2.

Young female in first year's plumage (October 27th, 1872). Very similar to the adults, and presenting few appreciable differences; many of the feathers of the upper surface glossed with silvery white, the light mottlings being larger on many of the dorsal feathers; middle tail-feathers more coarsely and distinctly mottled with whitish, affording indications of five ill-defined bars; upper wing-coverts much blacker than in the adult, with fewer transverse vermiculations. Total length 27 inches, wing 19·4, tail 11·5, tarsus 3·2.

Obs. The descriptions are taken from Swedish specimens, carefully sexed and dated; and other specimens from the same country show very little differences. On the other hand, the two examples mentioned below from Archangel are much blacker on the upper surface generally, the head and neck very bright orange buff, with the usual broad black mesial streaks; the under surface of the body is also much paler. (*Cf.* also Dresser, B. Eur. *l. c.*)

Hab. The whole of Europe and Northern Asia, ranging south into Northern China and the Himalayas, also occurring rarely in North-eastern Africa.

a. Ad. st.	Europe.	A. D. Bartlett, Esq.
b. Pull. sk.	England (bred in confinement).	J. Gould, Esq. [C.].
c. Ad. sk.	Hisinger, S. Sweden, Nov. 2nd, 1858.	Prof. A. W. Malm [P.].

d. ♂ ad. sk.	Upsala, Feb. 12th, 1872.	Purchased.
e. ♀ juv. sk.	Upsala, Dec. 30th, 1872.	Purchased.
f. ♀ juv. sk.	Upsala, Oct. 27th, 1872.	Purchased.
g. Ad. sk.	Archangel.	R. B. Sharpe, Esq. [P.].
h. Imm. sk.	Archangel.	R. B. Sharpe, Esq. [P.].
i. Ad. sk.	Xanthus, Asia Minor.	Sir C. Fellowes [P.].
k. Skeleton.	Russia.	Purchased.
l. Skull.	Rauhe Alp.	Dr. Günther [C.].
m. Sternum.		W. Yarrell, Esq. [C.].
n. Sclerotic bone.		W. Yarrell, Esq. [C.].

Subsp. *a.* **Bubo turcomanus.**

Strix turcomana, *Eversm. Add. Pall. Zoogr.* p. 3.
Strix sibirica, *Schl. & Susem. Vög. Eur.* Taf. 44.
Bubo sibiricus, *Gray, Cat. Accipitr. B. M.* 1844, p. 99; *id. Gen. B.* i. p. 37; *Bp. Consp.* i. p. 49; *Licht. Nomencl. Av.* p. 7; *Strickl. Orn. Syn.* p. 213.
Bubo cinereus, *Gray & Mitch. Gen. B.* i. pl. 13.
Bubo scandiacus, *Cab. J. f. O.* 1854, p. 367.
Bubo pallidus, *Brehm, Naum.* 1855, p. 270.
Bubo hemachalana, *Hume, Str. F.* i. p. 315; *Blanf. Ibis,* 1874, p. 81.
Bubo maximus, var. turcomanus, *Severtz. Turkest. Jevotn.* p. 111.
Bubo ignavus, *Dresser, Ibis,* 1875, p. 111.

Adult. General colour above pale tawny buff, many of the feathers inclining to whitish; head whitish, the feathers yellowish at base, all rather broadly centred with black, from which dark centres radiate irregular lines and spots of black; nape and hind neck light yellowish buff, many feathers whitish, all with broad black central streaks, the lateral lines or vermiculations very indistinct, in fact almost entirely absent; back blackish brown, yellowish buff at base, and shading off distinctly white in the centre of the feathers, all the lighter parts of the dorsal feathers narrowly waved and minutely barred or spotted with blackish; outer scapulars white on external webs, with only two or three narrow zigzag bars of blackish; wing-coverts tawny buff, obscured by brownish mottlings thickly distributed over all the least series and the innermost of the median and greater series, the outermost of the median row uniform brownish black towards the tips, which more or less incline to white, most of the greater series having a large white spot at the tip of the outer web, which has very few and narrow cross lines of black; primary coverts blackish brown, irregularly mottled with yellowish buff on the outer webs; quills for the most part clear tawny buff, almost orange in intensity, the primaries dusky brown at tips, all the quills distinctly barred with darker brown, not quite conterminous, the yellow interspaces with nothing more than a few minute dots of brown; the secondaries more dusky than the primaries, the light interspaces being clouded with coarse brown vermiculations, especially towards the tip, the innermost secondaries very much clouded with brown mottlings, and distinctly barred across with blackish brown; lower back, rump, and upper tail-coverts tawny buff, with a few indistinct wavy lines of blackish brown, rather more distinct and strongly characterized on the upper tail-coverts; tail yellowish buff, whitish at tip, and regularly barred with dark brown, of which about eight bars (some partly broken up) can be distinguished on the outer feathers, the two centre ones more dusky than the rest, and crossed with six or seven bands of dusky brown, often much dissolved and mingled with the interspaces, which are thickly mottled and vermiculated with brown; lores and feathers round the eye white, the former black at the tip of the shafts; just above the eye a patch of black feathers, continued to and appearing to form part of the ear-tufts, which are 3·2 inches long, and black with tawny bases, the long feathers uniform yellowish buff on the inner web, and barred with black near the base of the outer; ear-coverts dull fulvous, with indistinct brown cross-barrings; chin pure white, the throat encircled by a row of slightly recurved feathers, yellowish buff, streaked down the centre with black, and also laterally barred with narrow lines of the same, exactly resembling the plumes on the sides of the neck, of which this gular band seems a continuation; rest of under surface white, slightly washed here and there with yellowish, the breast-feathers broadly streaked down the centre with black, all these feathers slightly varied with lateral vermiculations of brown, the central streaks reduced to a narrow shaft-line on the abdominal feathers and flanks, these being, however, finely but regularly barred across with dark brown, disappearing on the under tail-coverts; crop covered with tawny down, succeeded by a patch of pure white feathers in the centre of the breast; leg-feathers buffy white, with a few remains of brown zigzag bars on the outer aspect; under wing-coverts pure white, here and there washed with yellowish, the outermost slightly marked with irregular lines of brown, the axillaries being also crossed with narrow lines of the same; lower series ashy brown, yellowish buff at base, forming a bar across the wing; inner lining of quills for the most part orange-buff, with a few distinct brown bars on the inner web, disappearing towards the tip of the quills, which are there dusky brown. Total length 26 inches, wing 19, tail 11·2, tarsus 3·4.

Hab. Central Asia, from the Ural to Turkestan and Thibet; also said to occur in Siberia.

a. Ad. st.	Siberia.	
b. Skeleton.	River Volga.	Purchased.

Obs. This bird, if not quite specifically distinct from *B. ignavus*, undoubtedly constitutes a well-marked race of that species. It is a very much paler bird both above and below, being especially white on the under surface, the cross-barrings on the abdominal plumes being much fewer and further apart; the legs also are covered with white feathers, which extend further on the foot and cover the junction of the toes and claws; the tail is different also—in *B. ignavus* the two central feathers being brown, with faintly indicated bands of fulvous vermiculations, whereas in *B. turcomanus* the prevailing colour of these feathers is tawny buff, about eight cross bands of brown being distinguishable, while, instead of the lateral feathers being barred with deep tawny and brown in about equal proportions, as in *B. ignavus*, in *B. turcomanus* these feathers are orange-buff, narrowly crossed with about eight bars of brown. The same differences are seen in the wing, the prevailing colour being orange-buff in the quill-lining of *B. turcomanus*; in the under wing-coverts there is also a difference, these being barred across in *B. ignavus* like the breast, whereas in the Siberian Owl they are nearly uniform. The differences in the upper surface of the two birds chiefly consist in the entirely paler colour of *B. turcomanus*, the tawny buff colour predominating.

2. Bubo virginianus.

Great Horned Owl, *Edwards, Birds*, pl. 60.
Le Grand Duc de Virginie, *Briss. Orn.* i. p. 484.
Wapacuthu Owl, *Penn. Arctic Zool.* p. 231.
Strix virginiana, *Gm. S. N.* i. p. 287; *Wilson, Am. Orn.* vi. p. 52, pl. 50. fig. 1; *Audub. Am. B.* pl. 61; *id. Orn. Biogr.* i. p. 313.
Strix wapacuthu, *Gm. S. N.* i. p. 290.
Strix maximus, *Bartr. Trav. Carol.* p. 285.
Bubo ludovicianus, *Daud. Traité*, ii. p. 210.
Bubo pinicola, *Vieill. Ois. Am. Sept.* pl. 19.
Strix crassirostris, *Vieill. N. Dict. d'Hist. Nat.* vii. p. 44.
Strix macrorhyncha, *Temm. Pl. Col.* ii. pl. 62.
Otus virginianus, *Steph. Gen. Zool.* xiii. pt. 2, p. 57.
Otus macrorhynchus, *Steph. Gen. Zool.* xiii. pt. 2. p. 59; *Cuv. Règn. An.* i. p. 341 (1829); *Less. Traité*, p. 109.
Nyctea wapacuthu, *Steph. Gen. Zool.* xiii. pt. 2, p. 63.
Asio macrorhyncha, *Less. Man. d'Orn.* i. p. 117.
Bubo virginianus, *Sw. & Rich. Faun. Bor.-Am. Birds*, p. 82; *Bp. Comp. List B. Eur. & N. Am.* p. 6; *Audub. B. N. Am.* i. p. 143, pl. xxxix. (1839); *Gray, Gen. B.* i. p. 37; *Bp. Consp.* i. p. 48; *Kaup, Tr. Z. S.* iv. p. 241, et *Contr. Orn.* 1852, p. 116; *Strickl. Orn. Syn.* p. 213; *Cass. B. Calif. & Texas*, p. 177; *id. B. N. Am.* p. 49; *Gray, Hand-l. B.* i. p. 43; *Coues, Key N. Am. B.* p. 202; *Allen, Bull. Harv. Coll.* ii. p. 337; *Dall & Bann. Tr. Chic. Acad.* i. p. 272; *Finsch, Abh. Nat. Ver. Bremen*, iii. p. 26; *Coues, B. N. W. Am.* p. 300.
Bubo arcticus, *Swains. Faun. Bor.-Am. Birds*, p. 86, pl. 30.
Ulula virginiana, *James. ed. Wils. Am. Orn.* i. p. 100.
Heliaptex arcticus, *Swains. Classif. B.* i. p. 328.
Asio virginianus, *Swains. Classif. B.* ii. p. 217.
Otus crassirostris, *Gray, Gen. B.* i. p. 40.
Bubo subarcticus, *Hoy, Pr. Acad. N. Sci. Philad.* vi. p. 211.
Asio crassirostris, *Strickl. Orn. Syn.* p. 208.
Bubo virginianus atlanticus, *Cass. B. N. A.* p. 49.
Bubo virginianus pacificus, *Cass. l. c.* p. 49.
Bubo virginianus arcticus, *Cass. l. c.* p. 50.
Bubo virginianus, var. arcticus, *Coues, Key N. Am. B.* p. 202; *Ridgw. B. N. Am.* iii. p. 64.
Bubo virginianus, var. pacificus, *Coues, Key N. Am. B.* p. 202; *Ridgw. B. N. Am.* iii. p. 65.
Bubo virginianus, var. virginianus, *Ridgw. B. N. Am.* iii. p. 62.

Adult. Above dark brown, varied with tawny or orange-buff, all the bases to the feathers being of this colour, and showing conspicuously on the head and neck, all the brown dorsal feathers profusely vermiculated with ashy white in the form of zigzag cross markings; scapulars nearly entirely orange-buff on the outer web, inclining to white near the tip, and regularly crossed with rather narrow bars of brown, the tips being mottled with brown; head and neck, as mentioned above, more orange-buff than the back, the bases to the feathers showing very conspicuously, all of them blackish brown at tip, crossed with zigzag bars of ashy white; wing-coverts blackish, especially the lesser series and those near the edge of the wing, these being only slightly vermiculated and barred with ashy white, the median and greater series more largely mottled with ashy white, the orange-buff bases to the feathers being also rather conspicuous, the greater series distinctly whitish at their tips, with several zigzag cross lines of brown; primary coverts regularly barred with dark brown and tawny buff in alternate bands, like the quills, the light bars slightly mottled with brown; primaries dark brown barred with tawny buff, more broadly on the inner web, these bars slightly obscured with brown vermiculations, especially towards the tips of the feathers, the secondaries alternately barred with dark brown and tawny buff in about equal proportions, the light bars vermiculated with brown on the outer web, the innermost secondaries being brown strongly vermiculated with ashy white, and thus resembling the scapulars; lower back orange-buff, with a few zigzag cross markings of brown, increasing in extent on the rump and upper tail-coverts, which are brown at their tips, vermiculated with ashy white like the rest of the back; tail dark brown, tipped with whitish and crossed with several zigzag lines of buffy white, indicating about seven bars on the middle feathers, inclining to buff near the base, the other feathers tawny buff on the inner web barred with brown, the outer web also alternately barred with brown and tawny buff, the external bars of this colour pretty clear on the outer feathers, but more or less clouded towards the centre of the tail; lores and feathers over the eye white, the shafts blackish at tip; plumes below the eye and ear-coverts tawny buff, the latter black at tip, forming a tolerably broad circular line down the sides of the neck; behind this black line a ruff of tawny feathers, whitish at tip, with a brown subterminal bar on the sides of the throat and cheeks, the feathers being very long and stiff, tawny buff with broad central streaks, these plumes continued across the throat, and separating the chin from the fore neck, both of the latter being pure white; rest of under surface of body white, thickly crossed with zigzag bars of blackish brown, all the bases to the feathers orange-buff, showing conspicuously; the centre of the breast white; the chest and the sides of the upper breast with large spots of black; leg-feathers buffy whitish, washed with orange-buff; under tail-coverts whitish, narrowly barred across with brown; under wing-coverts white, scantily spotted with dark brown, those near the edge of the wing with rather larger brown spots, and inclining to bright orange-buff, the lowest series blackish with buffy white bases, resembling thus the inner lining of the quills, which are blackish brown barred with buffy white: ear-tufts 2·6 inches long, blackish brown, tawny buff at base and for the greater part of the inner web; "bill black; toes ashy; claws horn-colour, black at ends; iris gamboge-yellow" (*Ridgway*). Total length 23 inches, wing 14·9, tail 9, tarsus 2·3.

Obs. The specimens from North America in the Museum not being sexed, I borrow the measurements given by Mr. Ridgway (*t. c.* p. 63). He gives those of the male as follows:—Wing 14·50, tail 8·20, culmen 1·10, tarsus 2, middle toe 2.

Female. General appearance same as in the male. Black blotches on head above and nape less conspicuous, the surface being mottled like the back &c.; primary coverts with three well-defined, narrow, pure black bands; primaries with only six bands, these broader than in the male; secondaries with only five bands; tail with but six dark bands, these very much narrower than the light ones. Tibiæ and tarsi with sparse transverse bars of dusky. Wing 16, tail 9, culmen 1·20, tarsus 2·20, middle toe 2·10 (*Ridgway*).

Mr. Ridgway gives nine dusky bars on the primaries of the male and eight on the secondaries, seven being the number on the tail-feathers. From the sizes here given it appears that the example described by me (one of Mr. Doubleday's Delaware specimens) is a male. I can, however, only count seven dark bands on the primaries and six on the secondaries, while the tail has seven more or less distinct ones. Generally, however, as will be seen below, the bars are identical in number on both series of the quills, while not unfrequently the number on the tail-feathers corresponds also.

The American Horned Owl varies greatly, and sometimes apparently these variations are coincident with a different locality; but I do not think these differences are greater than those we are accustomed to see in *B. ignavus* of Europe. Generally speaking, specimens from British Columbia and Vancouver's Island are much darker than ordinary North-American examples, and belong to the race first separated by Mr. Cassin, and constituting the var. *pacificus* of Mr. Ridgway's recent work. He gives the range of this subspecies as the "Pacific coast north of the Columbia, Labrador," and he calls it a "northern littoral form." We have in the Museum six specimens from the above-mentioned localities, on which I have made the following notes *à propos* of Mr. Ridgway's account:—

a. ♂. West side of the Rocky Mountains (*J. K. Lord*). Total length 21 inches, wing 14·85, tail 9, tarsus 2·6. Bands on primaries and secondaries 7, bars on tail 7.

b. ♀. West side of the Rocky Mountains (*J. K. Lord*). Total length 20 inches, wing 14·85, tail 9, tarsus 2·45. Bands on primaries and secondaries 8, and on tail 8. The sex is probably an error in this specimen: they both agree with Mr. Ridgway's diagnosis as to the dark coloration of this race; but they have not the face "greyish;" for it is orange-buff barred with black. Both are very thickly barred with black below.

c. Vancouver's Island (*J. K. Lord*). Total length 21 inches, wing 15·4, tail 9·8, tarsus 2·75. Bars on primaries 8, and the same number on the secondaries and tail. Ear-coverts dull tawny shaded with grey and barred with black.

d. Vancouver's Island (*J. K. Lord*). Total length 20 inches, wing 15·2, tail 9·5, tarsus 2·6. Bars on primaries, secondaries, and tail 6.

e. Vancouver's Island (*Dr. Lyall*). Total length 21 inches, wing 14·7, tail 9, tarsus 2·45. Bars on secondaries 6, and on tail 7. It is impossible to count the bars on the primaries, as they are nearly all merged, producing a uniform surface below.

f. Vancouver's Island (*Dr. Lyall*). Total length 21 inches, wing 14·8, tail 9, tarsus 2·55. Bars on primaries and secondaries 6, and on tail 6. This bird is quite light-coloured compared with the foregoing specimens, and has a greyish face with a few black markings. In tone of colour it is not very different from Mr. B. Ross's bird from Fort Simpson, but does not show the orange-buff bases to the feathers quite so much.

These Pacific-coast birds, as a rule, have the lower surface of the primaries nearly uniform brown; but these are also, perhaps in older birds, often as distinctly barred as in true *B. virginianus*.

Mr. Ridgway separates as a race, equal in value to the subspecies he calls *pacificus*, the pale form which Swainson named *arcticus*; and of this he gives the habitat as the "western region of North America, from the interior Arctic districts to the tablelands of Mexico."

On this subspecies Dr. Coues's remarks should be studied (B. N.W. Am. p. 301).

I consider the type specimen to be nothing but a white variety of *B. virginianus*; but should it constitute a *race*, as Mr. Ridgway thinks, it will hold the same relation to *Bubo virginianus* that *B. turcomanus* does to *B. ignavus*; but it is whiter in proportion. The type in the British Museum may be briefly described as follows:—General colour white, slightly tinged here and there with light orange-buff; all the upper surface mottled with zigzag cross markings of brown, rather broad and distantly vermiculated, the greater coverts white at tip of outer web; quills buffy white, with about six bars of dark brown on the outer web, the interspaces very slightly mottled with brown vermiculations, excepting at the tips of the primaries, which are very closely vermiculated with sandy brown, as also are the primary coverts at their extremities; tail-feathers white, inclining to orange-buff on the inner web, and crossed with six regular though narrow bands of dark brown; lores and general facial aspect white, with silvery-white shaft-lines, the hinder margin of the ear-coverts black; chin and fore neck pure white, separated from each other by a line of stiff feathers, which are white, with a broad mesial blackish streak; rest of under surface pure white, the leg-feathers and under tail-coverts immaculate, the breast irregularly spotted with blackish brown, the rest of the body very narrowly barred with the same; under wing-coverts pure white, the outermost of the lower series black at the tips, and forming a wing-spot which resembles the inner lining of the quills; bill and claws bluish black; iris yellow. Total length 20 inches, wing 14·9, tail 9·1, tarsus 2·6.

Mr. Ridgway thinks it likely that *Bubo arcticus* is a Mexican bird, on the strength of Mr. Sclater's paper in Proc. Zool. Soc. for 1860, where a specimen is recorded from Orizaba and named *B. virginianus*. I have seen two specimens from the valley of Mexico, whence they were received by M. A. Bouvier. The following notes refer to these birds:—

a. Ad. Valley of Mexico (*Bouvier*). Total length 20 inches, wing 14·5, tail 8·4, tarsus 2·55. This bird is very grey, but does not seem to me to be more than an old *B. virginianus*. Ear-coverts dull buff, narrowly barred with blackish, and relieved with silvery white shaft-streaks on the fore part; bars on the primaries 8, on the secondaries 7, and on the tail also 7.

b. [Jun.]. Valley of Mexico (*Bouvier*). Differs from all the Horned Owls of America which I have seen as yet by the broad character of the white markings on the under surface, so that the black bars are very distant. The character of the upper surface is also peculiar, the general colour being blackish, and the ashy white markings on the back, scapulars, and wing-coverts very broad and coarse; the tail is very dark, the bands eight in number, and much broader than usual; bands on primaries 8, on secondaries 7; ear-coverts dull tawny buff, with a few indications of blackish cross lines. Total length 20 inches, wing 14·5, tail 9, tarsus 2·6.

Mr. Salvin has kindly shown me a series of Central-American Horned Owls. They are chiefly from Guatemala, and belong to the ordinary North-American type, viz. *B. virginianus*. But in the Salvin-Godman Collection there is a remarkable bird from Ecuador, which is so black as to suggest the possibility of its being a case of melanism. It has so little left of the tawny colour and markings that I am unable to say whether it is a specimen of *B. virginianus* or *B. magellanicus*, though I should suspect that it would belong to the former species, if a permanent dark race or a distinct species should not be some day found to inhabit Ecuador.

Hab. The whole of North America, ranging south into Central America, as far as Guatemala.

a, *b*. Ad. st.	North America.	
c, *d*. Ad. st.	Arctic America.	Admiral Sir George Back [P.].
e. Ad. st.	Delaware, U. S.	E. Doubleday, Esq. [P.].
f. Ad. st.	North America.	Hudson's Bay Co. [P.].
g. Ad. sk.	Fort Simpson, June 1861.	Bernard R. Ross, Esq. [P.].
h, *i*. Ad. sk.	Vancouver's Island.	Dr. Lyall [P.].
k, *l*. Ad. sk.	Vancouver's Island.	J. K. Lord, Esq. [C.].
m, *n*. ♂ ♀ sk.	West side of the Rocky Mountains.	J. K. Lord, Esq. [C.].
o. Ad. sk.	Valley of Mexico.	M. A. Bouvier.
p. Jun. sk.	Valley of Mexico.	M. A. Bouvier.
q, *r*, *s*. Skeletons.	N. America.	Purchased.

3. Bubo ascalaphus.

Bubo ascalaphus, *Savign. Descr. Egypte*, p. 295, pl. 3. fig. 2; *Gould, B. Eur.* i. pl. 37; *Rüpp. Syst. Uebers.* p. 12; *Gray, Gen. B.* i. p. 37; *Cass. Cat. Strigidæ Philad. Mus.* p. 8; *Kaup, Tr. Z. S.* iv. p. 239, et *Contr. Orn.* 1852, p. 115; *Strickl. Orn. Syn.* p. 215; *Fritsch, Vög. Eur.* tab. 12. fig. 3; *Schl. Mus. P.-B.* Oti, p. 8; *Newt. Ooth. Wolley.* p. 164; *Heugl. Orn. N.O.-Afr.* i. p. 111; *Gray, Hand-l. B.* i. p. 44; *Shelley, B. Egypt*, p. 180; *Bree, B. Eur.* 2nd ed. i. p. 145.

Strix ascalaphus, *Temm. Pl. Col.* ii. pl. 57; *Schl. & Susem. Vög. Eur.* Taf. 49; *Bree, B. Eur.* i. p. 130 (1863).

Otus ascalaphus, *Steph. Gen. Zool.* xiii. pt. 2, p. 56; *Cuv. Règn. An.* i. p. 341 (1829); *Less. Traité*, p. 109; *Degl. & Gerbe, Orn. Eur.* i. p. 139.

Asio ascalaphus, *Less. Man. d'Orn.* i. p. 115.

Ascalaphia savignii, *Is. Geoff. St.-Hilaire, teste G. R. Gray, List Gen. B.* p. 7 (1841); *Bp. Consp.* i. p. 18; *id. Rev. et Mag. de Zool.* 1854, p. 542; *Loche, Expl. Sci. d'Algér. Ois.* i. p. 102; *Souza, Cat. Lisb. Mus. Acciptr.* p. 58.

Adult. General colour above clear tawny buff, varied with black markings occupying the centres and expanding towards the tips of the feathers, the tawny portion of which is distributed in lateral spots or notches, many of which incline to white, the bases of all the feathers being nearly entirely tawny buff; scapulars almost all tawny, with a large rounded white spot near the tip of the exterior feathers; the tips of the feathers blackish brown, with a few small bars of the same on both webs; head and neck tawny buff like the back, many of the feathers inclining to whitish at tip, with broad dark brown centres, less characterized on the hind neck, which therefore appears more tawny than the rest of the upper surface; wing-coverts tawny buff like the back, with the same central markings of brown, the tawny colour more distributed in spots than on the back, many of them inclining to white, especially on the outer webs of the median and greater series, which have quite an ocellated appearance, all the least wing-coverts and those bordering the outside of the wing darker brown and less spotted with tawny than the rest of the coverts; wing-coverts nearly uniform brown, inclining to tawny buff only at base and on the lower half of the inner web; quills tawny buff, tipped narrowly with buffy white, the primaries brown at tip and broadly barred with brown, the tawny interspaces nearly three times the width of the brown cross bars, the secondaries much more dusky, being more numerously barred with brown, the tawny buff interspaces about double the width of the brown bands, and much obscured with vermiculations of brown, the innermost mottled with tawny buff and brown and resembling the back; lower back, rump, and upper tail-coverts nearly uniform tawny buff, with a few brownish mottlings, in the form of bands near the extremity of the latter; tail-feathers tawny buff, tipped with buffy white, and crossed with about five distinct bands of darker brown on the outer feathers, the basal bands obsolete on these, but more distinct on the others, which have about seven distinguishable, the two centre feathers much more dusky, the dark bands being broader and about nine in number, the tawny buff interspaces obscured by mottlings of brown; lores and feathers in front of the eye white, with dusky blackish tips; feathers over the eye black, forming a small spot; ear-tufts about 1·75 inch long, tawny buff, with a black mesial mark towards the tip; sides of face and ear-coverts uniform tawny buff, the latter blackish at their tips, forming a circular line of black down the sides of the head, succeeded by a ruff of tawny buff feathers, inclining to whitish on the lower margin of the ear-coverts, these feathers mesially streaked with black, a small patch of which is also seen behind the lower part of the ear-coverts on the sides of the neck; cheeks and chin white, as also a patch on the fore neck, this latter separated from the chin by a patch of stiff feathers tawny buff with black central streaks; under surface of body bright tawny buff, broadly streaked with brown on the chest, the breast less distinctly streaked, the markings being mere longitudinal spots, many of the feathers being laterally white; belly crossed with numerous narrow zigzag bars of rufous brown, many of the interspaces being white, these narrow bars disappearing on the under tail-coverts; leg-feathers uniform buffy white; under wing-coverts tawny, more or less washed with white, the inner ones especially whitish, those near the edge of the wing rather minutely spotted with black, the lower series dark brown, white at the base, thus resembling the inner lining of the quills, which are whitish on the inner web, with broad brown bars; bill black; iris deep yellow. Total length 21 inches, wing 15·8, tail 9·5, tarsus 3·15.

Obs. A male collected by Mr. E. Bartlett at the Pyramids measures as follows:—Total length 19 inches, wing 13·8, tail 8·5, tarsus 2·9.

Hab. Northern Africa, from Egypt to Algeria.

a. Ad. st.	Egypt.	W. B. D. Turnbull, Esq. [P.].
b. ♂ imm. sk.	Egypt.	E. Bartlett, Esq. [C.]
c. Skeleton.	N. Africa.	Zoological Society.
d. Sternum.		

4. Bubo bengalensis*.

Otus bengalensis, *Frankl. P. Z. S.* 1831, p. 115; *Gould, Cent. B. Himal. Mts.* pl. 3.

Urrua cavearia, *Hodgs. J. A. S. B.* vi. p. 372.

Urrua bengalensis, *Jerd. Madr. Journ.* x. p. 87; *id. B. Ind.* i. p. 128; *id. Ibis*, 1871, p. 345; *Blyth, Ibis*, 1863, p. 26.

Mesomorpha cavearia, *Hodgs. J. A. S. B.* x. p. 28.

Bubo bengalensis, *Gray, Gen. B.* i. p. 37; *Cass. Cat. Strigidæ Philad. Mus.* p. 9; *Blyth, Cat. B. Mus. A. S. B.* p. 35; *Bp. Consp.* i. p. 49; *Kaup, Tr. Z. S.* iv. p. 239, et *Contr. Orn.* 1852, p. 115; *Horsf. & Moore, Cat. B. Mus. E. I. Co.* i. p. 73; *Bp. Rev. et Mag. de Zool.*

* Professor Sundevall considers that the title of *Strix capensis* of Daudin, founded on Levaillant's Grand Duc (Ois. d'Afr. pl. 40), is referable to this species; but the figure is too bad to allow of its exact determination.

1854, p. 542; *Strickl. Orn. Syn.* p. 216; *Schl. Mus. P.-B.* Oti, p. 8; *Gray, Hand-l. B.* i. p. 44; *Hume, Str. F.* i. p. 163.
Bubo cavearius, *Hodgs. J. A. S. B.* xix. p. 169.
Ascalaphia bengalensis, *Blyth, Ibis,* 1866, p. 252; *Hume, Rough Notes,* ii. p. 366; *Nests & Eggs Ind. B.* p. 62.

Adult. Upper surface blackish on the back, much mottled with light yellowish buff, distributed in lateral spots on most of the feathers, some of these spots whitish and slightly obscured by zigzag markings of dark brown, the outer scapulars almost entirely whitish, especially on their external web, forming a conspicuous shoulder-patch; wing-coverts blackish like the back, the least series mottled with a few whitish or yellowish spots, the median and greater coverts with large subterminal spots of white on both webs, very distinct and clear on the outer, but slightly marked with brown mottlings on the inner one; primary coverts dark brown at tip, slightly vermiculated, the rump-feathers also with a central arrow-shaped spot of brown; tail clear tawny buff, whitish at tip, crossed with rather narrow bands of dark brown, of which there are about seven on the outer web, the centre feathers dark brown, crossed with about five bands of whitish buff, all mottled with zigzag brown markings; face very light-coloured, the lores and plumes in front of the eye white, with black tips to the shafts; above the hinder part of the eye a small black patch; cheeks white; ear-coverts dull tawny buff with dusky shaft-lines, the hinder margin black, forming a lateral line down the sides of the face, this line being succeeded by a ruff of pale tawny buff feathers, with broad central streaks of dark brown, these showing very distinctly on the lower part; head and neck much lighter than the back, the general colour being bright tawny buff, streaked broadly down the centre of each feather with blackish brown, many of the plumes of the wing inclining to whitish buff near their tips; ear-tufts 2·25 inches long, black, narrowly margined on the outer web, but broadly on the inner one, with yellowish buff; chin whitish, the throat crossed with a bar of stiff recurved fulvous feathers, streaked down the centre with brown, in continuation of the facial ruff; breast whitish buff, with broad mesial streaks of dark brown with fulvous mottling, the basal half tawny buff, barred across with brown; primaries tawny buff, brown at tip, and barred across with dark brown, the secondaries marked in the same manner, but rather duller, the buff interspaces inclining to whitish externally, and all more or less mottled with brown vermiculations, the innermost almost entirely brown, largely spotted with white on the outer and with buff on the inner web, all these spots more or less obscured by brown zigzag bars or vermiculations; lower back, rump, and upper tail-coverts clear tawny buff, with a few irregular and nearly obsolete bars of brown, some of these streaks dilating towards the tips; belly light yellowish buff, with narrow and indistinct transverse bars of brown, a little plainer on the feathers of the lower breast, which have also tolerably distinct central streaks of brown, the under tail-coverts fulvous, less distinctly barred; leg-feathers uniform fulvous; under wing-coverts tawny buff, slightly barred with narrow markings of brown, the outermost rather largely spotted with brown, the lower series fulvous at base, brown at tip, forming a wing-bar, thus resembling the inner lining of the quills, which are fulvous, barred with brown on the inner web; bill horny black; irides intense orange-yellow. Total length 20 inches, wing 15·6, tail 8·8, tarsus 2·95.

Obs. On looking over the series of *Bubo bengalensis,* I was struck with the variation in colour which is exhibited by the species. Some specimens are very black in comparison with others; and these I cannot make out to be immature. Himalayan examples are generally more dusky about the face than those from other parts of India; and one presented by Mrs. Hillier is remarkable for its tawny appearance. In this stage it comes exceedingly close to *Bubo ascalaphus*; and indeed these two species are scarcely separable; a comparison of a large series might even prove their absolute identity; but we have in the Museum only two specimens from Egypt. They are more ocellated in appearance than the Indian birds, especially below, where the light markings are very broad and arranged in the form of spots, while the transverse bars are much narrower in *B. bengalensis.*

Hab. The Peninsula of India.

a. Ad. sk.	India.	Mrs. Hillier [P.].
b. Ad. st.	Nepal.	B. H. Hodgson, Esq. [P.].
c. Imm. st.	Nepal.	B. H. Hodgson, Esq. [P.].
d. Jun. sk.	Nepal.	B. H. Hodgson, Esq. [P.].
e. Ad. sk.	Nepal.	B. H. Hodgson, Esq. [P.].
f, g. Ad. sk.	Behar.	B. H. Hodgson, Esq. [P.].
h. Ad. sk.	Rajkote, Kattiawar, June 4th, 1871.	Major J. Hayes Lloyd [P.].
i. Ad. sk.	Jaitpore, Kattiawar, Dec. 12th, 1871.	Major J. Hayes Lloyd [P.].
k. Ad. sk.	Kamptee, C. India.	Dr. B. Hinde [P.].
l. Jun. sk.	Kamptee, C. India.	Dr. B. Hinde [P.].
m. Skeleton.		Zoological Society.

5. Bubo capensis*.

Bubo capensis, *Smith, S. Afr. Q. Journ.* ii. no. 4, pt. 1, p. 317; *Gray, Gen. B.* i. p. 37; *Rüpp. Syst. Uebers.* p. 49; *Smith, Ill. Zool. S. Afr.* Aves, pl. lxx.; *Cass. Cat. Strigidæ Philad. Mus.* p. 8; *Bp. Consp.* i. p. 49; *Kaup, Contr. Orn.* 1852, p. 116; *Bp. Rev. et Mag. de Zool.* 1854, p. 542; *Strickl. Orn. Syn.* p. 217; *Schl. Mus. P.-B.* Oti, p. 9; *Layard, B. S. Afr.* p. 39; *Heugl. Orn. N.O.-Afr.* i.

---

* This specific name is generally credited to Daudin, who founded his *Strix capensis* on Levaillant's "Grand Duc." This Professor Sundevall refers to *Bubo bengalensis*; but the question is too obscure to admit of our substituting for the latter well-established title the older, but totally inapplicable, one of *capensis* for an Indian bird. The synonymy of this doubtful Cape species is as follows:—

Le Grand Duc, *Levaill. Ois. d'Afr.* i. pl. 40.
Strix capensis, *Daud. Traité.* ii. p. 209 (*ex Levaill.*).
Bubo africanus, *Stephens, Gen. Zool.* xiii. pl. 2, p. 55.

p. 110; *Gray, Hand-l. B.* i. p. 43; *Gurney, P. Z. S.* 1871, p. 148; *Schl. Revue Accipitr.* p. 4; *Sharpe, ed. Layard, B. S. Afr.* p. 70.
Bubo dilloni, *Prevost et Des Murs, Rev. Zool.* 1846, p. 242; *iid. in Lefebvr. Voy. Abyss. Zool.* p. 73, pl. 3.
Bubo maximus, var. capensis, *Kaup, Isis*, 1848, p. 770.
Nisuella dilloni, *Bp. Rev. et Mag. de Zool.* 1854, p. 542.
Bubo ascalabotes, *Licht. Nomencl. Av.* p. 7.
Bubo ascalaphus, *Finsch, Tr. Z. S.* vii. p. 210.

Not quite adult (type of species). Above brown, everywhere strongly mottled with orange-tawny, the feathers of the head and neck brown, with large spots of tawny colour on each web of the feather, larger and forming bands on those of the hind neck, the tawny colour predominating on this portion of the upper surface; back brown, mottled with irregular markings of tawny, which forms large spots and bars on the scapulars, more distinct on the outer webs of these feathers, which are also tipped with whitish; least wing-coverts almost entirely uniform brown, with only a few small rounded spots of tawny buff, the median series mottled with larger spots of the same, inclining to white on the outer web of some of these coverts, the greater series brown like the secondary quills, and, as in the latter, barred with rows of tawny buff markings, most regular in appearance but difficult to describe, these tawny markings not being actually bars, but enclosing a more or less dissolving spot of brown, and giving a generally barred appearance to the wing; the primary coverts more entirely brown, the cross markings, which are like those on the secondaries, very indistinct; the primaries brown, regularly barred across with tawny buff, these cross bars only slightly obscured by remains of brown mottling, increasing towards the tips of these quills, which are almost entirely brown; lower back, rump, and upper tail-coverts alternately barred with brown and tawny buff in regular succession; tail brown, slightly tipped with fulvous and crossed with seven other bands of buff on the centre feathers, these bands being exactly of the same character as those on the wings, enclosing brown mottlings, the outer feathers banded with dark brown and dull tawny in about equal proportions, the light inter-spaces inclining to white on the outermost rectrix, especially near the base; lores and feathers in front of the eye white, the shafts black near the tip and elongated into hair-like bristles; feathers under the eye whitish; sides of the face dull tawny, shading into dark brown at their tips, the ear-coverts broadly brown on their hinder margin; above the eye a small patch of blackish feathers, reaching to the base of the ear-tufts, which are black, spotted with tawny buff on the inner margin; behind the ear-coverts runs a tolerably conspicuous ruff of bright tawny buff feathers, with brown bases and indistinct cross bars of the same; chin white, as also the fore neck, separated from each other by a bar across the throat, consisting of tawny feathers mesially streaked with brown; crop-patch composed of fluffy dull white feathers, inconspicuously barred with dull brown; rest of under surface tawny buff, barred across with brown, the light portions of the feather, however, being the broadest, and often nearly white; under tail-coverts whitish, barred across with brown, much more narrowly than the breast; leg-feathers whitish, the tibial plumes tawny, with very few zigzag markings of brown; under wing-coverts pure white, plainly crossed with zigzag bars of brown, the lower series yellowish white for basal half, barred with brown, the apical half being entirely brown, forming a wing-bar, and resembling the inner lining of the quills, which are yellowish white on inner web, broadly barred with dark brown; bill dark livid brown; claws horn-colour, lightest towards base; iris bright orange-yellow. Total length 24 inches, wing 15·3, tail 9·2, tarsus 2·6.

Hab. Africa: Cape colony and Abyssinia: not known on the west coast.

a. Imm. st.	Cape of Good Hope.	Sir A. Smith [C.].

6. Bubo magellanicus.

Hibou des Terres Magellaniques, *Buff. Pl. Enl.* i. pl. 385.
Bubo magellanicus, *Gm. S. N.* i. p. 286; *Gray, Cat. Accipitr.* p. 46 (1844); *D'Orb. Voy. Am. Mérid. Ois.* p. 137; *Bp. Consp.* i. p. 48; *Pelz. Reise Novara, Vög.* p. 26; *Gray, Hand-l. B.* i. p. 43.
Nacurutu, *Azara, Apunt.* ii. p. 192.
Strix nacurutu, *Vieill. N. Dict.* vii. p. 44.
Asio magellanicus, *Less. Man. d'Orn.* i. p. 116.
Bubo crassirostris, *Burm. Th. Bras.* ii. p. 121; *id. La Plata Reis.* ii. p. 439.
Bubo virginianus, *Scl. & Salv. P. Z. S.* 1869, p. 155; *iid. Nomencl. Av. Neotr.* p. 116 (pt.).
Bubo virginianus, var. magellanicus, *Ridg. B. N. Amer.* iii. p. 61; *Coues, B. N.W. Am.* p. 301.

Adult male. Above dull tawny buff, more or less mottled with blackish brown over the whole of the upper surface, the buff colour prevailing on the head and neck, these barred crosswise with blackish brown, with a tolerably broad central line of the same dilating towards the apex of the plume, the light markings paler and more whitish on the feathers of the crown; ear-tufts 1·75 inch long, blackish, inclining to pale tawny buff on the inner webs; feathers of the back and scapulars barred near the base, and thickly vermiculated towards the tips with blackish, giving a darker appearance to this part of the upper surface; wing-coverts coloured like the back, the least series thickly mottled with blackish, the tawny-buff colour predominating in the centre of the shoulder, the greater coverts also being tawny buff, slightly mottled with brown, and distinctly barred with the latter colour; primary coverts deep reddish buff, obscurely mottled and tipped with blackish brown, and distinctly barred across with the same colour; primaries for the most part tawny or orange buff, inclining to dusky brown at the tips of the primaries, the light interspaces also being slightly mottled with the same colour, all the quills crossed with seven bands of darker brown, these bars much more indistinct on the innermost secondaries, where they form coarse irregular zigzag brown markings; lower back, rump, and upper tail-coverts dull tawny buff, all regularly barred across with dark brown; tail tawny buff, tipped with whitish and crossed with seven blackish bars, less distinct on the centre feathers, where the interspaces are also mottled with dark brown; lores and feathers in front of the eye white, with hair-like shaft-lines of black; feathers above the eye black; sides of face dull tawny buff, narrowly barred with blackish on the lower part of the ear-coverts, the hinder part black, forming a circular band down the sides of the head; this black bar followed by an indistinct ruff of fulvous feathers, blackish at the base; cheeks whitish, as also the chin and fore neck, the latter separated by a band of stiff tawny buff feathers mesially streaked with blackish; centre of the body dull ochraceous buff, with only a few faint indications of dusky brown cross lines on the abdomen and feathers of the legs, the latter being, with this exception, nearly uniform buff; all the sides of the body ochraceous buff, closely barred with dull brown, more widely on the flanks and more thickly on the breast, many of the feathers of which have patches or spots of dull brown in the centres of the feathers; under tail-coverts resembling the flanks, but more widely barred; the under wing-coverts also ochraceous buff, narrowly but distinctly barred with brown, the axillars similarly coloured, the outermost of the lower series blackish at tip and forming a wing-spot, thus resembling the inner lining of the quills, which are for the most part dull brown, barred on the inner web with fulvous; horny sheath of beak and toes black, the cere somewhat greyer; iris reddish yellow. Total length 19·5 inches, wing 14·5, tail 8·5, tarsus 2·55.

Adult female (Peru). Total length 20 inches, wing 14·85, tail 9·2, tarsus, 2·65.

Hab. The southern portion of S. America, ranging on the western side up to Peru, and on the eastern side to Southern Brazil.

a. ♀ ad. sk.	Tinta, Peru, April 23rd, 1869.	Henry Whitely, Esq. [C.].
b. Ad. st.	Chili.	Mr. Brydges [C.].
c. ♂ ad. sk.	Prov. Santiago, Chili, May 1872.	E. C. Reed, Esq. [C.].
d, e. Ad. st.	Straits of Magellan.	The Admiralty [P.].
f. Skeleton.	Chili.	Purchased.

7. Bubo maculosus*.

Strix maculosa, *Vieill. N. Dict. d'Hist. Nat.* vii. p. 44; *id. et Oud. Gal. Ois.* i. pl. 23 bis; *Grill, Zool. Anteckn.* p. 47.

---

* Bubo maculosus.

Le Choucouhou, *Levaill. Ois. d'Afr.* i. pl. 39.

Strix nisuella, *Daud. Traité,* ii. p. 187 (*ex Levaill.*).

M. Jules Verreaux, in a note communicated to Dr. Hartlaub (Orn. W.-Afr. p. 19) says that this species is the same as *B. maculosus* (V.); and Professor Sundevall, in his critique on Levaillant (p. 27), agrees with this, and says he has seen the type in the Leyden Museum. Prof. Schlegel, however, does not mention this in his 'Catalogue;' and I do not consider the plate in Levaillant's work sufficiently decisive to upset the better-known title of Vieillot's.

Strix africana, *Temm. Pl. Col.* ii. pl. 50.
Otus africanus, *Steph. Gen. Zool.* xiii. pt. 2, p. 58; *Cuv. Règn. An.* 1829, p. 341.
Bubo africanus, *Boie, Isis,* 1826, p. 976; *Kaup, Tr. Z. S.* iv. p. 240, et *Contr. Orn.* 1852, p. 116.
Asio africana, *Less. Man. d'Orn.* i. p. 116.
Otus maculosus, *Less. Traité,* p. 109; *Smith, S. Afr. Q. Journ.* ii. p. 315; *Gray, Gen. B.* i. p. 40; *Cass. Cat. Strigidæ Phil. Mus.* p. 8.
Bubo maculosus, *Bp. Consp.* i. p. 49; *Strickl. Orn. Syn.* p. 215; *Gurney, Ibis,* 1859, p. 242; *Schl. Mus. P.-B.* Oti, p. 14; *Sclater, Ibis,* 1864, p. 307; *Pelz. Reis. Novara, Vög.* p. 26, pl. 6. fig. 1; *Layard, B. S. Afr.* p. 40; *Gray, Hand-l. B.* i. p. 44; *Bocage, Jorn. Lisb.* 1870, p. 338, 1871, p. 4, 1873, p. 19; *Gurney, P. Z. S.* 1871, p. 148, et in *Anderss. B. Dam. Ld.* p. 42; *Sharpe, ed. Layard, B. S. Afr.* p. 73.
Nisuella maculosa, *Bp. Rev. et Mag. de Zool.* 1854, p. 542.

Nestling. Covered with long and fluffy feathers, of a dull brownish white colour, with indistinct bars of dull brown on the head and under surface of the body; feathers of the upper surface light brown, barred and mottled with darker brown, the wing-coverts ocellated on both webs with yellowish buff; these as well as the scapulars broadly tipped with whitish, causing a distinctly mottled appearance.

Adult female. Above dark brown, strongly mottled and vermiculated with fulvous, the markings somewhat coarser on the scapulars, which have on their outer webs large spots of white, separated from each other by a narrow brown bar, and giving a distinctly ocellated appearance; wing-coverts light brown, slightly mottled with fulvous, the median and greater series distinctly ocellated with white on outer webs, the primary coverts also marked with white externally; quills pale brown, narrowly tipped and rather broadly barred with buffy white, all the bars more or less vermiculated with brown, the bars on the outer primaries white and very broad, all of them very distinct below; upper tail-coverts brown, irregularly barred with white; tail brown, tipped with whitish, and crossed with four bands of light fulvous, all the interspaces and the tips more or less mottled with brown vermiculations, the centre feathers with no distinct cross bars, but much vermiculated with fulvous on the outer margins; crown of head and ear-tufts deep brown spotted with white, and producing a barred appearance on the latter and on the hind neck, very conspicuous on the neck, which has also large ocellated spots of white; lores white, tipped with black; sides of face and ear-coverts white, barred with brown, the latter tipped with black, forming a ring; feathers of facial ruff white at base, dark brown at tip; sides of face white, narrowly barred with brown; chin whitish; throat brown, notched on both webs with fulvous; on the lower throat a conspicuous patch of white; rest of under surface of body white barred narrowly with brown and freckled (especially on the sides of the body) with pale rufous brown; the tarsal feathers whitish, with very few indications of brown cross bars, which are, however, very distinct on the under tail-coverts; under wing-coverts white, narrowly barred with brown, as also are the axillaries; bill black; iris yellow. Total length 21 inches, culmen 1·8, auricular tufts 2·4, wing 13·4, tail 7·7, tarsus 3·1.

Obs. Some specimens are not mottled with white but with rufous buff, and have consequently a more rufescent appearance; but it appears that the species gets darker and more uniform with age, especially on the upper surface and the sides of the face.

Hab. The whole of South Africa, ranging as high as the Zambesi on the east coast, and to Benguela on the west.

a. Pull. st.	South Africa.	South-African Museum.
b. Juv. st.	South Africa.	South-African Museum.
c. Ad. st.	South Africa.	
d. Ad. st.	South Africa.	Earl of Derby [P.].
e. Ad. st.	South Africa.	Zoological Society.
f. Juv. sk.	Benguela.	J. J. Monteiro, Esq. [C.].
g, h, i. Skeletons.	South Africa.	Zoological Society.
k. Sternum.		Zoological Society.

8. **Bubo cinerascens.**

Otus africanus, *Rüpp. Neue Wirb.* p. 45 (*nec Temm.*); *id. Syst. Uebers.* p. 12.

Bubo cinerascens, *Guér. Rev. Zool.* 1843, p. 321; *Gray, Gen. B.* i. p. 37; *Cass. Cat. Strigidæ,* p. 8; *Bp. Consp.* i. p. 49, et *Rev. et Mag. de Zool.* 1854, p. 542; *Strickl. Orn. Syn.* p. 217; *Heugl. in Peterm. Mitth.* 1861, p. 20; *Schl. Mus. P.-B.* Oti, p. 15; *Blanf. Geol. & Zool. Abyss.* p. 302; *Antin. & Salvad. Viagg. Bogos,* p. 39.

Otus cinerascens, *Des Murs, in Lefebvr. Voy. Abyss. Zool.* p. 74, pl. 4.

Bubo cinerascens, *Guér. & Lafr. in Ferret et Galin. Voy. Abyss. Zool.* iii. p. 181, pl. 2.

Otus maculosus, *Horsf. & Moore, Cat. B. Mus. E. I. Co.* i. p. 80; *Heugl. J. f. O.* 1863, p. 13.

Ægolius africanus, *Heugl. Syst. Uebers.* p. 14.

Bubo maculosus (*nec V.*), *Hartl. Orn. Westafr.* p. 19; *id. J. f. O.* 1861, p. 101; *Brehm, Reis. n. Habesch,* pp. 208, 270; *Heugl. Orn. N.O.-Afr.* p. 114; *Finsch & Hartl. Vög. Ostafr.* p. 103.

Bubo africanus, *Scl. Rep. Speke coll. Somali Ld.* p. 9; *Speke, Ibis,* 1860, p. 244; *Heugl. in Peterm. Mitth.* 1861, p. 20.

Otus melanotus, *Heugl. J. f. O.* 1863, p. 12.

Bubo selenotis, *Heugl. J. f. O.* 1867, p. 293 (*ex Pr. Würt. MS.*).

Huhua cinerascens, *Gurney, Ibis,* 1868, p. 149; *id. in Anderss. B. Dam. Ld.* p. 42.

Adult. Above brown, mottled and transversely vermiculated with darker brown, and crossed with zigzag lines of whitish on the scapulars and inner secondaries; the outer scapulars white, narrowly barred across with brown, forming a very conspicuous shoulder-patch, the bases of the feathers on this part inclining to orange-buff; wing-coverts brown, with less numerous zigzag bars, the greater coverts mottled with white, the median ones with a large round spot of white at the tips, forming an indistinct band; quills brown, barred across with light fulvous brown, very distinct on the inner web, all the bars more or less mottled with brown vermiculations, especially near the tips, which incline to whitish; tail brown, whitish at tips, and crossed with six fulvescent bands, all of which are mottled with brown, and incline more or less to white on the outermost feather; the under surface of the wings and tail ashy brown, crossed with broad and distinct bands of yellowish near the base, whitish near the tip; crown of head brown, spotted on either web with white inclining to irregular bars on the ear-tufts; the nape and hind neck, forming the wig, whitish, inclining more or less to yellowish, and barred with light brown; lores and feathers over the eye pure white, the sides of the face dusky brown, indistinctly barred with white, inclining to brown at the tips of the ear-coverts, and forming a circular frill; facial ruff white, narrowly barred with brown, and touched here and there with orange-buff; under surface of body white, the chin and a patch on the lower throat pure white, the remainder of the body barred with brown, more broadly on the chest, which therefore assumes a dusky appearance; the flanks, sides of the body, and under tail-coverts washed with yellowish; under wing-coverts white, narrowly barred with brown, more coarsely on the outermost, the lower series white broadly barred with dark brown, thus resembling the inner lining of the wing; iris dark, clear blue; eyelids blood-red; feet pale leaden; bill blackish horn-colour, paler at the tip of upper mandible. Total length 19 inches, culmen 1·9, wing 12·8, tail 7·8, tarsus 2·4, ear-tufts 1·5.

Female. According to Von Heuglin, the female is smaller than the male.

Young. This plumage is described by Von Heuglin as being more washed with rufous, vermiculated and waved on the upper surface with darker grey; head and ear-tufts smoky, varied with pale rufous; iris rufous umber.

Hab. N.E. Africa.

a, b. Ad. sk.	Bogos Land.	Herr Esler [C.].
c. Ad. st.		Purchased.
d. Ad. st.	River Niger.	Purchased.

9. Bubo lacteus.

Strix lactea, *Temm. Pl. Col.* ii. pl. 4; *Chapm. Trav. S. Afr.* ii. p. 392.
Bubo lacteus, *Steph. Gen. Zool.* xiii. pt. 2, p. 55; *Cuv. Règne An.* i. p. 344 (1829); *Smith, S. Afr.-Journ.* ii. p. 317; *Rüpp. Syst. Uebers.* p. 12; *Gray, Gen. B.* i. p. 37; *Cass. Cat. Strigidæ Philad. Mus.* p. 17; *Bp. Consp.* i. p. 49; *Horsf. & Moore, Cat. B. Mus. E. I. Co.* i. p. 73; *Strickl. Orn. Syn.* p. 217; *Hartl. Orn. W.-Afr.* p. 19; *Schl. Mus. P.-B.* Oti, p. 11; *Gurn. P. Z. S.* 1864, p. 2; *Heugl. Orn. N.O.-Afr.* i. p. 112; *Gray, Hand-l. B.* i. p. 44; *Blanf. Geol. & Zool. Abyss.* p. 302; *Finsch & Hartl. Vög. Ostafr.* p. 101; *Gurney, P. Z. S.* 1871, p. 148; *Antin. & Salvad. Viagg. Bogos,* p. 28; *Sharpe ed. Layard B. S. Afr.* p. 71.
Scops lactea, *Less. Man. d'Orn.* i. p. 118.
Otus lacteus, *Less. Traité,* p. 111.

Buto sultaneus, *Less. Traité*, p. 115; *Pucher. Rev. et Mag. de Zool.* 1849, p. 27; *Gray, Gen. B.* i. p. 37.
Bubo verreauxii, *Bp. Consp.* i. p. 49; *Layard, B. S. Afr.* p. 38; *Gray, Hand-l. B.* i. p. 44; *Buckley, Ibis*, 1874, p. 362.
Urrua lacteus, *Kaup, Tr. Z. S.* iv. p. 244, et *Contr. Orn.* 1852, p. 117.
Huhua verreauxi, *Gurney, Ibis*, 1868, p. 147; *id. in Anderss. B. Dam. Ld.* p. 41.

Adult female. General colour greyish brown; the back brown, scantily freckled with greyish white, more coarsely on the scapulars, the outermost of which are more or less white externally, more or less vermiculated with brown zigzag lines; head and neck lighter than the back, the crown crossed with broader lines of white; ear-tufts 2·5 inches long, brown freckled with white lines, the tip uniform brown; wing-coverts coloured like the back, and freckled in the same way, many of the feathers uniform brown on the inner web, the greater series white at the tips of the feathers, forming an irregular bar down the wing, these spots slightly varied with brown frecklings; primary coverts nearly uniform dark brown, with faintly indicated bars of lighter brown; quills brown, tipped with whitish and barred with darker brown, the lighter interspaces slightly freckled with brown, especially on the inner web of the secondaries, which incline to whitish, the innermost secondaries brown freckled with whitish lines, and resembling the back; upper tail-coverts rather lighter than the back, the white vermiculations rather broader; tail-feathers ashy brown, tipped with whitish and crossed with four bars of darker brown, the light interspaces being freckled with greyish white on the centre feathers, and whitish on the inner webs of the outer ones; lores whitish, obscured by blackish bristles; sides of face white, the hinder margin of the ear-coverts black, forming a lateral line down the sides of the head, behind which is an indistinct ruff of white feathers, barred narrowly with greyish black, resembling the plumes of the sides of the neck; chin and throat white, as well as the cheeks, these feathers broadly streaked with black, the lower throat narrowly barred with greyish black; fore neck white, with scarcely any cross lines; rest of under surface of body ashy grey, everywhere vermiculated with whitish, more thickly on the chest, the breast-feathers as well as those of the flanks and sides of the body white, barred across with brown, the lower flank-feathers widely barred; centre of abdomen white; leg-feathers and under tail-coverts white, with more or less indistinct zigzag cross lines of greyish brown; under wing-coverts whitish, with greyish transvermiculations, exactly like the breast, the lower series black, white at base, forming a wing-bar, and resembling the inner lining of the wing, which is blackish brown, broadly barred with white on the inner web; cere bluish grey; bill pale ash-colour; iris very dark hazel; eyelid a conspicuous pink.

Obs. Specimens of this Owl from South Africa are often regarded as belonging to a different species, which is called *Bubo*, or *Huhua, verreauxi*. Mr. Gurney has argued in favour of their recognition as distinct species; and he gives the following measurements in support of his plea (Ibis, 1868, p. 148):—

	Ulna.	Carpus to tip of 4th primary.	Tarsus.	Tail.
	in.	in.	in.	in.
B. verreauxi ♀	8·0	18·5	3·0	11·0
B. lactea ♀	7·5	16·0	2·5	9·5

I cannot, however, see more in these differences than the usual slight increase of size exhibited by South-African birds over examples of the same species from more northern localities. I therefore unite these two titles under one species, and give the measurements of the series now before me:—

		Total length.	Wing.	Tail.	Tarsus.
		in.	in.	in.	in.
a. [♀] ad.	Bogos Land (*Esler*)	28	19·3	11·0	3·3
b. [♂] ad.	Bogos Land (*Esler*)	—	17·6*	[missing]	2·9
c. [♂] ad.	Sooroo (*Blanford*)	24	18·6	10·0	2·9
d. ♂ ad.	Bissao (*Beaudouin*)	26	17·5	9·5	2·85
e. Ad.	Zambesi (*Livingstone*) ..	22	18·6	9·8	2·95
f. ♂ ad.	Cape colony (*Verreaux*) .	24	18·0	10·5	2·85
g. Ad.	Cape colony (*Verreaux*) .	24	19·3	10·5	3·0

Hab. The whole of Africa, excepting the forest districts of the west coast.

a. ♂ ad. sk.	Bissao (*Beaudouin*).	R. B. Sharpe, Esq. [P.].
b. ♂ ad. sk.	Bogos Land.	Herr Esler [C.].
c. ♀ ad. sk.	Bogos Land.	Herr Esler [C.].
d. ♂ ad. sk.	Sooroo.	W. T. Blanford, Esq. [C.].
e. Ad. sk.	Zambesi.	Dr. Livingstone [C.].
f, g. Ad. st.	Cape colony.	M. J. Verreaux [C.].

10. **Bubo coromandus.**

Petit Hibou de Coromandel, *Sonn. Voy. Ind.* ii. p. 186.
Coromandel Eared Owl, *Lath. Gen. Syn. Suppl.* p. 44.
Strix coromanda, *Lath. Ind. Orn.* i. p. 53.
Otus coromandus, *Steph. Gen. Zool.* xiii. pt. 2, p. 58.
Bubo coromanda, *Gray, Gen. B.* i. p. 37; *Cass. Cat. Strigidæ Philad. Mus.* p. 9; *Bp. Consp.* i. p. 49; *Horsf. & Moore, Cat. B. Mus. E. I. Co.* i. p. 75; *Schl. Mus. P.-B.* Oti, p. 9; *Gray, Hand-l. B.* i. p. 44; *Hume, Str. F.* i. p. 164.
Urrua coromanda, *Hodgs. J. A. S. B.* vi. p. 373; *Kaup, Tr. Z. S.* iv. p. 243; *id. Contr. Orn.* 1852, p. 117; *Bp. Rev. et Mag. de Zool.* 1854, p. 542; *Strickl. Orn. Syn.* p. 218; *Jerd. B. Ind.* i. p. 130; *id. Ibis*, 1871, p. 345; *Hume, Nests and Eggs Ind. B.* p. 63.
Mesomorpha coromanda, *Hodgs. J. A. S. B.* x. p. 28.
Urrua umbrata, *Blyth, J. A. S. B.* xiv. p. 180.
Bubo umbratus, *Blyth, Cat. B. Mus. A. S. B.* p. 35.
Ascalaphia coromanda, *Blyth, Ibis*, 1866, p. 253; *Hume, Rough Notes*, ii. p. 371.
Bubo sinensis, *Heude, Ann. Sci. Nat.* (5) xx. art. 2; *Sharpe, Ibis*, 1875, p. 265.

Adult female. General colour greyish brown, with dark brown centres to the feathers, which are obscurely vermiculated with darker brown; head rather lighter than the back, the ground-

* This bird is moulting, and perhaps has not the wing perfectly grown.

colour being paler and more hoary; ear-tufts nearly uniform dark brown, lighter at the base and vermiculated with greyish on the inner web; scapulars rather browner than the back, the mesial brown streaks being very distinct, the outermost with large yellowish white markings, these more or less mottled with brown, but forming a tolerably distinct shoulder-patch; wing-coverts resembling the back and vermiculated in exactly the same manner, the centres to the feathers darker brown, the outer median and greater coverts with distinct rounded spots of yellowish white; primary coverts ashy brown, broadly banded with darker brown; quills ashy brown, the primaries tinged with fulvous, all crossed with broad bands of darker brown, the light interspaces minutely vermiculated and dotted with brown, the innermost secondaries thickly vermiculated and resembling the back; lower back, rump, and upper tail-coverts ashy brown, numerously crossed with rather narrow fulvous cross bars, the latter vermiculated with fulvous; tail dark brown, broadly tipped with buffy white, and crossed with four other broad bands of deeper buff, these bands very distinct on all excepting the two centre feathers, which have the light bands very much obscured with brownish vermiculations, even extending to the apical whitish band; lores and feathers in front of the eye white, with hair-like black shaft-streaks; sides of face greyish brown, lighter behind the eye, all the feathers with narrow mesial streaks of blackish, the hinder margin of the ear-coverts dark brown, forming a longitudinal patch, which is succeeded by a tolerably well-defined ruff of fulvous plumes slightly vermiculated with greyish brown and streaked down the centre with darker brown; chin whitish; rest of under surface of body greyish white, with distinct mesial streaks of darker brown to all the feathers, which are also faintly but regularly barred across with zigzag lines of brown, the bars rather further apart on the abdomen, which consequently appears lighter than the breast; lower throat crossed with an indistinct bar of long fulvous feathers with dark brown centres in continuation of the facial ruff; bases to the feathers of the fore neck whitish, giving an indistinct appearance of a throat-patch; leg-feathers fulvescent, very narrowly barred across with zigzag lines of brown; under wing-coverts fulvescent, with dark brown shaft-lines, and crossed with irregular narrow bars of brown, broader on the outermost, the lower series yellowish at base, dark brown at tip, forming a wing-bar and resembling the inner lining of the quills, which are dark brown below, broadly barred with yellowish buff on the web; "legs and feet feathered, the latter sparsely; terminal joint bare, pale grey, with one or two large, soft, transverse scales; claws black; irides deep yellow; bill greyish white or pale lavender, with the tips and culmen pale yellowish horny" (*Hume*). Total length 21 inches, wing 16·4, tail 8·5, tarsus 2·65.

Young. Like the adult, but paler and not showing any distinct ear-tufts, the vermiculations on the back much broader and lighter, the ear-coverts rather darker than the head and greyer; under surface of body lighter than the back, less distinctly vermiculated, and the mesial shaft-streaks much narrower; thighs and leg-feathers white, narrowly streaked with small brown lines; tail-feathers brown, with five lighter buff bands in addition to the pale tip.

Hab. India: common in the N.W. provinces and extending into Sindh; eastwards extending into Tipperah and British Burmah.

a. ♀ ad. sk.	Futtegurh, Jan. 10, 1871.	Andrew Anderson, Esq. [C.].
b, c. Juv. sk.	Nepaul.	B. H. Hodgson, Esq. [P.].
d. Ad. st.	India.	Dr. Royle [P.].
e. Juv. st.	India.	Dr. Royle [P.].
f. ♀ skeleton.		Zoological Society.

11. **Bubo shelleyi.** (Plate II.)

Huhua shelleyi, *Sharpe and Ussher, Ibis*, 1872, p. 182.

Adult. Above blackish chocolate, the crown and auricular tufts (3·5 inches long) perfectly uniform; the nape with a few fulvous cross bars, for the most part concealed on this part, but broad and distinct on the sides of the neck, outer scapulars, and upper tail-coverts; wing-coverts blackish like the back, the least ones perfectly uniform, and the primary coverts nearly uniform also, excepting as regards a few concealed indications of paler brown bars on the inner web; the greater and median wing-coverts distinctly barred across with fulvous, these bars inclining to ashy brown towards the tips of the feathers; quills ashy brown, regularly barred across with darker brown, the interspaces inclining to fulvous on the inner webs of the feathers, the lower surface of the wing brown, with very broad buffy white cross bars; tail light brown, fulvescent at tip, brown at base, and crossed with seven other bands of dark brown, these bars being much broader near the base and becoming gradually narrower in extent towards the tip; forehead and feathers of the face fulvous white, distinctly barred across with dark brown, broader at the tips, and forming a distinct fringe to the facial ruff; under surface of body whitish, tinged with yellowish on the flanks, and inclining to silvery white on many of the body- and flank-feathers; all the under surface broadly and distinctly barred with dark brown, becoming much narrower on the tarsal feathers; under wing-coverts tawny rufous, barred across with blackish brown, the greater series almost entirely fulvous like the wing-lining, and having only a small brown marking towards the tip; bill yellow. Total length 25·5 inches, culmen 2·8, wing 19·4, tail 10·5, tarsus 3·5.

Hab. The forest country in the interior of Fantee.

a. Ad. sk.	Denkera.	R. B. Sharpe, Esq. [P.]

12. **Bubo nipalensis***.

Bubo nipalensis, *Hodgs. As. Res.* xix. p. 172; *Gray, Hand-l. B.* i. p. 44.

* Mr. Holdsworth, in his excellent list of Ceylonese birds, refers the species of Ceylon to *Huhua pectoralis* of Jerdon. The latter species was described from Southern India, and is stated to hold an intermediate position between *B. nipalensis* and *B. orientalis*. Not having seen examples from South India,

Huhua nipalensis, *Hodgs. J. A. S. B.* vi. p. 362; *Jerd. B. Ind.* i. p. 131; *Blyth, Ibis*, 1866, p. 254; *Hume, Rough Notes*, p. 378; *id. Str. F.* i. p. 431.
Etoglaux nipalensis, *Hodgs. J. A. S. B.* x. p. 28.
Bubo orientalis, *Blyth, Cat. B. Mus. A. S. B.* p. 34.
Bubo pectoralis, *Cass. Cat. Strigidæ Philad. Mus.* p. 9.
Urrua nipalensis, *Kaup, Contr. Orn.* 1852, p. 117; *id. Tr. Z. S.* iv. p. 244.
Ptiloskelos amherstii, *Tickell, J. A. S. B.* xxviii. p. 448.
Huhua orientalis, *Blyth, J. A. S. B.* xxviii. p. 411.
Huhua pectoralis, *Holdsw. P. Z. S.* 1872, p. 416.

Young. General colour white, somewhat washed with yellow here and there, and barred with broad markings of dark brown, sometimes heart- and sometimes spade-shaped on the back, less prominent and narrower on the head; sides of face dull white, washed with brown; under surface of body whitish, barred across with pale brown, these bars much darker on the under wing-coverts; wings and tail coloured much as in the adult, the former with many white feathers barred with brown like the back (the remains of young plumage), the tail with remains of seven fulvous bars besides the white tip; toes bare, with a few bristly feathers sprouting.

Adult. Above brown, narrowly tipped and banded across with tawny buff over the whole of the upper surface, these bars less distinct on the crown, but broader and deeper-coloured on the hind neck; the light-coloured bars broader and coarser on the scapu-

---

I cannot speak positively on the species; but if distinct, the synonymy will be as follows:—

BUBO PECTORALIS.

Huhua pectoralis, *Jerd. Madr. Journ.* x. p. 89, pl. 1; *Holdsw. P. Z. S.* 1872, p. 416.
Bubo pectoralis, *Gray, Gen. B.* i. p. 37; *Bp. Consp.* i. p. 49; *Strickl. Orn. Syn.* p. 216.

Mr. Hume, however, speaks with great certainty on the subject of the Ceylon bird:—"Whether *H. pectoralis* (Jerdon) be, or be not, a good species I cannot yet positively affirm, because I have hitherto failed to procure a Nilghiri or Malabar specimen; but the Ceylon bird, of which I have a very fine specimen now before me, is no more distinct from *Huhua nipalensis* than *Ketupa ceylonensis* of Ceylon is distinct from that of Northern India. *H. nipalensis* is, I believe, a rare bird in European collections; but I have carefully examined some ten specimens, five of which are now in my collection, and I find that, even as regards size, there is no such marked difference between the Ceylonese and Nepaulese birds. In two males before me from Nepaul, the wings vary from 16 to 16·5 inches, in the females from 17·5 to 18·5 inches; in a supposed male from Ceylon the wing is 16·75 inches.

"As regards plumage, the bird is one that varies very greatly; if there is a difference, the Nepaulese birds are rather darker; as for the so-called pectoral band, which merely depends on the breadth of the subterminal bars on the breast-feathers, this apparently depends upon age, and I have a Nepaulese bird in which the so-called pectoral band is a great deal more marked than in the very fine Ceylon specimen before me; while I have another Nepaulese bird, a young male, I believe, in which there is scarcely a trace of this band. The Ceylon bird may, I think, be referred to *H. nipalensis*; and this being the fact, I think it extremely doubtful whether the Malabar and Nilghiri birds will prove distinct."

lars, the outermost of which are broadly tipped and spotted with yellowish buff on the outer web, forming a tolerably distinct shoulder-patch; wing-coverts dark brown, narrowly barred with fulvous like the back, the median and greater series broadly tipped and externally barred with yellowish buff, producing a chequered appearance; primary coverts nearly uniform dark brown, with faint indications of lighter brown bars; quills dark brown, barred across with light brown, inclining to fulvous on the inner web of the secondaries, which have the lighter bars freckled with fulvous even on the outer web; lower back dark brown, barred and tipped with fulvous; rump dark brown, narrowly tipped and barred with the same colour, the upper tail-coverts broadly barred with fulvous nearly as widely as the intervening brown bars; tail dark brown, broadly tipped with whitish and crossed with six other bands of fulvous; face dusky brownish, with whitish shaft-streaks; feathers above the eye blackish; ear-tufts (3·1 inches long) dark brown, notched and barred with fulvous or white on the inner web; sides of neck dark brown, broadly barred with white or tawny buff; cheeks white, all the feathers stiff and mesially streaked with dark brown; chin whitish; rest of under surface of body white, washed here and there with fulvous, and regularly barred across with broad markings of dark brown, these markings becoming much smaller on the leg-feathers and much wider apart on the under tail-coverts; under wing-coverts white, broadly barred with dark brown, the lower series brown, white at base, and resembling the inner lining of the quills, which are brown, broadly barred with whitish, especially near the base of the inner webs. Total length 25 inches, wing 18·1, tail 11, tarsus 3·2.

Obs. Another specimen, also collected by Mr. Hodgson in Nepaul, is very similar to the one described, but has a few indications of dark brown bars on the ear-coverts, and has the smaller ear-tufts white, barred with dark brown, these apparently being still remains of the immature plumage. Total length 25 inches, wing 18·6, tail 11, tarsus 3·2.

Hab. Himalaya mountains, ranging eastward into Tenasserim. Ceylon. ? Southern India.

a. Juv. sk.	India (*Jerdon*).	J. Gould, Esq. [P.].
b. Juv. st.	Nepaul.	B. H. Hodgson, Esq. [P.].
c. Ad. st.	Nepaul.	B. H. Hodgson, Esq. [P.]. (Type of species.)
d. Ad. st.	Nepaul.	B. H. Hodgson, Esq. [P.].
e. Imm. st.	Nepaul.	B. H. Hodgson, Esq. [P.].
f, g. Bones (various).	Nepaul.	B. H. Hodgson, Esq. [P.].

13. Bubo orientalis*.

Strix orientalis, *Horsf. Tr. Linn. Soc.* xiii. p. 140.
Strix sumatrana, *Raffl. t. c.* p. 279.
Strix strepitans, *Temm. Pl. Col.* ii. pls. 174, 229.

* Professor Schlegel keeps the Javan bird as a distinct race from the species inhabiting the Malayan islands; and Count Salvadori follows him.

Scops strepitans, *Steph. Gen. Zool.* xiii. pt. 2. p. 54; *Less. Man. d'Orn.* i. p. 118.
Bubo strepitans, *Cuv. Règn. An.* i. p. 344.
Syrnium strepitans, *Less. Traité*, p. 114.
Bubo orientalis, *Gray, Cat. Accipitr.* 1844, p. 100; *id. Gen. B.* i. p. 37; *Cass. Cat. Strigidæ Philad. Mus.* p. 9; *Bp. Consp.* i. p. 49; *Horsf. & Moore, Cat. B. Mus. E. I. Co.* i. p. 72; *Strickl. Orn. Syn.* p. 216; *Schl. Mus. P.-B.* Oti, p. 12; *Gray, Hand-l. B.* i. p. 44.
Urrua orientalis, *Kaup, Contr. Orn.* 1852, p. 117; *id. Tr. Z. S.* iv. p. 243.
Bubo orientalis minor, *Schl. Mus. P.-B.* Oti, p. 15; *id. Revue Accipitr.* p. 5.
Huhua orientalis, *Holdsw. P. Z. S.* 1872, p. 416.
Bubo sumatranus, *Salvad. Ucc. Born.* p. 19.

Young. Above white, everywhere banded with light brown, inclining to rufous on the greater upper wing-coverts; primary coverts blackish brown, mottled with rufous at extreme tips; quills dark brown, the primaries with faint indications of lighter brown bands on inner webs, the secondaries rather lighter, barred with rufous buff, the innermost white like the back, and barred with brown in the same manner, the outermost strongly mottled with rufous or irregularly barred with the same, the lower surface of the quills light brown, barred across with ashy white; lower back, rump, and upper tail-coverts white, barred with brown like the rest of the upper surface; tail dark brown, broadly tipped with white and barred across with three or four fulvous bands, by no means equidistant or of equal breadth, the lower surface of the tail light brown, plainly barred across with white; sides of face, eyebrow, lores, and throat white; remainder of under surface white, barred across with light brown.

Adult female. Above brown, everywhere crossed with numerous zigzag bars of tawny rufous, rather broader than the back, and mottled with tawny buff on the scapulars, some of the latter having the outer webs broadly white, forming a shoulder-stripe; wing-coverts coloured like the back, the greater and median series margined with white and mottled with the same; quills dark brown, the primaries indistinctly, but the secondaries very plainly, barred with light fulvous brown, the cross bars mottled with darker brown, the innermost thickly mottled and vermiculated with tawny buff and white; tail dark brown, whitish at tip and crossed with six dull tawny bands, of unequal breadth and distance from each other, and often more or less inclining to whitish; lower surface of wings and tail ashy brown, barred across with whitish, the inner webs of the feathers white at base; crown of head brown, barred with rufous like the back, the feathers over the eye uniform and forming a distinct eyebrow, continuous with the ear-tufts, which are dark brown, the smaller feathers barred on the inner web with white or with rufous; sides of face and lores dirty white; all the feathers on the sides of the neck and throat white, barred with brown, the bars somewhat tinged with rufous on the hinder part of the latter; under surface of body white, with scattered and irregularly shaped spots and bars of brown; the chest rufous brown, barred across with fulvous or dull white; under wing-coverts white, barred with brown like the breast, the lower series white at base, ashy brown at tip, resembling the inner lining of the wing. Total length 20·5 inches, culmen 2·3, wing 13·5, tail 7·2, tarsus 2·3, ear-tufts 2·6.

Adult male. Total length 20 inches, wing 13·5, tail 7·5, tarsus 2·1.

Hab. Malacca, Java, Sumatra, Borneo, Bangka.

a. Juv. st.	Java.	
b. ♀ ad. st.	Sumatra.	Purchased.
c. ♀ ad. st.	Borneo.	
d. Juv. sk.	Borneo.	J. Gould, Esq. [P.].
e. ♂ ad. sk.	Singapore.	A. R. Wallace, Esq. [P.].

14. Bubo leucostictus.

Bubo leucostictus, *Hartl. J. f. O.* 1855, p. 354 (*ex Temm. MS. in Mus. Lugd.*); *id. Orn. W.-Afr.* p. 387; *Cass. Pr. Phil. Acad.* 1859, p. 32; *Hartl. J. f. O.* 1861, p. 101; *Schl. Mus. P.-B.* Oti, p. 16; *Gray, Hand-l. B.* i. p. 44; *Schl. Revue Accipitr.* p. 6.
Huhua leucosticta, *Sharpe, Ibis,* 1869, p. 387.

Adult. Above dark brown, the feathers with rufous cross markings, very much larger and broader on the scapulars; the hind neck paler brown, with fulvous cross bars, probably the remains of immaturity; head blackish, with small rufous spots, larger and more ovate on the nape and hind neck, a few of the scapulars with white spots on the outer web; wing-coverts resembling the back, with very few markings on the least series and bastard wing, but having much more distinct and somewhat vermiculated bands of rufous buff on the median and greater coverts; primary coverts and quills dark brown, banded across with ashy brown, all the light bars slightly vermiculated with darker brown; the secondaries pale brown, banded with vermiculated bars of blackish, always more or less mottled with fulvous; tail dark brown, tipped with white and crossed with eight other fulvous bars, all more or less mottled with brown; forehead spotted and barred with white, the ear-tufts black, barred with white on the inner web; sides of face and ear-coverts dusky fulvous, barred with black; sides of neck white, barred with blackish and inclining to brown at the tip; throat-feathers white, barred across with dark brown; the breast more or less dark brown, washed with rufous and barred across with white, the latter colour increasing on the abdomen, where the feathers are rufous at the tip, but the brown cross bars are reduced to narrow transverse lines; leg-feathers dark brown, barred with fulvous; under tail-coverts white, with narrow rufous brown bars; under wing-coverts blackish, with a few fulvous bars, some of the inner ones barred with white, the lower series uniform blackish; inner lining of quills blackish, barred on the inner web with fulvous, more distinctly towards the tips of the secondaries; bill yellowish, darker at base. Total length 16·5 inches, wing 13, tail 7·7, tarsus 1·9. (*Mus. Norvicensi.*)

Hab. West Africa: Gold Coast to Gaboon.

15. Bubo poensis.

Bubo poensis, *Fraser, P. Z. S.* 1853, p. 13.
Bubo fasciolatus, *Hartl. J. f. O.* 1855, pp. 354, 360 (*ex Temm. MS. in Mus. Lugd.*); *id. Orn. W.-Afr.* p. 18; *Schl. Mus. P.-B.* Oti, p. 15; *Sclater, P. Z. S.* 1863, p. 376, pl. xxxiii.; *Gray, Hand-l.* i. p. 44.
Huhua poensis, *Sharpe, Ibis,* 1869, p. 194, pl. 4.

Adult. Above blackish brown, transversely banded with orange-tawny, these bars being broader than the blackish cross bands, but being more concealed do not give the general tone to the upper surface; scapulars silvery white for the greater part of their external webs, forming a conspicuous shoulder-patch; least wing-coverts blackish, those running along the upper bend of the wing white, varied with a few narrow black bars and washed with tawny; greater and median wing-coverts orange-tawny, the latter more or less overshaded with blackish brown, all of them transversely banded with black, and some of the outermost of the median series externally silvery white at tip, forming an indistinct bar; primary coverts entirely blackish; quills fulvous brown, numerously banded with black, the secondaries orange-tawny banded with black, the outermost primaries very dark brown, with black cross bands; under surface of wing dusky blackish, with fulvous cross bars, the secondaries tawny buff, narrowly banded below with black; upper tail-coverts and tail dingy tawny, with numerous black cross bands, about nine in number, on the latter; ear-tufts black, tawny at base; frontal and facial feathers rufous, with black shaft-lines, the bases of all fulvous, very distinct on the ear-coverts, the hindermost of which are broadly tipped with black, as also the cheeks; under surface of body silvery white, narrowly banded across with blackish, each of these black cross bars margined by a conterminous line of orange-tawny, imparting a shade of the latter colour to the lower surface; the sides of the chest bright tawny rufous, with a few black cross bars; tarsal feathers buffy white, thickly but narrowly barred with black; under wing-coverts orange-tawny, sparsely barred with black, some of the feathers inclining to silvery white, the lower series buff at base, blackish at tip, uniform with the lower surface of the wings; ear-tufts 2·1 inches long. Total length 18·5 inches, culmen 1·9, wing 12·8, tail 6·2, tarsus 2·2.

Young. Very much lighter than the adult, the general colour above being light tawny fulvous, narrowly barred across with black, the scapulars very pale and light fulvous; underneath whitish, with very narrow cross lines of brown.

Hab. West Africa, in the forests of the Gold Coast, Fernando Po, and Gaboon.

a.	Ad. sk.	Denkera, Gold Coast.	R. B. Sharpe, Esq. [P.].
b.	Ad. sk.	Interior of Fantee.	George Lyall, Esq. [C.].
c.	Juv. st.	West Africa.	L. Fraser, Esq. [C.].
d.	Ad. sk.	Lopé, Ogowé River, Gaboon.	MM. Marche and De Compiègne, [C.].

16. Bubo philippinensis.

Scops philippensis, *Gray, Cat. Accipitr.* 1844, p. 45.
Syrnium philippense, *Gray, Cat. Accipitr.* 1848, p. 105; *Bp. Consp.* i. p. 52; *Strickl. Orn. Syn.* p. 191.
Pseudoptynx philippensis, *Kaup, Contr. Orn.* 1852, p. 117; *id. Tr. Z. S.* iv. p. 244; *Walden, Tr. Z. S.* ix. p. 144, pl. xxv. fig. 2.
Bubo philippinensis, *Schl. Mus. P.-B.* Oti, p. 14; *id. Dierent.* p. 10, fig. 3; *Wall. Ibis,* 1868, p. 25; *Gray, Hand-l. B.* i. p. 45.

Adult (type of species). Above tawny rufous, the feathers mesially dark brown, giving a closely striped appearance; the wing-coverts rather darker, some of the scapulars and greater coverts externally creamy buff, forming a kind of longitudinal bar down the latter; entire head and auricular tufts coloured like the back, but somewhat more narrowly streaked; quills dull brown, tipped with fulvous, more plainly on the secondaries, the primaries spotted on the outer and barred on the inner web with pale tawny rufous, the bars more distinct on the secondaries; tail-feathers dull brown, tipped with fulvous and crossed with a few bars of tawny buff, the middle feathers nearly uniform, with only one bar and indications of a second, the bars gradually increasing in number towards the outer tail-feather, which has five bars on the inner web, and is externally tawny buff; lores white, the anterior bristles black towards tip; sides of face tawny, with white shafts; facial disk tawny, with central streaks of brown to the feathers, which are fulvous at base; under surface of body buffy white, longitudinally streaked with dark brown centres to the feathers, each mesial streak having a narrow edging of rufous on each side, and decreasing in width towards the abdomen and under tail-coverts, the tarsal feathers entirely uniform; under wing-coverts buffy white, the outermost longitudinally spotted with dark brown, the greater series whitish at base, brown at tip, resembling the inner lining of the wing; bill horn-blue at base, yellow at tip; feet fleshy brown, claws light horn-colour. Total length 21·5 inches, auricular tufts 1·3, culmen 2·3, wing 14, tail 7, tarsus 2·7.

Hab. Philippine Islands.

a. Ad. st. Philippine Islands. Hugh Cuming, Esq. [P.]. (Type of species.)

4. SCOPS*.

		Type.
Scops, *Savigny, Descr. de l'Egypte,* p. 291 (1809)		S. giu.
Lophostrix, *Less. Compl. Buff.* i. p. 261 (1837)		S. cristatus.
Ephialtes, *Keys. & Blas. Wirb. Eur.* p. xxxiii (1840)..		S. giu.

---

* The following species is, I believe, distinct; but I am not able from memory to place it in the synoptic table. However, the following description was made by me two years ago from the type specimen:—

SCOPS NOVÆ-ZEALANDIÆ.

Scops novæ-zealandiæ, *Bp. Consp.* i. p. 47; *Strickl. Orn. Syn.* p. 202; *Schl. Mus. P.-B.* Oti, p. 27; *Finsch, J. f. O.* 1867, p. 245; *Gray, Hand-l. B.* i. p. 46; *Buller, B. N. Zeal.* p. 25; *Schl. Revue Accipitr.*

Pisorhina, *Kaup, Isis,* 1848, p. 769 .... S. menadensis.
Megascops, *Kaup, Isis,* 1848, p. 769................ S. lempiji.
Acnemis, *Kaup, Isis,* 1848, p. 769 .... S. gymnopodus.
Ptilopsis, *Kaup, Isis,* 1848, p. 769.................. S. leucotis.
Lempijius, *Bp. Rev. et Mag. de Zool.* 1854, p. 542.... S. semitorques.

Range. The whole of the Old World excepting the extreme north; absent also in Australia and Oceania, but probably present in New Zealand. The whole of the New World, excepting the extreme southern portion of South America.

Obs. Difficult to understand as all Owls are, the species of the genus *Scops* are in every way the most difficult to identify. The impossibility of procuring series of some of the species to study at the same time, the absence of information as to the sequence of plumages from the young stage to that of the adult, and the puzzling way in which some species seem to possess rufous phases, while others do not—these are all problems which time alone can solve. Thanks to many kind friends, I have been able to compare with the series in the Museum the specimens of Scops Owls in the private collections of Lord Walden, Messrs. Salvin and Godman, Mr. Swinhoe, and Captain Shelley, as well as many of the rare species contained in the Norwich Museum. These advantages have, I trust, not been without some benefit in my studies of these difficult

---

p. 13; *Finsch, J. f. O.* 1874, p. 222; *Sharpe, Voy. Ereb. & Terror, Birds,* 2nd ed. p. .
Pisorhina novæ-zealandiæ, *Bp. Rev. et Mag. de Zool.* 1854, p. 543.
Strix parvissima, *Ellm. Zoologist,* 1861, p. 272.

Adult (type of species). Upper surface brown, very minutely vermiculated with darker brown, sometimes forming an irregular spot, and varied with a few wavy lines of dull fulvous, rather brighter and more ochraceous on the outer margin of the scapulars, where the alternate bars of dark brown and fulvous are a little more regular; greater wing-coverts mottled with ashy grey, especially towards the tips; primaries darker brown, externally notched with fulvous, these marks tolerably distinct, except towards the tips, where they are obscured by greyish mottling, the secondaries more ashy brown, mottled profusely with darker brown and with indications of five lighter and more fulvous bars across them; under surface of wing uniform brown, with slight ashy mottlings towards the tip, the under wing-coverts almost entirely ochraceous, this shade extending some way along the inner web of the quills, which are also barred with greyish brown; tail brown, mottled with darker brown, with indications of seven fulvescent bands; crown and hind neck, as well as sides of face and ear-tufts, darker brown than the back; lores and ear-coverts rufescent; an indistinct superciliary line indicated by fulvous mottling, which also appears on the inner webs of the ear-tufts; round the hind neck a narrow collar produced by fulvous mottling; under surface of body rufous ochre, becoming paler and more fulvous on the abdomen, thighs, and under tail-coverts; feathers of lower breast mesially streaked with blackish brown, a little varied with whitish, and mottled with darker brown; the throat and chest blacker, much mottled with irregular wavy lines of this colour, especially on the sides of the breast. Total length 10 inches, culmen 0·85, wing 7, tail 3·8, tarsus 1·2, middle toe 6·8, ear-tufts 1·1. (*Mus. Lugd.*)

Professor Schlegel ('Revue,' p. 13) says that it is to be distinguished from all the other Scops Owls by the "absence of light spots on the scapulars. The upper parts are very dark, the top of the head being brownish black; nape ornamented with a band of rufous feathers, vermiculated with black, and followed by a black collar."

birds; and I cannot sufficiently thank those gentlemen who have aided me by the loan of specimens, especially Mr. Allan Hume, who has sent over from India beautiful series of all the Indian Scops Owls as determined by him, thus enabling me to make out several species which must otherwise have been uncharacterized in the present work.

I can hardly expect that all ornithologists will acquiesce in my views as to the subspecies or races which I have believed it to be my duty to recognize. These races *do* exist in nature, and they may be called by whatever name naturalists please, "varieties," "races," "subspecies," "climatic forms," &c.; but it has seemed to me better to keep these forms, many of which are very well characterized, distinct from one another, than to merge them all as one species, and thus to obliterate all records of *natural facts*, which are plain enough to the practised eye of the ornithologist, though difficult to describe in words. As will be seen below, the series in the Museum is very fine, and nearly all the variations mentioned in the present work can be examined by any one wishing to test the validity of my conclusions.

I have also deemed it more suitable to separate in the diagnostic tables the species of the Old World from those of the New, as the facies of the Scops Owls from the American continent is very different from that of the Old-World species, and can hardly lead to their being confounded one with another. There appear to be only two representative species, as follows:—

Old World.	New World.
Scops giu.	*Scops flammeola.*
Scops semitorques.	*Scops kennicotti.*

These are, however, very easily distinguishable as species, though belonging to the same groups.

A. *Species of the Old World.*

a. Bill more or less dusky (in skin), horn-coloured, or brownish, always marked with some dark shade.
  a'. Sides of face greyish, or brown of some shade or other, never *pure white*.
    a''. Bands on head and neck never very distinct; collar round hind neck composed of feathers mottled with buff or white, never forming an ochraceous collar as in the other group; sides of neck generally coloured like hind neck, but not forming a continuation of neck-collar.
      a'''. Legs slender.
        *a*⁴. General colour grey or brown of various shades; upper surface distinctly vermiculated.
          *a*⁵. Tarsi feathered.
            *a*⁶. Upper and under surface distinctly

streaked with black shaft-lines ear-coverts greyish, the shaft-lines only slightly produced into hair-like bristles .... *giu* and allies, pp. 47–62.

b^6. Upper and under surface sandy brown, with more or less distinct subterminal bars or spots of blackish, but no *streaks*; ear-coverts sandy buff, the shafts enormously produced into hair-like bristles.................. { *spilocephalus*, p. 63. *hambroecki*, p. 64.

b^5. Tarsi bare for half their length .... *gymnopodus*, p. 65.

b^4. General colour bright bay, with few or no vermiculations at all.

c^5. Smaller: wing 5·3 inches; ear-coverts with small hair-like bristles *sunia*, p. 67.

d^5. Larger: wing 5·9–6·1 inches; ear-coverts enormously bristled ...... *rutilus* (red phase), p. 80.

b'''. Legs very stout; sides of crown and sides of neck uniform with the rest of the upper surface, the ear-coverts similarly coloured and indistinguishable from the crown; no very perceptible collar on hind neck.

c^4. No line of black down the sides of the neck; general colour sandy rufous, with the finest possible blackish cross vermiculations .... *megalotis*, p. 69.

d^4. A distinct crescentic line of black running along the hinder margin of the ear-coverts down the sides of the neck .... *magicus* and allies, pp. 69–79.

b''. On the head and hind neck a triple light-coloured band, that on the neck very broad and conspicuous, the one on the occiput indistinct, and the one on the nape plainer; sides of neck *ochraceous* like the collar round the hind neck; crown of head with a broad lateral fulvous or ashy buff band on each side from the forehead to the auricular tufts, which usually partake of the colour of this band.

c'''. Toes *plumed* for the greater part of their length .... { *semitorques*, p. 83. *plumipes*, p. 85.

d'''. Toes bare.

e^4. Tarsi plumed to the base of the toes, the feathering running some way down the middle and outer toes, so that the junction with the tarsus is always hidden .... *lettia* and allies, pp. 85–90.

f^4. Tarsi not plumed the whole way, the junction of the toes always bare .... *lempiji* and allies, pp. 91–95.

b'. Sides of face pure unspotted white .... *leucotis*, p. 97.

b. Bill bright yellow in skin; no collar on hind neck; frontal band very broad and conspicuous;

under surface not *streaked*, only *spotted* with black; vermiculations nearly or quite obsolete both above and below; centre tail-feathers nearly uniform; upper surface marked with arrow-shaped spots of buff or white.

c'. Size very large: wing 7·15 inches; sides of neck with a broad crescentic band of black, also crossing the throat.................. *sagittatus*, p. 98.

d'. Size small; wing not exceeding 5·6 inches; some of the ruff-feathers tipped with black, but not enough to form a band.

c''. Under surface very strongly vermiculated with fine zigzag lines of blackish; upper surface of a rufous chocolate tint, spotted with rufous ochre; frontal band vot very broad .... *balli*, p. 100.

d''. Under surface without vermiculations, or at most with a very faint trace; upper surface sandy buff in tint, spotted with fulvous or white.

e'''. Spots and bars on upper surface fulvous, as also frontal band; scapular- and wing-spots fulvous .... *rufescens*, p. 102.

f'''. Spots and bars on upper surface creamy white, as well as the frontal band, wing-spots, and scapular spots .... *icterorhyncha*, p. 103.

1. Scops giu.

Le Petit Duc, *Briss. Orn.* i. p. 495, pl. xxxvii. fig. 1; *Buff. Pl. Enl.* i. pl. 436.

Strix scops, *Linn. S. N.* i. p. 132; *Meyer & Wolf, Vög. Deutschl.* pl. 10, p. 76; *Temm. Man. d'Orn.* i. p. 103; *Naum. Vög. Deutschl.* i. p. 466, pl. 43. fig. 3; *Roux, Orn. Prov.* i. p. 73, pl. 48; *Werner, Atlas, Rapaces,* pl. 42; *Nordm. in Demid. Voy.* iii. p. 110; *Yarr. Brit. B.* i. p. 113; *Sundev. Sv. Fogl.* p. 187.

Strix giu, *Scop. Ann.* i. p. 19; *Lath. Ind. Orn.* i. p. 56.

Strix zorca, *Gm. S. N.* i. p. 289 (*ex Cetti*).

Strix carnioliaca, *Gm. S. N.* i. p. 290 (*ex Scop.*).

Strix pulchella, *Gm. S. N.* i. p. 290.

Scops ephialtes, *Savign. Descr. Egypte,* p. 291; *Kaup, Tr. Z. S.* iv. p. 222; *id. Contr. Orn.* 1852, p. 110.

Bubo scops, *Boie, Isis,* 1822, p. 549; *Jenyns, Man. Brit. Vertebr.* p. 91.

Scops asio, *Steph. Gen. Zool.* xiii. pt. 2, p. 51.

Scops aldrovandi, *Flem. Brit. An.* p. 57; *Selby, Ill. Brit. Orn.* i. p. 92, pl. 22; *Gould, B. Eur.* i. pl. 41; *Bp. Comp. List B. Eur. and N. Am.* p. 6; *Macgill. Brit. B.* iii. p. 422; *Hewits. Eggs Br. B.* i. p. 54, pl. xix. fig. 1; *Degl. & Gerbe, Orn. Eur.* i. p. 142.

Scops carnioliaca, *Brehm, Vög. Deutschl.* p. 126.

Scops europæus, *Less. Traité,* p. 106.

Scops senegalensis, *Sw. Classif. B.* ii. p. 217; *id. B. W. Afr.* i. p. 127; *Cass. Cat. Strigidæ Philad. Mus.* p. 10; *Kaup, Contr. Orn.* 1852, p. 142; *Bp. Rev. et Mag. de Zool.* 1854, p. 543; *Strickl. Orn. Syn.* p. 199; *Hartl. Orn. W.-Afr.* p. 20; *id. J. f. O.* 1861, p. 101.

Scops zorca, *Swains. Classif. B.* ii. p. 217; *Bp. Rev. et Mag. de Zool.* 1854, p. 543; *Strickl. Orn. Syn.* p. 197; *Fritsch, Vög. Eur.* tab.

13. fig. 5; *Schl. Mus. P.-B.* Oti, p. 19; *Loche, Expl. Sci. Algér.* i. p. 104; *Gould, B. Gt. Br.* i. pl. xxxiii.; *Bettoni, Ucc. nidif. Lomb.* tav. 17; *Heugl. Orn. N.O.-Afr.* i. p. 117; *Gray, Hand-l. B.* i. p. 45; *Schl. Revue Accipitr.* p. 13.

Ephialtes scops, *Keys. & Blas. Wirb. Eur.* p. xxxiii; *Gray, Gen. B.* i. p. 38 (1845); *Cass. Cat. Strigidæ Philad. Mus.* p. 10; *Borrgr. Vogelf. Norddeutschl.* p. 65; *Salvad. Faun. Ital. Ucc.* p. 34.

Otus scops, *Schl. Rev. Crit.* p. 16.

Scops longipennis, *Kaup, Tr. Z. S.* iv. p. 223; *id. Contr. Orn.* 1852, p. 110.

Scops kamtschatkensis, *Bp. Rev. et Mag. de Zool.* 1854, p. 543; *Gray, Hand-l. B.* i. p. 46.

Scops minor, *Brehm, Naum.* 1855, p. 270.

Scops rupestris, *Brehm, l. c.* p. 270.

Scops rufescens, *Brehm, l. c.* p. 270.

Scops pygmæa, *Brehm, l. c.* p. 270.

Scops vera, *Finsch, J. f. O.* 1859, p. 381.

Ephialtes zorca, *Jaub. & Barth. Lap. Rich. Orn.* p. 78.

Scops giu, *Newt. Ooth. Wolley.* p. 153; *id. ed. Yarr. Brit. B.* i. p. 173; *Shelley, B. Egypt*, p. 178; *Irby, B. Gibraltar*, p. 58.

Adult male. Above grey, washed here and there with orange- or sandy buff; on the crown the latter colour prevails, all the feathers being mesially streaked with black, and varied with very minute wavy lines of brown; the ear-tufts externally sandy brown, greyish on the inner webs, and varied with cross markings of white, so that when the tufts are erect there is a continuous line of grey from above the eye on each side of the forehead, the feathers of the latter situated above the eye being greyish; across the back of the head an indistinct greyish line of feathers, many of them being spotted with white on both webs; hind neck decidedly greyer than the back, most of the feathers being greyish white with very few wavy lines of brown, the dark shaft-streaks also less distinctly indicated; general aspect of the back grey, but washed with orange-buff on the interscapulary region, and with deep sandy buff on the inner line of the scapulars, all the feathers minutely vermiculated with dark brown, the blackish shaft-streaks also strongly pronounced, and many of the dorsal and scapular feathers varied with indications of whitish bars; the outer web of the scapulars white or fulvous, broadly tipped with black, and forming a tolerably distinct shoulder-patch; wing-coverts coloured like the back, the least ones darker, the brown centres of the feathers showing more distinctly, the median and greater series largely spotted with white on the outer web; primary coverts rather darker than the others, being brown, mottled with dull sandy buff, the outer ones externally notched with fulvous; quills brown on their concealed inner webs, but greyish externally, harmonizing with the general colour of the bird, being minutely vermiculated like the back; innermost secondaries greyish, with narrow blackish shaft-streaks, the outer ones indistinctly barred with fulvous, the primaries more distinctly, these light bars on the outer web shading into pure white externally, and producing a very marked chequered appearance, all the bars less distinct at the tips of the quills, which are, for the most part, greyish, thickly vermiculated with brown wavy markings; upper tail-coverts greyish, like the back, with a few indications of whitish cross markings, the outer ones inclining to sandy buff; tail-feathers brown, coarsely mottled with brownish vermiculations, and irregularly barred with fulvous, these bars being rather more distinct and whiter on the outer feathers; lores whitish, the shafts elongated and blackish at tip; over the fore part of the eye a few white feathers terminally margined with blackish; sides of face grey, with a few indistinct dusky cross lines, the ear-coverts slightly washed with sandy colour, more particularly below the eye; behind the ear-coverts a crescentic band of black feathers extending below on to the sides of the neck, all these plumes whitish or fulvous at base; chin whitish; throat and sides of neck clear grey, finely vermiculated with brown cross lines, and slightly washed with orange-buff, the shafts of the feathers distinctly black; all the rest of the under surface greyish, strongly washed with orange-buff on parting the feathers, all of which have black shaft-lines, broader on the flanks; the breast and sides of body lighter than the chest by reason of whitish bars on most of them; the lower flanks and under tail-coverts white, crossed with one or two sandy buff bars, the narrow linear vermiculations nearly absent; leg-feathers orange-buff, with a few minute bars of brown; under wing-coverts buffy white, slightly streaked and barred with brown, the lower series ashy brown, broadly barred with whitish on the inner web, and resembling the inner lining of the quills; bill black; toes brown; claws white at base, nearly black at tip; iris yellow. Total length 7·5 inches, wing 5·9, tail 2·9, tarsus 0·95.

Adult female (Safed, Tristram). Total length 8 inches, wing 6·15, tail 2·9, tarsus 1·1.

Hab. Central and Southern Europe, ranging occasionally into the British Isles, and migrating into N.E. Africa and Senegambia.

a, b. Ad. st.	Europe.	
c. Ad. st.	Great Britain.	John Gould, Esq.
d. ♂ ad. st.	Dauphiné, France.	R. B. Sharpe, Esq. [P.].
e. Ad. sk.	Genoa, Spring of 1871 (*A. B. Brooke*).	R. B. Sharpe, Esq. [P.].
f. Ad. sk.	Southern Spain.	Lord Lilford [P.].
g. ♂ ad. sk.	Granada, Spain (*H. Saunders*).	R. B. Sharpe, Esq. [P.].
h. ♀ ad. sk.	Valencia, Spain, April 2, 1872 (*H. Saunders*).	R. B. Sharpe, Esq. [P.].
i. Ad. sk.	Algeria.	R. B. Sharpe, Esq. [P.].
k, l, m. Ad. sk.	Tangier.	M. Olcese [C.].
n. Juv. sk.	Tangier.	M. Olcese [C.].
o. ♀ ad. sk.	Belgrade, Turkey, March 29, 1873 (*T. Robson*).	R. B. Sharpe, Esq. [P.].
p. Ad. sk.	Asia Minor (*Robson*).	J. Gould, Esq. [P.].
q. ♀ ad. sk.	Safed, Palestine.	Canon Tristram [C.].
r. Ad. sk.	M'bao, Senegal.	M. A. Marche [C.].
s. Juv. sk.	River Gambia.	H. Whitely, Esq. [P.].

Obs. Amongst specimens from various parts of Europe I perceive but little variation in colour; and the following are the measurements of the series now before me.

		Total length. in.	Wing. in.	Tail. in.	Tarsus. in.
a. ♂ ad.	Dauphiné, France	8·0	6·15	2·9	0·95
b. Ad.	France (*Mus. Cantabr.*).	8·0	6·1	2·9	1·15
c. Ad.	Spain (*Lilford*)	8·5	6·0	2·9	1·0
d. ♂ ad.	Granada (*H. Saunders*)..	7·5	5·8	2·9	1·05
e. Ad.	Morocco (*Olcese*)	7·0	5·75	2·6	1·05
f. Juv.	Morocco (*Olcese*)	7·0	5·85	2·75	1·05
g. Ad.	Morocco (*Olcese*)	7·0	5·9	2·5	1·1
h. Ad.	Morocco (*Olcese*)	7·0	5·55	2·7	1·1
i. Ad.	Asia Minor (*Robson*) ..	7·0	5·9	2·6	1·1
k. ♀ ad.	Safed (*Tristram*)	8·0	6·15	2·9	1·1

The Morocco birds are rather smaller than the French, and differ from the Dauphiné example described in having much broader dark central streaks to the breast-feathers; but the Spanish birds agree better with the North-African ones. One of the latter seems to be a younger bird, and is remarkable for the coarseness of its markings both above and below, and for the distinctness of the light bands on the tail, six being distinguishable on the centre rectrices besides the terminal band of buff; the same number can be traced on the tails of the adult ones, but they appear to get dissolved with age. Old birds may also apparently be detected by the great extent of white on the under surface, occupying the fore neck and chest, which are then very slightly barred with zigzag brown lines; the hind neck also is broadly barred with white. The oldest example before me appears to be that from Asia Minor, which exhibits in the strongest degree the above-mentioned characters, and has only five fulvous bars on the middle tail-feathers, besides the light terminal bar. The Dauphiné specimen has only four caudal bands distinguishable on these feathers, and is probably of about the same age as the one from Asia Minor. Both are males: and the small number of tail-bands is perhaps a character of this sex; for the male from Granada has only five, while the Safed bird has seven.

Great variation exists in the amount of rufous shading on specimens from the same locality; but I have never met with one which could be said to constitute a regular rufous phase, such as we know to exist in *Scops asio* of America. A French specimen collected by Mr. Swainson is in the Cambridge Museum, and nearly equals in intensity the type of *Scops kamtschatkensis* in the Paris Museum. I carefully compared this with ordinary European examples, and can affirm with certainty that it is nothing but a more than ordinarily rufous *Scops giu*; no authority for its Kamtschatkan habitat exists beyond the assertion of the late M. Jules Verreaux. In order to make sure of the identity of these birds, I give the description which I made from the Paris specimen.

Adult female (Kamtschatka; obtained from Verreaux, 1847). General colour orange-rufous, varied with blackish vermiculations, some of the feathers also having a central arrow-shaped streak of black, these black markings much more distinct on the crown, but less so on the ear-coverts (0·9 inch long), the latter being paler orange than the rest of the upper surface, and for the most part uniform on the inner web; facial aspect grey, washed with orange-rufous above the eye and behind the ear-coverts, vermiculated cross-wise with brown; the lores white with blackish shafts washed with orange; under surface of body whitish, the chest and sides washed with orange-rufous, all the feathers minutely vermiculated with ashy brown, a few of the chest-feathers with broad mesial streaks of black, narrower on the flanks, the markings disappearing on the under tail-coverts, which are almost entirely white, with only a few transverse vermiculations of greyish brown; legs whitish, inclining to orange at the summit of the tarsus, which has a few transverse markings of brown; under wing-coverts yellowish buff, as also the axillaries, the outermost washed with orange and spotted with brown, the lower series fulvous at base, greyish brown at tip, resembling the inner lining of the wing, which is ashy brown, barred with buffy white on the inner webs of the quills; upper wing-coverts a little duller than the back, the transverse vermiculations very indistinct, but the central streaks plain, the greater coverts with a few white spots on the outer web, in this respect resembling the scapulars, which are more coarsely transvermiculated with brown than the back, and have white spots on the outer web, which is otherwise orange at base; primary coverts dull orange-brown, vermiculated with greyish brown; quills brown, orange on the outer web, and irregularly vermiculated with greyish brown; the primaries barred with lighter brown on the inner web, this bar being represented on the outer one by a bar of white more or less tinged with orange, the interspaces orange-brown, with darker brown vermiculations; the secondaries decidedly greyer than the primaries and more minutely vermiculated; upper tail-coverts resembling the back; tail darker brown, barred with orange, of which about five bands can be distinguished, becoming more distinct towards the tip, where they are obscured by the greyish vermiculations of the interspaces; bill horn-colour. Total length 6·7 inches, wing 5·8, tail 3, tarsus 1·05.

In the Paris Museum there is another specimen marked as from Kamtschatka, and absolutely identical with ordinary European specimens.

Another supposed closely allied species is the *Scops senegalensis* of Swainson, which must next be inquired into. On this bird Dr. Finsch (Tr. Z. S. vii. p. 210) writes:—"The African *Scops* agrees in every respect with our European *S. zorca* [=*giu* of this work], except in the shorter wings, this being, as far as I could find out, a permanent character, common also to specimens from the west and north-east as well as to those from the south (*Strix latipennis*, Licht.). I do not esitate to regard this peculiarity as of specific value." I have before me specimens from Africa, the measurements of which I give, in order to test the above-mentioned character:—

			Total length. in.	Wing. in.	Tail. in.	Tarsus. in.
a.	Ad.	River Gambia (*Rendall*) ..	7·5	5·35	2·6	0·95
b.	Ad.	River Gambia (*Rendall*) ..	7·5	5·45	2·6	0·95
c.	Ad.	M'bao, Senegal (*Marche*) ..	7·5	5·35	2·7	1·0
d.	Ad.	Damara (*Andersson*)	6·5	5·0	2·6	0·9
e.	Ad.	Benguela (*Monteiro*)	7·5	5·3	2·6	0·95
f.	Ad.	Cape of Good Hope	7·0	5·35	2·6	0·9
g.	Ad.	Cape of Good Hope	7·5	5·4	2·8	0·9

It will be seen by the above measurements that Dr. Finsch was quite right in his facts about the smaller size of the African bird; but the plumage of the Senegal *Scops* is not distinguishable from that of European birds, and I cannot allow the distinctness of *S. senegalensis* as constituting any thing more than a slightly smaller race of *S. giu*. I am indebted to the kindness of Mr. Salvin for submitting to me the type of *Scops senegalensis* of Swainson, which is the first bird mentioned in the above table of measurements.

The South-African birds, however, are different from *S. giu*, being so extremely dark in colour, as well as of such a small size; but they can hardly be regarded as constituting more than a dark race, though the characters appear pretty constant. The true *S. capensis* occurs in N.E. Africa, as Lord Walden has kindly lent me the example collected by Mr. Jesse during the Abyssinian expedition, and referred to by Dr. Finsch (*l. c.*). Dr. von Heuglin also considers that both forms occur in N.E. Africa, *S. giu* being a winter visitor to Egypt, while *S. capensis* he believes to be a resident in the warmer parts of N.E. Africa (*cf.* Heugl. Orn. N.O.-Afr. i. p. 118).

Subsp. *a*. **Scops capensis***. (Plate III. fig. 1.)

Le Scops, *Levaill. Ois. d'Afr.* i. p. 162.
Scops europæus, *Smith, S.-Afr. Q. Journ.* 2nd ser. no. 4, pt. 1, p. 314.
Scops capensis, *Smith, l. c.* p. 314; *Bp. Rev. et Mag. de Zool.* 1854, p. 543; *Bocage, Jorn. Lisb.* ii. pp. 38, 338; *Gray, Hand-l. B.* i. p. 45;

---

* Mr. Cassin described a Scops Owl from Western Africa, characterized as follows:—

Scops hendersoni.

Ephialtes hendersoni, *Cass. Pr. Phil. Acad.* 1852, p. 186.
Scops hendersoni, *Hartl. Orn. W.-Afr.* p. 20; *Gray, Hand-l. B.* i. p. 46.

"Colours generally very similar to those of *E. capensis* (Smith) and *E. senegalensis* (Sw.), but darker and with the fine lines and points of brown throughout the plumage much more numerous and distinct. Entire plumage above cinereous, mixed on the neck, back, and rump with bright fulvous, every feather longitudinally striped and with numerous very irregular transverse lines and points of brownish black, most distinct on the head and back; frontal feathers nearly white, tufts and face very pale cinereous, striped longitudinally and minutely lined with dark brown; plumage below variegated with white and brown, with a tinge of fulvous, every feather having stripes and minute lines and points, as on the back, but less numerous, with the white predominating and nearly pure on the lower abdomen and under tail-coverts; quills brown,

Souza, Cat. Lisb. Mus. p. 59; *Gurney in Anderss. B. Dam. Ld.* p. 38; *Schl. Revue Accipitr.* p. 14; *Sharpe, ed. Layard, B. of S. Afr.* p. 75.
Scops latipennis, *Kaup, Tr. Z. S.* iv. p. 223; *id. Contr. Orn.* 1852, p. 110; *Gurney, P. Z. S.* 1871, p. 148.
Ephialtes latipennis, *Licht. Nomencl. Av.* p. 7.
Scops senegalensis (*nec Swains.*), *Strickl. & Sclater, Contr. Orn.* 1852, p. 142; *Mont. P. Z. S.* 1865, p. 91; *Bocage, Jorn. Lisb.* i. p. 132; *Blanf. Geol. & Zool. Abyss.* p. 303; *Finsch, Tr. Z. S.* vii. p. 210; *Antin. & Salvad. Viagg. Bogos*, p. 40.
Scops fazoglensis, *Würt. Naum.* 1857, p. 432; *Heugl. J. f. O.* 1867, p. 293.
Ephialtes capensis, *Gurney, Ibis*, 1859, p. 242.
Scops zorca africanus, *Schl. Mus. P.-B.* Oti, p. 20; *Heugl. Orn. N.O.-Afr.* i. p. 117.
Ephialtes senegalensis, *Layard, B. S. Afr.* p. 41.
Strix scops, *Chapm. Trav. S. Afr.* ii. App. p. 393.

Similar to *Scops giu*, but rather smaller and with shorter wings; general colour constantly of a dark grey; white spots on scapulars and wing-coverts very large and distinct; grey shade on under surface also very much darker, and not so much white apparent; tarsus scarcely feathered so far towards the joints of the toes. For measurements, *vide suprà*. "Bill and toes of a bluish horn-colour; iris bright yellow" (*Andersson*).

Hab. South Africa and North-eastern Africa, in both of which countries it is resident.

a. Ad. st.	Cape of Good Hope.	Sir A. Smith [C.].
b. Ad. st.	Cape of Good Hope.	Sir A. Smith [C.].
c. Ad. sk.	Benguela.	J. J. Monteiro, Esq. [C.]

Subsp. β. **Scops pennatus.**

Scops pennata, *Hodg. J. A. S. B.* vi. p. 369 (pt.); *Blyth, J. A. S. B.* xiv. p. 183; *Kaup, Contr. Orn.* 1852, p. 110; *id. Tr. Z. S.* iv. p. 223; *Bp. Rev. et Mag. de Zool.* 1854, p. 543.
Ephialtes pennatus, *Jerd. B. Ind.* i. p. 136.

Similar to *S. giu*, but of a very dark grey colour, and distinguished by its grey ear-coverts; the ear-tufts for the most part rufous ochre, and not mottled with white as in European examples. Three Nepaulese specimens in the Museum measure as follows:—

---

with quadrangular spots of white on the outer webs; tail pale cinereous, with irregular transverse bars of brown, and exhibiting a different style of colouring from either of the species above mentioned; tarsi nearly white, obscurely spotted with brown.

"Sexes similar.

"*Hab.* Angola, at sea off Novo Redondo.

"*Obs.* Resembling *E. capensis* and *E. senegalensis*, but smaller than either, and has the bill shorter and weaker; wing with the third quill longest, but only slightly exceeding the fourth; tail short."

The characters above given do not seem to me to distinguish the species from *E. capensis*; but as Mr. Cassin had adult and young specimens of the latter bird before him at the time he wrote the description of *S. hendersoni*, I have not united the two, preferring to wait for further evidence.

		Total length. in.	Wing. in.	Tail. in.	Tarsus. in.
a. Ad.	Nepaul (*Hodgson*)	7·5	5·55	2·6	0·95
b. Ad.	Nepaul (*Hodgson*)	7·5	6	2·75	1·05
c. Ad.	Nepaul (*Hodgson*)	7·8	5·8	2·9	1

Hab. Indian peninsula.

a, b. Ad. st.	Nepaul.	B. H. Hodgson, Esq. [P.].
c. Ad. sk.	Nepaul.	B. H. Hodgson, Esq. [P.].
d. Sternum.	Nepaul.	B. H. Hodgson, Esq. [P.].

Obs. Mr. Hodgson's drawings show, even if specimens bearing his numbers and original tickets did not exist to prove the fact, that his *Scops pennatus* embraced two very distinct species, viz. the bird now named by every one *S. pennatus* and the species afterwards called *S. spilocephalus* by Blyth. *S. pennatus* is therefore a composite species, the type of which would be difficult to determine; but as Blyth has separated the spotted bird it remains to allot to the grey Scops Owl of the Himalayas the title of *pennatus*. By its dark coloration the Indian bird approaches *S. capensis*, but differs from it and from *S. giu* by the possession of a very distinct rufous phase (not, however, to be confounded with *S. sunia*). Mr. Hume has kindly lent me the rufous bird figured by him in his new work on the Ornithology of India; and I think that every one will admit that such a rufous bird was never seen in Europe.

Subsp. γ. **Scops stictonotus.** (Plate III. fig. 2.)

Scops bakkamæna, *Swinh. Ibis,* 1860, p. 47.
Scops japonicus, *Swinh. Ibis,* 1863, p. 89; *David, N. Arch. Mus. Bull.* p. 4.
Scops sunia, *Swinh. P. Z. S.* 1871, p. 343; *id. Ibis,* 1874, p. 433.

Adult. General colour above ashy brown, mottled with numerous blackish cross-vermiculations, the whole of the upper surface varied with central spots or cross bars of pale yellowish buff, much broader on the hinder neck, and forming there a tolerably distinct collar; some of the feathers of the head blackish in the centre, and nearly all of them spotted with pale buff, most of the shafts being of the latter colour; ear-tufts (0·85 inch long) marked and spotted like the crown, the bases to the feathers pale yellowish buff, extending for the greater part of the inner web; scapulars as distinctly barred with yellowish buff as the hind neck, the outer webs of external ones white, with a distinct terminal bar of blackish brown, some of the feathers slightly tinged with yellowish buff; wing-coverts rather more rufous than the back and distinctly spotted with tawny buff, the median and greater series slightly washed with grey and less thickly vermiculated with brown, the median series distinctly barred with whitish, large spots of which are apparent at the tips of the outer median and greater coverts, the spurious quills externally notched with white; primary coverts brown, vermiculated with sandy buff, and distinctly barred across with the same colour, so as to exactly resemble the secondaries, which are barred in the same manner externally, the inner webs barred with paler brown, shading off into yellowish white on the interior margin of the quills; the primaries externally notched or barred with white, giving a very distinct chequered appearance to the outer aspect of the wing, all the tips of the feathers barred with sandy buff, many of the darker bars dissolving and breaking up into mottlings and vermiculations; the innermost secondaries coloured like the back with concealed bars of yellowish buff, and barred with whitish near the tips of the feathers; tail brown, the dark bars varied with sandy buff vermiculations, all the feathers very distinctly crossed with bars of the same colour, inclining to white near the tips, about six bands being distinguishable on the centre feathers, these being a little more distinct on the exterior rectrices, the outermost of which are notched with white; loral plumes whitish at base, black at tips, the shaft-lines produced into long black hair-like bristles; feathers over the eye buffy white, with narrow subterminal zigzag lines of brown, and the inner webs also brown, so that the superciliary line is not at all distinct; behind the upper part of the eye some dusky brown feathers; ear-coverts dusky grey, clouded with blackish cross markings; behind the ear-coverts an indistinct ruff of orange-buff feathers, tipped with brownish black, the plumes adjoining the throat being also orange-buff, but finely barred with dark brown; chin-feathers buffy white; rest of under surface dull greyish, washed here and there with pale orange-buff, some of the feathers distinctly streaked down the centre with blackish brown, and all more or less distinctly waved with narrow zigzag lines of brown; the chest crossed with many distinct bars of whitish, these bars predominating on the abdomen, where the cross lines are more scanty and the whitish colour prevails, the flank-feathers having irregular cross bars of orange-buff, and being only vermiculated with brown dots at the tip, these markings being still more scanty on the under tail-coverts, which are to a great extent white; leg-feathers buffy white, streaked with brown; under wing-coverts buffy whitish, those near the edge of the wing brown, mottled and barred with sandy buff, the lower series pale ashy brown, sandy buff at base, and resembling the inner lining of the wing, which is ashy brown, very broadly and distinctly banded with yellowish white on the inner web; bill horn-brown, the lower mandible yellowish; feet probably fleshy brown in life. Total length 7·5 inches, wing 5·5, tail 2·85, tarsus 0·95.

Obs. This little Owl belongs to the same group as *S. giu*, and is quite different in appearance, without giving very tangible characters for specific separation. Its general colour is ashy brown above, without any greyish shade; the ear-coverts dusky grey; the collar round the neck is very indistinct, and is represented by certain pale buff-coloured bars or spots without any appearance of white whatsoever; on the back, however, are several very distinct spots and bars of the same pale buff-colour, giving a very marked character to these peculiarities. The specimen which I have described is said to be from China, and presented to the Museum by Mr. J. R.

Reeves. The same subspecies is found in Camboja, and even extends to Nepaul, specimens from both these localities being in the British Museum.

The typical specimen is said to be from China, whence also I have seen two other specimens in Mr. Swinhoe's collection, which I am inclined to refer to *S. stictonotus*. One from Amoy (*S. bakkamœna*, Swinhoe, Ibis, 1860, p. 47) is certainly the same; and another from Chefoo (*S. sunia*, Swinhoe, Ibis, 1874, p. 433) must also be the same, I think, although there is rather more white on the eyebrow and under surface of the body, and the ear-coverts are paler grey than in the typical bird. The fulvous collar, however, is present, and the bars on the dorsal plumage are very distinct. I give measurements of the other specimens examined by me:—

	Total length.	Wing.	Tail.	Tarsus.
	in.	in.	in.	in.
a. ♂ ad. Chefoo (*Swinhoe*)	7·5	5·75	2·85	1 05
b. Ad. Amoy (*Swinhoe*)	7·5	5·5	2·7	1
c. Ad. Camboja (*Mouhot*)	8	5·75	2·8	1·1
d. Ad. Nepaul (*Hodgson*)	8	5·55	2·8	0·95

Hab. China, ranging westwards to Siam and the eastern Himalayas.

a. Ad. st.	China.	J. R. Reeves, Esq. [P.].
b. Ad. sk.	Camboja.	M. Mouhot [C.].
c. Ad. sk.	Nepaul.	B. H. Hodgson, Esq. [P.].

Subsp. δ. **Scops japonicus.**

Otus scops japonicus, *Schl. Faun. Japon. Aves*, p. 27, pl. 9.
Scops japonicus, *Bp. Rev. et Mag. de Zool.* 1854, p. 543; *Gray, Hand-l. B.* i. p. 45.

Adult. Above brown, all the feathers coarsely vermiculated with zigzag black lines, the interspaces often fulvescent and producing a strongly mottled appearance, the centres of the feathers being blackish; scapulars coloured like the back, the outermost white or slightly washed with buff, and mottled with some coarsely marked spots of black near the tip, forming an ill-defined terminal bar; head rather darker than the back, and washed with dull orange-buff, most of the feathers being laterally spotted with this colour and conspicuously black in the centre; on the nape a faintly indicated greyish band, some of the feathers being subterminally barred with white, the hind neck darker, but the interscapulary region varied with orange-buff bases to the feathers, slightly washed with grey, the plumes also much mottled with spots of white; ear-tufts 1·1 inch long, orange-buff for the greater part of the inner web and at the base, the outer webs obscured with blackish mottlings and zigzag lines; wing-coverts coloured and vermiculated exactly like the back, the greater series shaded with grey, but the rest of the coverts rather more dusky than the back, the spurious quills externally barred but not distinctly notched with sandy rufous; primary coverts dark brown, vermiculated with sandy rufous, forming a couple of bars towards the tips of the feathers; quills dark brown, barred with sandy rufous on both webs, these bars not being strictly conterminous, and always more or less mottled with dusky vermiculations, the light bars on the outer webs clearer and inclining to whitish on the primaries, giving a chequered appearance to the outer aspect of the wing, the innermost secondaries vermiculated and exactly resembling the back; tail dull brown, very coarsely mottled with blackish zig-zag lines, all the tips being vermiculated in this manner, the outer feathers crossed with about eight tolerably distinct lighter bars of sandy rufous, these cross bars very indistinct on the centre feathers, where not more than remains of seven can be traced; lores dusky white, barred across with black, the elongated shafts being black at the tips; above the fore part of the eye a few white feathers narrowly barred with blackish; sides of face ashy grey, indistinctly barred across with blackish; behind the ear-coverts an indistinct ruff of orange-buff plumes, broadly tipped with blackish and subterminally vermiculated with greyish; under surface of body hoary, the chin-feathers whitish, the breast and flanks slightly varied with orange-buff bases to the feathers; the plumes of the throat and chest distinctly streaked down the middle with black, these streaks much narrower on the lower breast and flanks, all the feathers more or less coarsely vermiculated with zigzag cross lines of dark brown, and irregularly spotted or barred with greyish white, these white bars increasing much in width on the flanks and abdomen; under tail-coverts white, slightly washed with rufous, and minutely freckled near the tip with brown vermiculations; leg-feathers tawny, with small cross bars of dark brown; under wing-coverts whitish, excepting those near the edge of the wing, which are rufescent, slightly marked with brown; the greater series ashy brown, yellowish white at base, resembling the inner lining of the quills, which are ashy brown, barred with fulvous near the base and with sandy buff towards the tips, which are thickly vermiculated with sandy buff. Total length 8·5 inches, wing 6·75, tail 3, tarsus 1·15.

Hab. Japan.

a. Ad. st.	Nagasaki, Japan.	Capt. St. John [C.].

Obs. Of the same group as *S. giu*, but a much larger and browner bird, so that, with the faintly indicated occipital bars and the buff variations of the hind neck, it somewhat approaches the *S. lettia* section. The entire aspect of the bird is much darker than *S. giu*, and especially on the ear-coverts, which are dusky as in *S. stictonotus*, and not silvery grey as in *S. giu*. White spots on the wing-coverts are nearly absent. The under surface is much more dusky than in the European bird; and the toes are not only bare at the joint, but the lower portion of the tarsus is also naked, as will be seen by the following cut:—

1. 2.

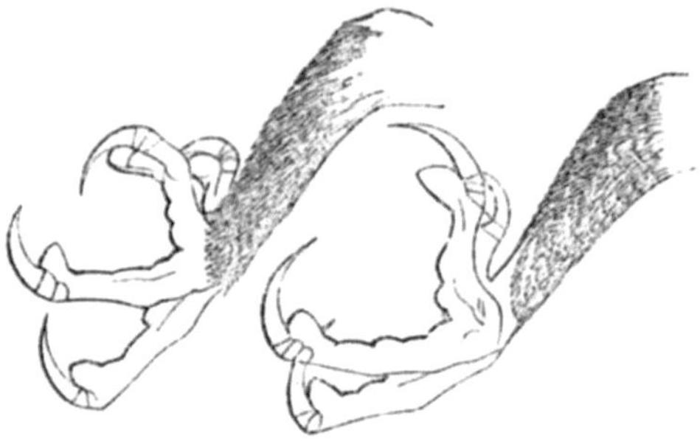

Fig. 1. Leg of *S. giu*, nat. size. Fig. 2. Leg of *S. japonicus*, nat. size, to show difference in size and extent of tarsal feathering.

Subsp. ε. **Scops malayanus.** (Plate IV. fig. 1.)

Scops malayanus, *A. Hay, Madr. Journ.* xiii. pt. 2, p. 147.

Adult. Above dark brown, freckled with very minute vermiculations of sandy buff, rather coarser on the scapulars; nearly all the plumes of the upper surface variegated with distinct spots or bars of white, in many cases concealed, but very broad and distinct on the hind part of the neck, where they form a tolerably pronounced collar; outer scapulars largely marked with sandy buff and white, forming a conspicuous shoulder-patch, the white predominating on the lower scapulars, which are broadly tipped with blackish on the outer web; head rather darker than the back, and slightly washed with greyish, most of the plumes blackish in the centre, and spotted on both webs with sandy buff or whitish; ear-tufts orange-buff, vermiculated with brown towards the tips, and carrying generally a subterminal white spot; wing-coverts rather darker than the back, but rather more rufescent, the mottlings very distinctly rufous sandy colour, the uppermost of the median series externally marked with white spots, the spurious quills also notched with white on the outer web; greater coverts slightly shaded with grey and largely marked with white on the outer web; primary coverts dark brown, barred indistinctly with frecklings of sandy buff; quills brown, externally barred with pale sandy buff, the tips of both primaries and secondaries thickly mottled with yellowish buff towards the tips, the primaries externally notched with white; lower back, rump, and upper tail-coverts dark brown, thickly vermiculated with sandy rufous, and varied with half-concealed sandy spots; tail dark brown, mottled with sandy-buff vermiculations, especially towards the tips of the feathers, and crossed with five sandy bars on the outer tail-feathers, four on the centre ones; lores whitish, the hair-like elongated shafts blackish at tip; feathers over the front of the eye also white, with narrow blackish tips; feathers round the eye sandy brown, mottled with blackish; ear-coverts greyish white, barred across with brown, and having the appearance of being streaked with the latter colour; behind the ear-coverts an indistinct ruff of orange-buff plumes tipped with black, those on the throat and cheeks white, more narrowly tipped with a line of black; chin white; sides of neck greyish, with very minute blackish vermiculations, and distinctly spotted or barred here and there with white; chest white, the feathers washed with orange-buff and crossed with several zigzag lines of dark brown, these increasing in number towards the tips of the plumes; all the cross lines disappearing on the rest of the underparts, which are nearly entirely white, some of them showing a dark shaft-streak, this being, however, more apparent on some of the feathers of the chest; under tail-coverts pure white, with only the faintest remains of zigzag lines at the tips of a few of them; thigh-feathers tawny, gradually shading off into white on the tarsus, and distinctly mottled with dark brown cross lines; under wing-coverts yellowish white, the outermost dark brown spotted with tawny buff, the lower series ashy brown, yellowish at the base, and thus resembling the inner lining of the quills, which are ashy brown barred with yellowish on the inner web, the outer web of the primaries notched with buff, and the tips of all mottled with dull sandy colour; bill dull horn-colour, the under mandible yellowish; feet apparently yellowish flesh-colour in life. Total length 7·5 inches, wing 5·5, tail 2·4, tarsus 1.

Rufous phase. Above tawny rufous, with very fine (almost indistinguishable) vermiculations, the hind neck mottled with fulvous bars, some of them inclining to white; scapulars externally white, with a blackish terminal bar, some of the greater wing-coverts tipped, and the bastard wing externally notched with white, the vermiculations rather more coarsely vermiculated with blackish; quills sandy rufous externally barred with brown, the primaries notched with white, giving the wing a chequered appearance, the inner webs brown, notched with fulvous; tail sandy brown, paler at tip, and crossed with five bands of pale rufous; head deeper tawny rufous than the back, more strongly mottled with black; lores and feathers over the fore part of the eye whitish, narrowly tipped with dusky, the lores shaded with rufous; sides of face and ear-coverts greyish, indistinctly barred across with black; chin whitish; sides of neck and chest sandy orange, mottled with brown, and with more or less concealed bars of white; the breast orange, slightly vermiculated and barred with white, the black streaks to the centres of the feathers very broad, less so on the abdomen and flanks, the lower parts of the body being white as in the brown phase, the vermiculations not numerous on the belly, and disappearing on the under tail-coverts; bill dusky, yellowish on the under mandible. Total length 7 inches, wing 5·35, tail 2·6, tarsus 0·95. (Malacca, Maingay: *Mus. Walden.*)

Hab. Malayan peninsula, ranging northwards into Southern China*.

a. Ad. sk.	Malacca.	Hugh Cuming, Esq. [C.].
b. Ad. sk.	Malacca.	A. R. Wallace, Esq. [C.].
c. Ad. sk.	Singapore.	Earl of Ellenborough [P.].

Obs. This is a very well-marked form of the *Scops giu* group, and is by no means so very different in appearance from *S. capensis*; but it is chiefly remarkable for the ocellated appearance of the head and hind neck, the latter having very broad white bars, so as to form a striking contrast to the head and back; the ear-coverts are dusky as in *S. stictonotus*. The character of the under surface is peculiar, and has been well commented on by Lord Walden in his original description:—"The under surface is distinctly divided into two equal portions, the first, including the chin, throat, and breast, being wood-brown, mottled with a little white, light rufous, and black, irregularly distributed; the lower division, including the belly, vent, thigh-coverts, and under tail-coverts, is white, speckled with deep brown and light rufous."

Considerable differences are exhibited in the three specimens possessed by the Museum, but thay all agree in one character, viz. the large white spotting on the hind neck, which gives an ocellated appearance. Mr. Cuming's specimen is greyish like the one described, but is much more obscurely coloured below by reason of the numerous greyish vermiculations which crowd the belly; everywhere on the upper surface the vermiculations are rather finer. Mr. Wallace's bird is much more rufous, but shows the white ocellation on the hind neck, and is also a good deal clouded on the belly with cross lines.

Subsp. ζ. Scops rufipennis.

Scops pennatus, *Jerd. Madr. Journ.* xiii. pl. 2, p. 119.

Of the *Scops giu* group, and very closely allied to *S. malayanus*, and resembling it in the dusky grey ear-coverts, but distinguished by the absence of the white ocellations on the hind neck and of the bars on the centre tail-feathers, and more especially by its rufous quills. The following is a description of the type:—

Adult. General aspect of upper surface more uniform than is usual in species of this genus, being of a dusky greyish brown, the feathers being blackish in the centre, but scarcely to be called streaked, excepting on the fore part of the crown, where the black shafts are very broad and distinct, all the feathers of the upper surface so finely pencilled with dark brown as to appear almost uniform, with here and there a few sandy-coloured mottlings, more distinct on the head, to which they impart a slightly spotted appearance; the collar on the hind neck very indistinct, some of the feathers being barred with

* Lord Walden has a specimen from Fokien, procured by Mr. Swinhoe in Sept. 1866, which agrees with typical Malayan examples, and not with *S. stictonotus*.

fulvous, and crossed with narrow bars of blackish; on the scapulars the blackish cross lines a little more coarsely defined than on the back, washed with orange-buff, and having the outer web pure white tipped with black, forming a conspicuous shoulder-patch; wing-coverts greyish like the back, the vermiculations very faint and often obsolete on the greater series, which have rather large white spots on the outer web, the median coverts coarsely vermiculated with sandy buff, the feathers with blackish shaft-streaks, the least series rufous with obscure blackish cross vermiculations, the outermost of the greater series and the primary coverts strongly rufescent, almost chestnut in tone, the latter finely vermiculated with blackish; innermost secondary quills coloured like the back, and finely vermiculated in the same manner, their centres streaked with dusky blackish along the shaft, the rest of the quills rufous, barred with dusky brown, these bars more or less vermiculated, the inner webs almost entirely dusky brown, barred with pale rufous inclining to yellowish on the inner web, the rufous bars on the outer web of the primaries inclining to white, and producing a somewhat chequered appearance; upper tail-coverts exactly like the back, the centre tail-feathers likewise strongly resembling the upper surface, inasmuch as they are without any distinct trace of cross bars, the outer feathers dark brown, vermiculated with sandy rufous, and crossed with seven bars of rufous, the subterminal one very indistinct, and lost in the vermiculations at the tips, the light bars inclining to white on the outer edge of the external web; loral plumes whitish, the shafts black, and produced into long hair-like bristles; feathers over the eye buffy white, tipped with blackish; sides of face dusky grey, indistinctly varied with fine cross lines of dull brown; behind the ear-coverts a tolerably distinct ruff of orange-buff feathers, broadly tipped with black, this ruff extending across the throat, but the feathers here finely barred with blackish; chin-feathers dull white; chest dull orange-buff, the feathers broadly centred with black and crossed with a few narrow zigzag lines of brown and vermiculated with the same at the tips, many of the feathers inclining to white; on the breast and the rest of the lower parts the white predominates, many of the feathers only having a few zigzag markings of brown, on many of them a strong tinge of rufous with broad black central streaks; some of the flank-feathers slightly washed with grey; under tail-coverts almost entirely white, excepting an arrow-shaped mark of rufous or brown near the tip; leg-feathers buffy white, the tarsus slightly streaked with brown; under wing-coverts fulvous, those near the edge of the wing mottled with brown, the lower series ashy brown with yellowish white bases, the quills being also ashy brown below, but inclining to rufous near the tips, the bands being entirely of this colour, and fulvous only near the base of the inner webs. Total length 8 inches, wing 5·1, tail 2·7, tarsus 0·85.

Hab. Eastern Ghauts.

a. Ad. sk.	Madras: Eastern Ghauts.	Dr. Jerdon [C.].

Subsp. η. **Scops brucii.**

Ephialtes brucei, *Hume, Str. F.* i. p. 8.

Adult male. Above ochraceous grey, with nearly obsolete dusky cross vermiculations in the form of widely separated hair-like lines, every feather distinctly streaked with black down the shaft; the head likewise streaked in the same manner as the back; the hind neck very slightly paler than the back, scarcely sufficient to be perceptible; scapulars a little browner than the back, the outermost buffy white, tipped with black on the outer web; wing-coverts coloured like the back, and mesially streaked in the same manner, the vermiculations rather plainer on the greater series, which are more or less barred with fulvous on the outer web; primary coverts alternately barred with dark brown and sandy buff, the latter somewhat mottled with ashy brown; quills ochraceous grey like the back and streaked in the same way, the primaries tipped with this colour but plainly vermiculated with dark brown, all the quills barred with darker brown on the inner webs, the secondaries with pale fulvescent bars on the outer web, the primaries externally barred with buffy white, giving a chequered appearance to the outer aspect of the quills; tail ochraceous grey, with tolerably plain blackish vermiculations, crossed with five (or six on some feathers) faintly indicated fulvous bars, margined above and below with blackish zigzag lines; loral plumes and feathers over the fore part of the eye whitish, narrowly tipped with dusky brown; ear-tufts similar to the crown, and streaked in exactly the same way; ear-coverts greyish, with white shaft-lines and indistinct cross bars of dusky brown, the hindermost tipped with blackish and merging in the ruff, the feathers of which are ochraceous, finely vermiculated and broadly tipped with blackish, the gular feathers mesially streaked but not barred with blackish; chin-feathers whitish; rest of under surface of body ochraceous grey, much resembling the back, and similarly streaked with black, the streaks on the chest much broader, the flanks and abdomen inclining to whitish, the vermiculations rather scantier and almost disappearing on the under tail-coverts, which are for the greater part white; leg-feathers buffy white, broadly and distinctly streaked with dark brown for their whole extent; under wing-coverts yellowish white, with a few streaks of dark brown near the outer margin, the lower series ashy brown at tip, forming a wing-bar, and resembling the inner lining of the quills, which are distinctly barred below with ashy brown; bill dusky; feet slaty; claws black; iris bright yellow. Total length 9 inches, wing 6·35, tail 3·3, tarsus 1·35. (*Mus. A. O. Hume.*)

Obs. Mr. Hume has been so kind as to lend me the type of this bird, which forms an extremely well-marked race of *S. giu*; and it seem likely that, although future researches and larger series may destroy the validity of many of the other races by interposing connecting links, a further acquisition of specimens will only lead to a more definite recognition of this one. Simply from want of material, which may show variation in plumage (though Mr. Hume says that he has seen six examples from Ahmednuggur all similar in coloration), I keep *Scops brucii* as a race of *S. giu*; but it is impossible to overlook the fact that its colouring is quite peculiar, none of the other subspecies being any thing like it in the uniformity of its general colouring, which I have described as ochraceous grey.

2. Scops spilocephalus.

Scops pennata, *Hodgs. J. A. S. B.* vi. p. 369 (*partim*).
Ephialtes spilocephalus, *Blyth, J. A. S. B.* xv. p. 8.
Phodilus nepalensis, *Gray, Hand-l. B.* i. p. 53.
Ephialtes gymnopodus, *Hume, Rough Notes,* ii. p. 390; *Jerd. Ibis,* 1871, p. 347.
Scops spilocephala, *Hume, Nests & Eggs Ind. B.* i. p. 66.

Adult (type of species). Above sandy brown, thickly vermiculated with wavy cross lines of blackish, disappearing to a great extent on the lower scapulars, and still more on the innermost secondaries, which are consequently nearly uniform sandy or fawn-colour; the outer scapulars fulvous or buffy white, with a broad terminal bar of blackish brown; round the hind neck a tolerably distinct collaret, most of the feathers having a broad subterminal bar of buffy white; head sandy brown, lighter than the back, the wavy black cross lines very distinct, nearly all the feathers with a broad subterminal spot or bar of buffy white, giving the crown a distinctly spotted appearance; wing-coverts rather deeper sandy brown than the back, coarsely vermiculated with wavy lines of blackish, the median series with ovate spots of buffy white on some of their outer webs, the spurious quills also slightly notched externally with buffy white; primary coverts deep sandy rufous, barred with blackish brown, and minutely vermiculated with lines of the same towards the tips; quills brown, indistinctly barred with lighter brown on the inner webs, increasing in distinctness towards the tips, which are sandy-coloured, vermiculated with lines of blackish; the outer webs of the quills sandy or fawn-coloured, more or less distinctly barred with blackish, the interspaces on the outer webs of the primaries inclining to white, and giving a chequered appearance to the external aspect of the wing; rump and upper tail-coverts deep sandy brown like the back, and vermiculated with blackish lines in the same manner, as well as crossed with many fulvescent bars, producing a somewhat spotted appearance; tail fawn-coloured, crossed with eight or nine bars of blackish brown, of which about eight can be traced on the centre feathers, and remains of nine on the exterior ones, all the bars being distinct near the base, but less strongly characterized towards the tip; forehead whitish, the plumes mesially streaked and terminally spotted with blackish brown; lores whitish, the feathers brownish at tip, the shafts developed into blackish hair-like bristles; a broad eyebrow of buffy white narrowly tipped with brown; ear-coverts sandy buff, barred with blackish, the hinder ones more fulvescent; behind the ear-coverts an indistinct ruff of buffy white plumes, narrowly barred and broadly tipped with deep brown; chin buffy white; under surface light sandy buff, with a few wavy lines and minute vermiculations of blackish, most of the feathers barred with white, these white bars margined with a conspicuous irregular spot or bar of blackish; thighs bright chestnut, shading off into buff on the tarsi; under wing-coverts yellowish white, with a broad longitudinal patch of dark-brown feathers near the edge of the wing, which is whitish, the lower series sepia-brown, yellowish white at base, thus resembling the inner lining of the quills, which are brown below, barred with fulvous on the inner web; bill yellowish; feet fleshy brown; iris yellow. Total length 9 inches, wing 5·75, tail 3·1, tarsus 1·25.

Obs. The above specimen, which I believe to be one of those figured by Mr. Hodgson as *S. pennatus* (*vide suprà*), is in bad condition; but the description agrees in the main with a better-preserved Hodgsonian skin, the type of *Phodilus nepalensis*, Gray. The Cambridge Museum likewise contains a specimen from Kumaon, procured by Captain Boys, which is in better preservation, and shows the centre of the abdomen to be very conspicuously white.

Hab. Himalaya Mountains.

a. Ad. sk.	Nepal.	B. H. Hodgson, Esq. [P.].
b. Ad. st.	Nepal.	B. H. Hodgson, Esq. [P.]. Type of *Phodilus nepalensis*.
c. Ad. sk.	India (*Jerdon*).	J. Gould, Esq. [P.].

Subsp. *a*. **Scops hambroecki.**

Scops japonicus, *Swinh. Ibis*, 1865, p. 348, et *Ibis*, 1866, p. 307.
Ephialtes hambroecki, *Swinh. Ann. N. H.* (4) vi. p. 153.
Lempijius hambroecki, *Swinh. P. Z. S.* 1871, p. 344.

Adult male (type of species). Above rufous chocolate, very finely vermiculated with black, these markings rather coarser near the ends of the feathers, where they sometimes form a subterminal spot; the head rather darker than the back, the black centres to the feathers broader; round the hind neck a very distinct white collar, the feathers being tipped with dusky black, and having a broad subterminal band of white; forehead distinctly buffy white, slightly spotted with brown, extending backwards so as to form an eyebrow as far as the inner web of the ear-tufts, the outer aspect of the latter being sandy rufous with blackish cross lines; on the occiput a few light rufous spots, scarcely forming a band; on the nape, however, a distinct indication of a band, the feathers being light fulvous; scapulars externally white, tipped with blackish, and forming a distinct shoulder-patch; wing-coverts a little darker than the back, the blackish centres larger and the vermiculations coarser, the greater series having indications of a whitish spot, the bastard wing externally notched with buffy white; primary coverts rufous-chocolate, with irregular cross bars of blackish; quills rufescent, barred with dusky blackish, the inner webs entirely blackish brown, but the outer webs barred with this colour, each bar more or less inclining to rufous on the outer margin, the interspaces on the primaries buffy white, giving a chequered appearance to the outer aspect of the wing; upper tail-coverts rufous, banded across with blackish; tail-feathers rufous, with about six bands of blackish, becoming very indistinct and breaking up into vermiculations towards the tip of the feather; lores very full and buffy white in colour, tinged with rufous near the tip, before which is a dusky black bar; ear-coverts bright rufous, with a few faint indications of brown cross lines, the hindermost tipped with black; sides of neck strongly mottled with white; chin buffy white, with a few brown spots; ruff composed of whitish feathers, barred with brown, these bars very distinct and somewhat tinged with rufous on the gular plumes; rest of under surface of body sandy white, speckled with fine brown vermiculations, some of the feathers with longitudinal streaks of dark brown down the centre; the sides of the chest rufous, with distinct vermiculations; leg-feathers sandy white, mottled with spots and bars of dark brown; under tail-coverts white, barred across with rufous brown; under wing-coverts brown, washed with rufous, the edge of the wing white, the innermost coverts yellowish buff, the lower series uniform ashy brown; bill greyish brown, lower mandible ochreous; feet bluish grey, tinged with brown; claws light brown at base; iris fine clear yellow. Total length 7·5 inches, wing 5·7, tail 3·5, tarsus 1·2. (*Mus. Norvic.*)

Hab. Tamsuy, Formosa (*Swinhoe*).

Obs. The authorities of the Norwich Museum having kindly lent me the type of this species, I have been enabled to describe it and compare it with *S. spilocephalus*, which it so closely resembles as to be almost inseparable. The type is in rufous plumage; and on comparing it with an Indian bird in rufous phase, lent me by Mr. Hume, the only differences that I can find are the slightly whiter face, eyebrow, and collar of the Formosan bird and the less spotted character of the head. The leg-feathers are also not so tawny; but I find approaches to all the characteristics of *S. hambroecki* in other Himalayan specimens, and therefore the comparison of a series may cause it to be united to *S. spilocephalus*. At present the type is unique.

3. **Scops gymnopodus.** (Plate IV. fig. 2.)

Scops gymnopodus, *Gray, Cat. Accipitr. B. M.* 1844, p. 44 (descr. nulla); *Strickl. Orn. Syn.* p. 203; *Gray, Hand-l. B.* i. p. 46.
Ephialtes gymnopodus, *Gray, Cat. Accipitr.* 1848, p. 96.
Acnemis gymnopodus, *Kaup, Isis*, 1848, p. 771; *id. Contr. Orn.* 1852, p. 111 (descr. orig.); *id. Tr. Z. S.* iv. p. 225.

Adult (type of species). Above dull sandy brown, everywhere minutely and almost imperceptibly vermiculated with wavy blackish hair-lines; the hind neck with an indistinct collar of orange-buff feathers, mottled at the tips with the same colouring as the back, and marked with blackish in various manners, sometimes as a subterminal bar, sometimes as a tolerably broad mesial streak, many of the feathers white in the centre; feathers of the crown varied with blackish mesial streaks, the cross vermiculations being also rather coarser than on the back, all with concealed tawny buff bases, but very few with any indications of a subterminal buff bar, so that the general appearance of the head is very uniform; ear-tufts 1 inch long, of the same colour as the head, but appearing rather lighter by reason of the orange-buff bases showing more plainly and extending for two thirds of the inner web; upper scapulars rather more blackish, the lower ones inclining to rufous sandy colour, the vermiculations less distinct and wider apart, the outermost for the greater part white, tipped with a bar of black, forming a very conspicuous shoulder-patch; wing-coverts darker brown than the back, the greater and median coverts paler and rather more rufescent, the vermiculations, as on the lower scapulars, being less distinct; the coverts near the edge of the wing notched with white, the median series with large ovate white spots on the outer webs, the greater coverts less distinctly spotted with white near the tips; quills dark brown on the inner webs, sandy brown on the outer, with tolerably distinct bars of blackish brown, more or less dissolving into vermiculations, the lighter interspaces becoming notches of white on the outer web of the primaries, and giving a chequered appearance to the external aspect of the wing; the innermost secondaries light sandy buff, coarsely vermiculated with blackish wavy lines; upper tail-coverts rufous sandy colour, with wavy linear vermiculations as on the back; tail dark brown, barred with sandy buff, the interspaces more or less mottled with the latter colour towards the tips of the feathers, these, like the extremities of the primaries, being thickly mottled with sandy buff; about seven light bars distinguishable on the tail, those on the outer rectrices inclining to whitish towards the margins of both webs; loral plumes whitish at base, the shafts developing into long hair-like bristles; over the front of the eye a patch of white feathers, each tipped with a terminal bar of blackish, the feathers above the eye with half-concealed white bases, forming an inconspicuous superciliary mark; ear-coverts greyish white, the feathers round the eye deep sandy brown, with narrow little shaft-streaks of fulvous; behind the ear-coverts an indistinct ruff, composed of orange-buff feathers, terminally barred with black, the plumes of the cheeks white, with orange-buff centres and a narrow blackish subterminal line; chin whitish; across the throat a continuation of the facial ruff equally indistinct, the feathers being sandy-coloured at tip, with whitish bases and wavy cross lines of black; chest sandy buff, the centres of the feathers longitudinally blackish, all the plumes with distinct blackish lateral lines (excepting at the tips, which are minutely vermiculated) and more or less barred with white, this colour gradually extending on the breast and flanks, which are almost entirely white, with blackish or sandy rufous centres to the feathers, all of which are more or less mottled with sandy and minutely vermiculated with wavy lines near the tips; the under tail-coverts like the belly, but having even less distinct cross vermiculations; thighs and feathered part of tarsus deep sandy buff, with blackish cross lines; under wing-coverts for the most part whitish, with a few blackish cross lines, those near the outer edge of the wing blackish, slightly mottled with fulvous, the lower series ashy brown, fulvous at base, like the inner lining of the quills, which is ashy brown, notched on the outer and barred on the inner web with fulvous; bill horn-colour, yellowish at base of lower mandible; feet apparently yellow in life. Total length 6·7 inches, wing 5·1, tail 2·5; tarsus 0·85, the hinder aspect entirely bare, feathered only for 0·45 in front, the bare part being 0·4 inch in length.

Hab. India.

a. Ad. st. India. J. R. Reeves, Esq. [P.]. Type of species.

Obs. I have given a very careful description and a figure of this obscure species, as hitherto it has not been correctly identified by Indian naturalists. Indeed its Indian habitat is by no means certain, as some of Mr. Reeves's birds came from Malacca and China, and it is just possible that it may have had a wrong locality affixed. Were it not for the exact correspondence of the feathering on both tarsi, the species looks like *Scops malayanus* with its tarsal feathers rubbed off. The following cut exhibits the characters of the leg, and will distinguish the species in future.

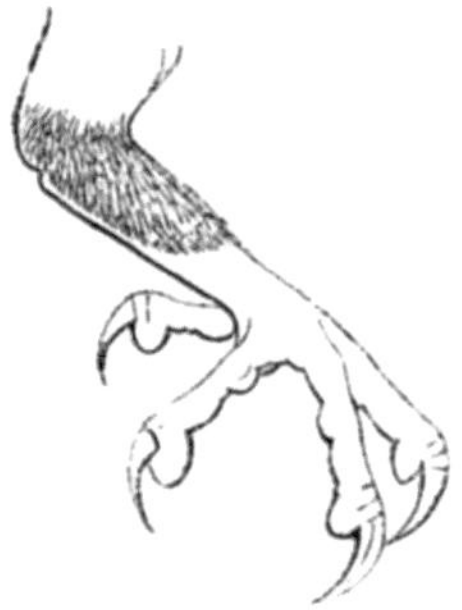

Leg of *Scops gymnopodus*, nat. size.

4. Scops sunia.

Scops sunia, *Hodgs. As. Research.* xix. p. 175; *id. in Gray's Zool. Misc.* p. 82; *Blyth, J. A. S. B.* xiv. p. 182; *Gray, Gen. B.* i. p. 38; *Jerd. Ill. Ind. Zool.* pl. xli.; *Bp. Consp.* i. p. 48; *Bp. Rev. et Mag. de Zool.* 1854, p. 543; *Gray, Hand-l. B.* i. p. 45; *Schl. Revue Accipitr.* p. 14.

Ephialtes sunia, *Gray, Cat. Mamm. &c. Nepal, coll. Hodgs.* p. 51; *Cass. Cat. Strigidæ,* p. 10; *Horsf. & Moore, Cat. B. Mus. E. I. Co.* i. p. 70; *Holdsw. P. Z. S.* 1872, p. 418.

Ephialtes bakkamæna, *Blyth, Ibis,* 1866, p. 255; *Jerd. Ibis,* 1871, p. 347.

Scops pennatus, *Gould, B. Asia,* part 22 (pt.); *Hume, Nests and Eggs Ind. B.* i. p. 65.

Ephialtes pennatus, *Hume, Rough Notes,* ii. p. 386; *Jerd. Ibis,* 1871, p. 347.

Adult. Above clear cinnamon rufous, nearly uniform, most of the feathers with a narrow central shaft-line of black, a little broader on the crown; scapulars externally yellowish buff or pure white, with a black subterminal bar; wing-coverts rufous like the back, with the same dusky black shaft-lines, broader on the greater series, which are white at the tips of their external webs; the spurious quills externally whitish, broadly barred with dark brown; primary coverts cinnamon, with a spot of yellowish white near the tip of the outer web, the inner webs blackish brown; quills dull cinnamon rufous, not so bright as the back, with faint cross bars of brown, almost obsolete on the innermost secondaries, less distinct on the outer ones, but plainer on the primaries, many of the light interspaces on the outer webs of the latter inclining to whitish; tail cinnamon, the centre feathers nearly uniform, with only faint indications of narrow blackish cross lines, these broader and more distinct on the outer feathers, the exterior ones having slight indications of whitish spots on the light interspaces; lores white, the elongated bristly shafts blackish at tip; over the eye a patch of white feathers, narrowly margined with black at their tips; sides of face cinnamon, with silvery white shaft-lines to the feathers; behind the ear-coverts an indistinct ruff of deep cinnamon feathers, broadly terminated with black; plumes on the chin whitish; chest cinnamon, some of the feathers buffy white at the tips, with broad mesial streaks of black; rest of under surface of body for the most part white, broadly streaked down the centre with black, many of the feathers barred with cinnamon, these light bars conterminously freckled with zigzag lines of blackish, the flank-feathers more inclining to cinnamon, especially on the inner webs; under tail-coverts nearly entirely white, washed with pale cinnamon near the base, and at the tips with a few indications of a central streak of dark brown; leg-feathers whitish on the outer face of the tarsus, the thighs and upper part of the latter cinnamon; under wing-coverts yellowish, those near the margin of the wing rufous, mottled with brown, the edge of the wing white; lower series of coverts ashy brown, yellowish at base; inner lining of quills light cinnamon, indistinctly barred with ashy brown, plainly near the base, but disappearing towards the tips of the quills; bill dusky horn-colour; feet fleshy brown; iris bright yellow. Total length 6·7 inches, wing 5·3, tail 2·5, tarsus 0·9.

Obs. The description is from one of Hodgson's skins; but some specimens show considerable differences, principally on the breast and abdomen, both the Penang and Madras specimens having the central black streaks very indistinct on the underparts, while they vary *inter se* in the mottling of the abdomen; the leg-feathers also are white, distinctly striated with blackish, whereas the Nepal bird scarcely shows any streaks on the tarsus. The Penang specimen shows slight indications of blackish frecklings on the upper parts; but this appearance is figured by Hodgson, who appears to have regarded this specimen as a more adult bird. The greatest difference of all is observable in the tail of the Penang *Scops*, this being dark brown, transversely banded with cinnamon, the dark bars breaking up into frecklings towards the tips of the feathers. This would appear almost to suggest that the specimen was the rufous phase of *S. malayensis*; but this I think unlikely, as I have before me what appears to be the rufous phase of that species. The general uniform appearance peculiar to *S. sunia* is not absent in the Penang example; and it appears to me to be a good species, especially considering the fact of rufous young having been found in the nest by Capt. Hutton (*cf.* Hume, *l. c.*).

Hab. India generally, ranging down into the Malayan peninsula as far as Penang.

a. Ad. sk.	Nepal.	B. H. Hodgson, Esq. [P.].
b. Ad. sk.	Nepal.	B. H. Hodgson, Esq. [P.].
c. Ad. sk.	Madras.	Dr. Jerdon [P.].
d. Ad. sk.	Penang.	J. Gould, Esq. [P.].
e. Juv. st.	?	E. Wilson, Esq. [P.].
f. Sternum.	Nepal.	B. H. Hodgson, Esq. [P.].

5. Scops megalotis.

Scops megalotis, *Gray, Cat. Accipitr.* 1844, p. 45; *Strickl. Orn. Syn.* p. 202; *Gray, Hand-l. B.* i. p. 46.

Ephialtes megalotis, *Gray, Cat. Accipitr.* 1848, p. 98; *Bp. Consp.* i. p. 45.

Ptilopsis megalotis, *Bp. Rev. et Mag. de Zool.* 1854, p. 543.

Lempijius megalotis, *Walden, Tr. Z. S.* ix. p. 145, pl. xxv. fig. 3.

Adult (type of species). Everywhere light rufous fawn, with separate vermiculations of dull blackish, a little coarser on the auricular tufts and on the wing-coverts, the vermiculations almost obsolete on the under surface of the body, the tarsal feathers fulvous, slightly mottled with rufous; greater and primary wing-coverts rufous fawn-colour, very coarsely and thickly vermiculated with blackish, forming five or six indistinct bars; quills blackish, with six or seven bars of fawn-colour, very distinct on the outer primaries, but gradually becoming more and more obscured with blackish frecklings till the bands are obsolete on the secondaries; tail blackish, with six or seven bars of fawn-colour, more or less obscured with frecklings of blackish, which are plentifully distributed over the whole tail; under wing-coverts uniform fulvous, the under surface of the wing ashy brown, with bands of fulvous; bill pale horn-colour, yellowish on under mandible; feet yellowish brown. Total length 11 inches, culmen 1·3, auricular tufts 1·45, wing 6·5, tail 3·5, tarsus 1·45.

Hab. Philippine Islands.

a. Ad. st.	Manilla.	Hugh Cuming, Esq. [C.].	Type of species.

6. Scops magicus.

Professor Schlegel unites several species of *Scops* under this one head; but I am inclined to consider with Mr. Wallace that they vary markedly, according to their localities, although, as in the case of

Scops giu and its allies, it is difficult to assign specific characters. Nevertheless I shall divide them into subspecies with distinctive titles, as follows; but the plates will give a still better idea of the birds than any descriptions.

Subsp. *a*. **Scops magicus.** (Plate V.)

Strix magica, *Müll. Natuurl. Gesch. Land- en Vogelk.* p. 110.
Otus magicus, *Schl. Fauna Jap. Aves*, p. 25.
Scops magicus, *Bp. Consp.* i. p. 46; *Strickl. Orn. Syn.* p. 203; *Schl. Mus. P.-B.* Oti, p. 22; *Gray, Hand-l. B.* i. p. 46; *Schl. Revue Accipitr.* p. 11.
Lempijius magicus, *Bp. Rev. et Mag. de Zool.* 1854, p. 543.
Ephialtes magicus, *Wall. Ibis,* 1868, p. 24; *Wald. Tr. Z. S.* viii. p. 39.

Adult female. General colour above dark sandy buff, everywhere on the upper surface barred with blackish, these dark cross markings being little, if any, narrower than the intervening light spaces; so that this species has a very strongly barred appearance, all the feathers being at the same time broadly streaked with black down the centre; hind neck rather lighter than the back, the interspaces rather paler and broader; scapulars externally yellowish buff, marked with blackish only towards or at the tip; wing-coverts decidedly darker than the back, the least ones spotted with sandy buff, the median and greater series barred with this colour, many of the cross bars paler and inclining to yellowish externally, some of the inner greater coverts also barred with yellowish near the tips of the feathers; primary coverts sandy buff, with dark brown bases, all of them also mottled with dark brown vermiculations in the form of irregular bars; quills dull brown on the inner webs, which, however, are indistinctly crossed with paler bands of sandy buff, especially near the tips, the light bands represented on the outer webs by sandy buff bars more or less inclining to white externally, and producing a distinctly chequered appearance, the brown interspaces very indistinct, and broken up by reason of extensive vermiculations of sandy buff; the secondaries very distinctly barred across with dark brown and sandy buff, exactly in the same manner as the primaries, but without the whitish spots on the outer web; tail-feathers very distinctly barred for their whole extent, the bars strongly characterized on the outer feathers, the dark brown and sandy buff cross markings being almost equal in width, the brown bars all more or less broken up with minute sandy mottlings, the centre feathers strongly barred with dark brown near the base, the bars somewhat dissolved towards the tips of the feathers; loral plumes whitish at base, dark brown at tip, the shafts blackish and elongated into hair-like bristles; feathers over the fore part of the eye whitish, narrowly tipped with dark brown; ear-coverts sandy buff, faintly barred with brown, the shafts distinctly whitish, and giving a narrowly streaked appearance; cheeks sandy buff, the feathers tipped broadly with black, and very stiff in character, extending backwards behind the ear-coverts and forming an indistinct ruff; chin buffy white, some of the lateral feathers with narrow linear cross markings of brown, the plumes of the lower throat sandy buff, with central markings and narrow lateral bars of black, the rest of the under surface entirely sandy buff, the chest-feathers very broadly centred with black, the abdominal plumes rather more narrowly; all the feathers crossed with narrow blackish lines, the interspaces becoming gradually paler on the lower breast and flanks, which have thus the appearance of being regularly barred with light buff; under tail-coverts yellowish buff, with blackish shaft-streaks and a few lateral bars towards the tips; under wing-coverts sandy buff, the outermost strongly mottled with brown, the lowest series ashy brown, fulvous at base, and thus resembling the inner lining of the quills, which are ashy brown below, more or less distinctly barred with sandy buff for the whole extent of the inner webs; bill dusky; feet pale; iris yellow. Total length 9 inches, wing 6·65, tail 3·5, tarsus 1·3.

Adult male. Different from the female in not being so universally barred, the upper surface being more vermiculated, while the under surface of the body has lost many traces of the blackish cross lines, and the lighter fulvous bars are not nearly so well characterized, though many of the bars on the lower breast, abdomen, and flanks are pure white. Total length 9 inches, wing 7, tail 3·7, tarsus 1·3.

Hab. Ceram; Amboina.

a. ♀ ad. st.	Ceram.	A. R. Wallace, Esq. [C.].
b. ♂ ad. st.	Ceram.	A. R. Wallace, Esq. [C.].
c. Ad. sk.	Ceram.	A. R. Wallace, Esq. [C.].
d. Ad. sk.	E. Ceram.	A. R. Wallace, Esq. [C.].

Obs. I follow Mr. Wallace in keeping the Ceram specimens as typical *S. magicus*, as they are doubtless the same as Amboyna birds. The characteristic feature of this Scops-Owl is its yellow plumage, which pervades the entire aspect of the bird both above and below. The ruff and ear-coverts are ochraceous buff; and there is no sign of white on the scapulars, which are externally yellowish. There is no perceptible collar round the hind neck; and the entire upper surface is very strongly banded with fulvous.

Leg of *S. magicus.*

Subsp. β. Scops leucospilus. (Plate VI.)

Ephialtes leucospilus, *Gray, P. Z. S.* 1860, p. 344; *Wall. Ibis*, 1868, p. 25.
Scops magicus (pt.), *Schl. Mus. P.-B.* Oti, p. 22; *id. Revue Accipitr.* p. 11.
Scops leucospila, *Gray, Hand-l. B.* i. p. 46.

Adult female (type of species). General colour above rather rufous brown, the feathers blackish in the centre, with distinct lateral lines of the same, the interspaces being rather lighter and more fulvous, so that a generally barred appearance is produced on the upper surface; besides these bars all the feathers are obscurely vermiculated with dusky brown, the central black markings to the feathers of the crown rather broader, many of them with nothing but a few lateral bars or spots of sandy buff, the latter rather larger on the sides of the crown; scapulars coloured like the back, the external ones white on the outer web, with a broad black bar at the tip; wing-coverts rather darker and more blackish than the back, the bases and centres to the feathers being blackish, spotted sparingly or barred with sandy buff, the greater series appearing more uniform than the other coverts by reason of the extreme minuteness of the vermiculations, the centres of these feathers streaked with black, some of the outermost largely spotted or half-barred with white; primary coverts dusky brown, with two indistinctly mottled bars of sandy buff; quills dull brown, barred on the inner web with sandy buff, the corresponding bars very distinct on the outer webs, especially near the base, the tips of all being rather obscured and the bars merged in the sandy vermiculations, the primaries externally notched with white, giving a distinctly chequered appearance to the outer aspect of the wing; upper tail-coverts rather more rufous than the back, the cross-barring less distinct; tail dull brown, crossed with about ten bands of sandy buff, these bands very distinct near the base of the feathers, but merging towards the tips, which are entirely sandy brown, vermiculated with blackish spots and lines, the outermost rectrix externally notched with white; lores white, becoming dusky brown, the shaft-lines produced into blackish bristles; feathers in front of and just over the fore part of the eye pure white, narrowly tipped with black, forming a distinct eye-patch, though scarcely an eyebrow; feathers round the eye reddish brown, the ear-coverts dull white, obscurely barred with blackish; behind the latter a ruff of white feathers, broadly barred with black at the tip, and narrowly barred with brown near the base, the throat-feathers more thickly lined but not so broadly tipped; chin white; chest sandy rufous, broadly streaked with black down the centre of the feathers, many of which are also barred and freckled with white, the white increasing on the breast and abdomen, the brown vermiculations decreasing in proportion, so that these parts are much lighter than the chest; under tail-coverts white, barred with sandy buff and streaked down the shaft with black; leg-feathers fulvous, rufescent on the tibia, all of them distinctly streaked with black; under wing-coverts yellowish white, those near the edge of the wing dark brown, mottled with sandy rufous, the under wing-coverts ashy brown, fulvous at base, and resembling the inner lining of the quills, which are ashy brown, barred with fulvous; bill horny black; feet pale; iris yellow. Total length 9 inches, wing 6·25, tail 3·2, tarsus 1·3.

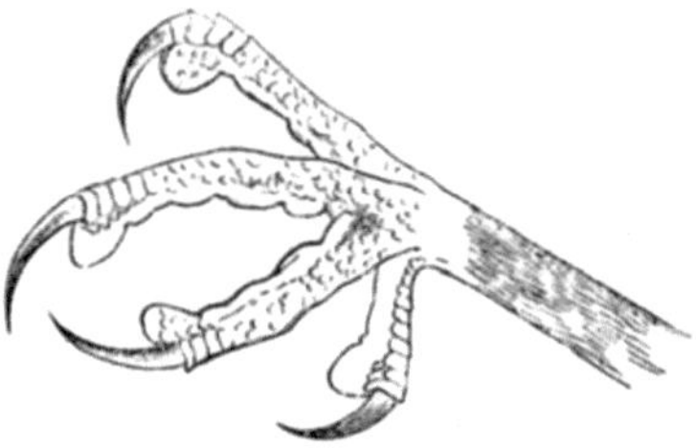

Leg of *S. leucospilus.*

Hab. Batchian; Gilolo.

a. ♀ ad. sk.	Batchian.	A. R. Wallace, Esq. [C.].	Type of species.
b. ♂ ad.	Batchian.	A. R. Wallace, Esq. [C.].	
c. Ad. sk.	Batchian.	Purchased.	
d. Juv. sk.	Gilolo.	A. R. Wallace, Esq. [C.].	

Obs. Scops leucospilus is distinguished by its dark brown plumage and large white shoulder-spots, but especially by its greyish white cheeks and ear-coverts: the under surface is whiter than in any of the allied races, this being the case even in a nestling specimen.

Subsp. γ. **Scops bouruensis.** (Plate VII. fig. 2.)

Ephialtes leucospila (pt.), *Wall. Ibis,* 1868, p. 25; *Wald. Tr. Z. S.* viii. p. 39.

Scops magicus (pt.), *Schl. Mus. P.-B.* Oti, p. 22; *id. Revue Accipitr.* p. 11.

Adult male. General colour above sandy brown, all the feathers coarsely vermiculated with blackish cross lines and broadly centred with black, especially on the head; on the nape a few feathers barred with white, but not sufficiently to cause even an indication of a band, the hind neck, however, being so broadly and distinctly banded with white on many feathers, and washed with ochraceous buff near their bases, that a tolerably distinct neck-band is seen; on many of the dorsal feathers appear a few sandy buff bars; the scapulars distinctly barred across with sandy buff near the tips of the feathers, the outer web pure white, with a tolerably broad black tip; wing-coverts coloured like the back and vermiculated in the same manner, the median and greater series broadly barred with white on the outer web, which is nearly entirely white on the external coverts belonging to these series; spurious quills externally notched with white; primary coverts dark brown, mottled on the outer web with sandy buff vermiculations; quills dark brown, all mottled with sandy buff at the tips, the inner web nearly uniform, with only here and there a faint indication of a lighter sandy-coloured bar, these represented on the outer web by a very distinct whitish notch, even on the secondaries, the innermost of which are vermiculated like the back, with indications of four or five cross bars of sandy buff; rump and upper tail-coverts more rufous than the back, the vermiculations being very indistinct, except towards the tips of the latter, which are also narrowly barred across with white; tail-feathers dark brown, mottled with sandy buff vermiculations, and having indications of about ten irregular bands of the same colour, disappearing gradually towards the tips of the feathers, which are entirely vermiculated with sandy colour, and inclining to white on the outer web of the external rectrices; loral plumes white, with narrow terminal black margins, the shafts also black, and produced into hair-like bristles; the frontal plumes also white, tipped and centred with black, these light feathers extending backwards towards the ear-tufts, which are sandy buff on the inner web, and barred with black on the outer; feathers round the eye rufous; ear-coverts and sides of face greyish white, with a few indications of obscure brown bars, more distinct on the hinder ear-coverts, and merging in the line of black-tipped feathers that mark the facial ruff, the plumes composing which are buff on the sides of the head, but become whiter on the throat, where they are not so broadly tipped or barred with black; sides of neck dark brown, but very plainly mottled with bars of greyish white; chin-feathers whitish; rest of under surface of body for the most part white, washed with sandy buff on the chest, most of the feathers with broad black centres, those on the chest with lateral zigzag lines of brown, increasing in number and blacker in proportion on the abdomen- and flank-feathers, many of which have concealed sandy buff bars, excepting those on the side of the vent and the under tail-coverts, which are very plainly barred with the latter colour, each bar having two conterminous zigzag lines of blackish; leg-feathers bright tawny, with narrow bars of blackish, the tarsal plumes buffy white, narrowly streaked with black; under wing-coverts fulvous, the edge of the wing white, the feathers adjoining being dark brown, mottled with sandy buff spots, the lower

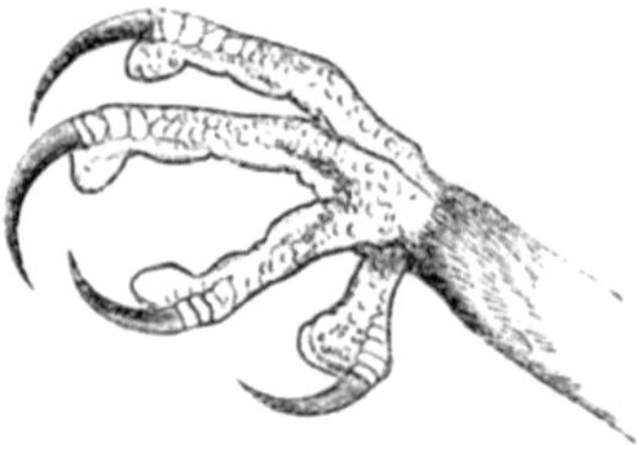

Leg of *S. bouruensis.*

series ashy brown, fulvous at base, and resembling the inner lining of the quills, which are ashy brown, barred with fulvous on the inner web; bill yellowish white; feet apparently yellow in life. Total length 10 inches, wing 7·3, tail 4, tarsus 1·45.

Hab. Bouru.

a. ♂ ad. sk. Bouru. A. R. Wallace, Esq. [C.]. Type of species.

Obs. *Scops bouruensis* is distinguished at once from *S. leucospilus* by its distinctly mottled forehead, which is largely spotted with white, and by its having the sides of the neck also whitish, mottled with black. The legs are more thickly feathered than in either *S. leucospilus* or *S. magicus*; the nape and hind neck are also mottled with white, forming two indistinct bands. This character goes far to indicate the position of the bird as intermediate between the *S.-magicus* group and *S. lettia* and allies. On the under surface the bird is very white, but the broad black centres give it a distinctive appearance.

Subsp. δ. **Scops morotensis.** (Plate VII. fig. 1.)

Ephialtes leucospilus (pt.), *Wall. ut suprà.*
Scops magicus (pt.), *Schl. ut suprà.*

Adult. Above dull brown, varied here and there with a few bars of black and sandy buff, these light bars a little more distinct on the hinder neck, but not sufficiently plain to form a collar; many of the dorsal feathers centred with black, but more especially those on the centre of the crown, which are black, with tiny lateral spots or notches of sandy rufous; the bases to the feathers on the sides of the crown pale sandy buff, but scarcely forming lateral stripes, the ear-tufts sandy buff on inner web, barred with brown, the outer web nearly uniform dark brown; the scapulars externally sandy buff, inclining to white, with a few narrow bars of dark brown, some of them barred with white and tipped with black; wing-coverts darker than the back, the least ones blackish brown, with a few bars and vermiculations of sandy buff, larger and paler on the outer web of the greater series, some of the latter broadly barred with fulvous and tipped with white on the outer web; primary coverts blackish brown, with a couple of sandy buff vermiculated bars near the tip; quills dark brown, almost uniform on the inner web, excepting at the tips, which are closely vermiculated with sandy buff, the outer webs also vermiculated with the same and barred with pale sandy buff, inclining to white on the outer web of the primaries, the innermost secondaries almost entirely sandy buff, paler and more rufous than the back, and mottled with black in the centres of the feathers; upper tail-coverts duller than the back and barred with blackish and sandy brown; tail dark brown, coarsely vermiculated with sandy buff on the edges of the feathers, which are likewise crossed with about nine bars of pale sandy colour, inclining to yellowish near the bases of the feathers, and to white on the margins of the outermost; loral plumes and feathers in front of the eye white, narrowly margined with black at the tip, the shafts of the former black and produced to a great length; sides of face and ear-coverts rufous sandy, indistinctly barred across with blackish, the hindermost tipped with black and merging in the ruff which runs down the sides of the head, the feathers composing which are yellowish tipped with black, becoming paler and more fulvescent on the throat, more narrowly tipped and scantily barred with black, the shaft-streaks being also of the latter colour; sides of neck exactly resembling the back, and descending somewhat on to the sides of the chest, which are plainly barred with sandy buff; the rest of the under surface sandy rufous, the feathers widely vermiculated with zigzag blackish lines, the centres to the chest-feathers very broadly black, with a few lateral bars of sandy buff or white; the abdominal plumes not so broadly streaked down the centre with black, but much more broadly notched and barred with white, which, however, never prevails to any great extent anywhere on the under surface, though the lower flanks are barred with white and sandy buff; under tail-coverts white, tinged with sand-colour near the tip, which has also some remains of blackish shaft-streaks; leg-feathers ochraceous, the tibial plumes barred and the tarsus streaked with dark brown; under wing-coverts yellowish white, as also the edge of the wing, the feathers adjoining the latter dark brown, mottled with sandy rufous, the lower series ashy brown, yellowish white at base, resembling the inner lining of the quills, which are ashy brown below, barred with yellowish white on the inner web. Total length 10·5 inches, wing 6·6, tail 3·8, tarsus 1·35.

Hab. Morty Island (Morotai).

a. Ad. sk.	Morty Island.	A. R. Wallace, Esq. [C.].
b, c. Ad. sk.	Molucca Islands.	Purchased.
d. Ad. sk.	Molucca Islands.	Purchased.

Obs. This is a very well-characterized Scops Owl. It is strictly of the same form and general coloration as *S. leucospilus*, but is a darker-looking bird, and is at once recognizable from that race by its buff-coloured ruff and ear-coverts, in these characters resembling true *S. magicus*. It is further like the latter bird in its rufescent under surface, on which there are very few white bars.

Subsp. ε. **Scops menadensis.** (Plate VIII. fig. 2.)

Scops menadensis, *Quoy et Gaim. Voy. de l'Astr.* i. p. 170, pl. 2. fig. 2; *Bp. Consp.* i. p. 47; *Kaup, Contr. Orn.* 1852, p. 112; *Strickl. Orn. Syn.* p. 202; *Schl. Mus. P.-B.* Oti, p. 20; *Gray, Hand-l. B.* i. p. 46; *Schl. Revue Accipitr.* p. 12.

Ephialtes menadensis, *Gray, Gen. B.* i. p. 38; *Wall. Ibis*, 1868, p. 25; *Wald. Tr. Z. S.* viii. p. 40.

Pisorhina menadensis, *Kaup, Isis*, 1848, p. 769; *Bp. Rev. et Mag. de Zool.* 1854, p. 543.

Otus manadensis, *Schl. Faun. Japon.* Aves, p. 26.

Megascops menadensis, *Kaup, Tr. Z. S.* iv. p. 230.

Adult. Above sandy brown, with numerous central streaks of black, more or less broken up into bars, each feather with conterminous fulvescent bars, deeper and more sandy rufous on the head, this giving a somewhat distinct appearance of bars on many of the feathers of the upper surface, every feather of which is minutely and almost invisibly vermiculated with blackish lines; scapulars more or less ochraceous buff on the outer web, arranged in broad bars, inclining to white on the lower scapulars; wing-coverts sandy brown like the back, and with bars of ochraceous buff as on the dorsal plumes, the outer coverts near the edge of the wing very distinctly notched with buff; primary coverts dull sandy brown, vermiculated with blackish, forming indications of cross bars; quills blackish, nearly uniform on the inner web of the primaries, excepting for a few sandy brown bars at the tip, these quills notched on the outer web with white or rufous, the secondaries not notched with these colours externally, but being generally of a sandy brown tint, vermiculated with black, giving the appearance of irregular bars, which are more plainly characterized on the inner web, especially towards the base, which is wholly blackish; tail sandy brown, vermiculated with zigzag blackish lines, and having tolerably distinct bars of black on the inner web of the external feathers, about ten being distinguishable on the outer tail-feathers; loral plumes sandy brown, whitish at base, the shafts elongated into black hair-like bristles; feathers over the fore part of the eye white, washed with rufous, and terminally tipped with brown; ear-coverts sandy brown, barred with blackish cross lines; behind the ear-coverts an indistinct ruff of sandy buff feathers, terminally barred with black; chin buffy white; throat sandy buff, with mesial streaks of black and lateral zigzag lines to each feather; rest of under surface greyish sandy, all the feathers minutely waved with linear transvermiculations and varied with white bars, these being narrow and indistinct on the chest, much broader on the flanks and sides of the body; long plumes at the sides of the vent broadly banded with buffy white, much broader than the sandy bars; under tail-coverts yellowish white, with one or two sandy bars near the tips of the feathers; leg-feathers fulvous, with distinct linear streaks of black; under wing-coverts sandy buff, more yellowish on the inner ones and on the edge of the wing, the plumes near the margin marked with wavy lines and vermiculations of blackish; the lowest series ashy brown, fulvous at base, resembling the inner lining of the

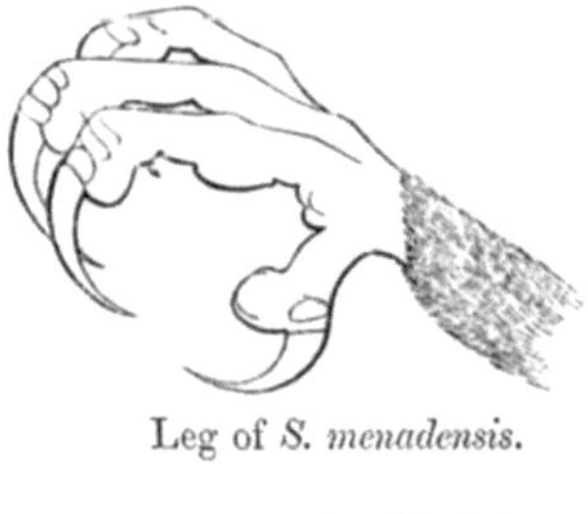

Leg of *S. menadensis*.

quills, which are ashy brown, barred with fulvous for the whole extent of the inner web, the lower apex of the quills numerously barred with sandy buff like the upper. Total length 8·5 inches, wing 5·8, tail 3·4, tarsus 1.

Another specimen in the collection measures as follows—total length 8 inches, wing 5·8, tail 3·1, tarsus 1.

Hab. Celebes and adjacent islands (Sanghir &c.).

a.	Ad. st.	Celebes.	A. R. Wallace, Esq. [P.].
b.	Ad. sk.	Menado, Celebes.	A. R. Wallace, Esq. [C.].
c.	Pull. sk.	Menado, Celebes.	J. Gould, Esq.
d.	♂ ad. sk.	Makassar, Celebes.	A. R. Wallace, Esq. [C.].

Obs. Scops menadensis is in reality a small form of *S. magicus*, to which it bears considerable resemblance; but the more slender legs and lesser size distinguish it from that and the allied subspecies.

Notwithstanding the oft-repeated assertion of the identity of *Scops menadensis* of Celebes and *S. rutilus* of Madagascar, I think that the two can well be distinguished; for, in addition to minor and less important characters, the aspect of the inner lining of the quills is, in my opinion, sufficient to separate them, in the brown phase at all events. In the Madagascar birds this is nearly uniform, with only a few broad bars of yellowish white near the base of the feathers; but in *S. menadensis* the whole wing is narrowly barred with fulvous for its entire extent, and on the outer web of the first primary the spots are eight or nine in number; in *S. rutilus* they are not more than six. The upper surface of *S. menadensis* is more stellated, the spots being more yellowish in appearance.

Subsp. ζ. **Scops siaoensis.**

Scops siaoensis, *Schl. Mus. P.-B. Revue Accipitr.* p. 13.

"Similar to *Scops menadensis*, but of very small size. Total length 7 inches, wing 4·9, tail 2·1." (*Schlegel, l. c.*)

Hab. Siao-oudang, between Celebes and Sanghir.

Subsp. η. **Scops albiventris.** (Plate VIII. fig. 1.)

Scops menadensis (pt.), *Schl. & Wall. l. c.* (ex Flores).

Adult. Above pale sandy brown, minutely vermiculated with blackish zigzag lines, the feathers mesially streaked with black, the plumes of the hind neck light buff, with distinct cross lines of black; the outer scapulars for the greater part pure white, with broad dark brown tips, forming a small shoulder-patch; wing-coverts coloured and marked exactly like the back, the light mottlings rather larger on the median and greater series, the inner webs of these being nearly uniform dark brown, the outer coverts adjoining the edge of the wing distinctly notched with white or buffy white; primary coverts sandy brown, mottled and barred across with blackish; quills blackish, mottled and barred with sandy brown towards the extreme tips of both webs of the primaries, these being notched or barred with sandy buff, more or less inclining to white, and giving a chequered appearance to two thirds of the primaries; secondaries sandy brown, tipped with greyish white, the innermost profusely mottled with blackish vermiculations resembling the scapulars, the outer secondaries blackish on the inner web, the outer webs narrowly barred with light sandy buff, the intermediate blackish bars being much broader and somewhat vermiculated with sandy lines; tail-feathers sandy brown, the central feathers mottled with blackish vermiculations in the form of irregular bars, these bars more distinct on the outer feathers, where about nine can be counted, the outermost tail-feather with distinct bars of blackish and sandy buff, many of the latter inclining to whitish on their outer margin; loral plumes whitish, washed with brown, with elongated blackish shaft-bristles; frontal plumes and those over the fore part of the eye white, with a few remains of blackish bars; ear-coverts greyish white, slightly washed with sandy rufous, and indistinctly crossed with narrow blackish bars; behind the ear-coverts an indistinct ruff of orange-buff feathers, narrowly barred with brown; the long stiff plumes reaching below the above-mentioned coverts of the same orange-buff colour, with a narrow tip of blackish, preceded by a distinct subterminal bar of white; chin-feathers whitish; throat and chest sandy buff, with a few indications of blackish zigzag lines, varied with irregular bars of white near the tips of the chest-feathers, the lower throat also slightly washed with greyish; rest of under surface white, with here and there a tinge of sandy or orange-buff, the blackish shaft-stripes very distinctly indicated on most of the feathers, which are varied with narrow wavy lines of blackish brown; the under tail-coverts almost entirely white, with only a few tiny lines of blackish near the tips of the feathers, which are also slightly washed with sandy buff; tibiæ distinctly washed with orange-buff, the tarsal plumes uniform buffy white; under wing-coverts yellowish white, as also the edge of the wing, the coverts nearest the latter sandy brown, indistinctly barred with blackish, the lowest series dusky brown, the bases to the feathers yellowish white; inner lining of quills dusky brown, regularly barred with fulvous on the inner web for the whole extent, the tips mottled with sandy buff, as they are on the upper surface. Total length 8·5 inches, wing 6·25, tail 3·3, tarsus 1·1.

Hab. Flores.

a. Ad. sk. Flores. A. R. Wallace, Esq. [C.]. Type of species.

Obs. On comparing this Scops Owl with the true *S. menadensis*, to which it has been joined by Mr. Wallace (Ibis, 1868, p. 25; *cf.* Lord Walden, Tr. Z. S. viii. p. 40), it is certain that it is closely allied, but differs in several particulars. It has a distinct wash of grey on the upper surface, and has the cheeks greyish white; but the principal difference lies in the white belly, where it is the prevailing colour, the cross markings and lines being very scanty.

Subsp. θ. **Scops rutilus.**

Scops rutilus, *Pucher. Rev. et Mag. de Zool.* 1849, p. 29; *id. Arch. du Mus.* iv. p. 326; *Bp. Consp.* i. p. 48; *Strickl. Orn. Syn.* p. 204; *Hartl. Faun. Madag.* p. 22; *Gray, Hand-l. B.* i. p. 46; *Gurney, Ibis*, 1869, p. 452.
Ephialtes rutilus, *Gray, Gen. B.* iii. App. p. 30 *a*.
Megascops rutilus, *Kaup, Contr. Orn.* 1852, p. 111; *id. Tr. Z. S.* iv. p. 229.
Pisorhina rutilus, *Bp. Rev. et Mag. de Zool.* 1854, p. 543.
Scops menadensis, *Hartl. Faun. Madag.* p. 23; *Schl. P. Z. S.* 1866, p. 41; *Grand. Rev. et Mag. de Zool.* 1867, p. 521; *Schl. & Poll. Faun. Madag.* p. 28; *Schl. Revue Accipitr.* p. 41.

Rufous phase. General colour above bright foxy rufous, very scantily varied with black markings, which are generally confined to indistinct shaft-stripes and small lateral cross markings, rather broader and more distinct on the crown of the head; the scapulars externally white towards the tip of the feather, contrasting markedly with the back, but not forming a continuous linear streak; upper wing-coverts duller rufous than the back, and more coarsely vermiculated with blackish cross markings, especially at the base of the median series, which, however, are almost uniform foxy rufous at the tips, the greater series nearly uniform externally, but barred with lighter buff, and mottled with blackish on the inner web; primary coverts blackish, only slightly varied with rufous externally and at the tips; quills for the most part dusky brown, especially on the inner webs, which are almost uniform, with but faint indications of lighter or more rufescent bars, the tips more distinctly rufescent, the corresponding light spots or bars on the primaries rufous, some of them whitish; secondaries dusky brown on the inner web, the light bars very indistinct, even on the outer webs, which are vermiculated with indistinct lines and spots of brown, the innermost secondaries almost entirely dull foxy rufous, with more or less distinct transvermiculations of blackish, forming here and there indications of bars; upper tail-coverts resembling the back, but a little more dingy; tail dull foxy rufous, barred with dull blackish, of which bars nine or ten can be distinguished on the outer feathers, where they are broader than the rufescent interspaces, the centre feathers more mottled with blackish, neither the dark nor the light bars distinctly indicated; lores white, with elongated black hair-like shafts; an indistinct eyebrow whitish, the feathers margined with foxy red, and tipped with dark brown; ear-tufts exactly resembling the crown; sides of face bright foxy red, with a few indistinct streaks of silvery white on the ear-coverts, behind which runs a slightly developed ruff of buff plumes, narrowly tipped with black; chin whitish; rest of under surface of body bright foxy red, like the back, many of the feathers mesially streaked with blackish, these streaks narrower on the flanks, many of the lateral and abdominal feathers being mottled and tipped with white; the feathers on each side of the vent white, barred with foxy rufous; on the fore neck an appearance of a white patch, centre of abdomen also whitish, the under tail-coverts slightly tipped with pale foxy red; leg-feathers whitish on the thighs, inclining to buff on the tarsal plumes; edge of wing and inner under wing-coverts whitish, those near the outer edge of the wing dark brown, narrowly barred with rufous, the lower series brown, buffy white at base. Total length 8·5 inches, wing 6·1, tail 3·3, tarsus 1·15.

Brown phase. Above very dark brown, almost blackish, varied with rufous buff markings, forming one or two rows of spots or bars of this colour near the tips of the feathers, which are also vermiculated with rufous buff, especially on the dorsal plumes and scapulars, the outer web of the latter having large white spots extending on to the inner web in some of them; head coloured like the back, but rather more minutely spotted with rufous buff, excepting on the auricular tufts, where the markings are larger, especially on the inner webs; wing-coverts blackish, the least series minutely spotted with rufous buff, the greater series more vermiculated with the latter colour, especially on the inner web, the outermost of this series being irregularly but conspicuously marbled with white, the greater series also white near the tip of the outer web, but all of these coverts more largely barred and mottled with rufous buff than the rest of the wing-coverts; primary coverts blackish, slightly barred on the outer web with rufous buff; quills blackish, nearly uniform on the inner webs of the primaries, externally notched with white near the base, these inclining to rufous towards the tips, which have indistinct remains of rufescent barring on the inner web also; the secondaries very faintly barred with rufous on the inner web, but more plainly on the outer one; the innermost secondaries blackish, very distinctly mottled with rufous buff vermiculations, and resembling the back; tail-feathers blackish, tipped with rufous and crossed with about five other narrow bars of rufous buff, the interspaces slightly freckled with dull rufescent colour near the tips of the feathers; lores dull whitish, washed with brown, the shafts black and elongated so as to form hair-like bristles; plumes at the base of the forehead and over the fore part of the eye white, with narrow rufous brown tips; ear-coverts dull rufous brown, transversely barred with blackish, the lower part varied with a slight indication of silvery white bars; behind the ear coverts a faintly indicated ruff of rufous buff plumes narrowly barred with black; chin and fore part of cheeks whitish; remainder of under surface rufous buff, the feathers broadly streaked down the centre with black, these markings wider on the chest, which has consequently a more dingy appearance; most of the feathers obscured with minute narrow vermiculations traversing the feathers, many of which incline to white on their outer margin; this white colour amounts only to a few lateral spots on the chest-feathers, gradually widening and taking the form of bars on the flanks and sides of the body, the feathers on the sides of the vent and under tail-coverts being white barred with rufous; leg-feathers buff, crossed with numerous small brown bars; inner under wing-coverts yellowish white, the edge of the wing also whitish, the coverts near the edge dark brown mottled with fulvous spots, the lowest series blackish, the bases yellowish white, resembling the inner lining of the quills, which are for the most part blackish, with a few broad bars of yellowish white near the base of the inner web, the tips of the quills with a few slightly indicated pale rufous bars. Total length 8 inches, wing 5·9, tail 2·9, tarsus 1·15.

Obs. The differences exhibited by the series of specimens in the Museum are not very considerable, although few are quite so dark as the one described. Two of these rather lighter examples have the auricular tufts barred across with rufous buff, but differ more conspicuously in the greater amount of variegation on the under parts, which are very prettily mottled with the white bars; these are fully developed as narrow bars on the breast-feathers; and the flanks and abdominal plumes have, in many instances, bars right across the feathers, both near the tip and near the base.

Hab. Madagascar.

a. Ad. sk.	Madagascar.	Purchased.
b. Ad. sk.	N.E. Madagascar.	Heer F. Pollen [C.].
c. Ad. sk.	S.W. Madagascar.	Mr. A. Crossley [C.]. (Brown phase.)
d. Ad. sk.	S.W. Madagascar.	Mr. A. Crossley [C.]. (Rufous phase.)
e. Ad. sk.	S.E. Madagascar.	Mr. A. Crossley [C.].
f. Juv. sk.	S.E. Madagascar.	Mr. A. Crossley [C.].
g. Ad. sk.	N. Madagascar.	Mr. A. Crossley [C.].
h, i. Skeletons.	N. Madagascar.	Mr. A. Crossley [C.].

7. Scops sylvicola.

Scops sylvicola, *Wall. P. Z. S.* 1863, p. 487; *Gray, Hand-l. B.* i. p. 45.
Ephialtes sylvicola, *Wall. Ibis*, 1868, p. 25.

Immature female (type of species). General colour sandy brown, freckled all over with a profusion of wavy lines of blackish, nowhere very strongly defined; the dorsal feathers very broadly centred with black, from which radiate some distinct wavy lines of the same colour, forming tolerably broad cross bars of black on the hinder neck; the head and interscapulary region paler, being of an ashy fulvous colour, with dusky brown cross lines; ear-tufts also pale, and resembling the pale buff colour of the head, 1 inch long; sides of face and throat fulvous, the dusky brown cross lines very indistinct; the rest of the under surface paler fulvous, the cross bars dusky and indistinct, but rather broader than on the throat, a few feathers white, mesially streaked with black, with lateral zigzag lines of rufous brown; leg-feathers white, with a few obsolete brown cross bars; wing-coverts sandy brown, with broad central markings of black, all the feathers pencilled with blackish cross lines, and most of them varied with distinct spots of sandy buff; the spurious quills externally notched with buffy white, the edge of the wing being also of the latter colour; primary coverts dull sandy, with numerous fine cross vermiculations of black, and indications of three cross bars of sandy colour; quills dark brown on the inner web, barred across with ashy brown; these bars nearly obsolete, but represented on the outer web with distinct square spots of sandy buff, sometimes slightly mottled with brown vermiculations, the darker brown cross bands being also sometimes mottled with ashy buff; the secondaries more rufous, the dark bands not so distinctly indicated, and the whole of the inner quills more vermiculated, so as to resemble the back in a greater degree; tail dark brown, crossed with nine bars of sandy buff on the middle feathers, and with ten or eleven on the outer ones; these bars less distinct towards the end of the tail-feathers, which, like the tips of the wings, are coarsely vermiculated, and thus the bars become merged. Total length 12·5 inches, wing 8·6, tail 4·9, tarsus 1·4.

Hab. Flores.

a. ♀ imm. sk.	Flores, 1862.	A. R. Wallace, Esq. [P.].	Type of species.

Obs. This bird is distinguished by its large size; but, from the immaturity of the type specimen, at present the only one known, I am unable to place it in its true position in the synopsis.

8. Scops semitorques.

Otus semitorques, *Schl. Faun. Jap. Aves*, p. 25, pl. 8.
Ephialtes semitorques, *Gray, Gen. B.* i. p. 38; *Cass. Cat. Strigidæ Philad. Mus.* p. 11.
Scops semitorques, *Bp. Consp.* i. p. 46; *Schl. Mus. P.-B.* Oti, p. 23; *Gray, Hand-l. B.* i. p. 46; *Sch. Rev. Accipitr.* p. 9.
Megascops semitorques, *Kaup, Tr. Z. S.* iv. p. 230; *id. Contr. Orn.* 1852, p. 112.
Asio semitorques, *Strickl. Orn. Syn.* p. 209.
Lempijius semitorques, *Bp. Rev. et Mag. de Zool.* 1854, p. 543; *Swinh. P. Z. S.* 1871, p. 343.
Scops asio, var. semitorques, *Ridgw. B. N. Am.* iii. p. 56.

Adult. Above dull sandy brown, obscurely vermiculated with blackish cross lines, most of the feathers distinctly blackish in the centre, and laterally barred with the same, especially on the scapulars, which are very distinctly marked with black and spotted with ochraceous buff, the outer feathers externally clear ochraceous buff, with a distinct shaft-streak and narrow edging of blackish; head and neck coloured like the back, most of the feathers blackish in the centre, with a double lateral spot of rufous buff; on the hinder head and neck three bars of buff, the first on the occiput, very small, and indicated only by a few bars of buff or greyish white on the feathers; on the nape a second one, of buff-barred plumes, forming a more distinct band than the previous one, but not so distinct as the one on the hind neck, which is very broad and composed of ochraceous-buff feathers, every one of which is tipped with blackish; wing-coverts coloured like the back, the least ones dusky and nearly uniform brown, the greater ones vermiculated and marked in the centre with blackish, the feathers more or less distinctly barred with fulvous, and inclining to buffy white near the tips of the outer webs, the spurious quills also distinctly notched on the outer web with buffy white; primary coverts blackish brown, very minutely vermiculated with sandy buff, forming three tolerably distinct bars on the outer webs; quills dark brown, banded with ashy brown on the inner webs, these lighter bars being replaced on the outer webs by others of sandy buff slightly vermiculated with brown, the primaries externally chequered with white bars, the interspaces also slightly vermiculated with sandy buff; the innermost secondaries vermiculated like the back, which they closely resemble; tail-feathers dark brown, coarsely vermiculated with zigzag lines of sandy buff, and crossed with seven or eight bars of the latter colour on the outer feathers, only seven being traceable on the centre ones, the outermost externally notched with indistinct white spots; lores white, slightly barred across with black, the feathers rufescent towards the tips, which are blackish; a very distinct eyebrow of greyish white feathers, narrowly tipped with brown; the sides of the crown rather greyer than the vertex, and so forming a light band on each side of the head continuous with the ear-coverts, which are whitish on the inner web, slightly barred with brown, and vermiculated near the tip with minute lines of the same, the outer webs blackish, mottled with a few spots of sandy buff, and vermiculated with brown on the outer margin; behind the upper

Leg of *Scops semitorques*, to show feathering on toes.

part of the eye a patch of rufescent feathers vermiculated with black, and reaching to the base of the ear-tufts; sides of face dull greyish, crossed with numerous faint bars of brown, washed with rufous brown near the eye, and relieved by some of the shaft-lines being silvery white; behind the ear-coverts a distinct ruff of ochraceous buff feathers, broadly tipped with black, this ruff descending on the gorget, and forming a collar across the throat, the plumes here being rather paler, mesially streaked and narrowly barred with black; chin whitish, the plumes of the throat buffy white, with fine transverse cross lines of blackish; below the gular ruff a line of white feathers across the fore neck, the centre of the body also pure white; rest of under surface sandy buff, with distinct blackish centres, all the feathers marked with wavy cross lines of dark brown, many of them barred with white, more broadly on the flanks; the under tail-coverts white, with central streaks of brown washed with fulvous near the tips, which have a few wavy lines of brown; tibial plumes bright tawny, the tarsal ones buffy white, with scanty cross lines of dark brown, the plumes extending for two thirds of the middle and outer digits, and for half the inner and hinder one; under wing-coverts whitish, the outer ones washed with buff, and marked with brown, the lower series ashy brown, yellowish white at base, like the inner lining of the quills, which are ashy brown, barred with yellowish white near the base of the inner web. Total length 10 inches, wing 7·25, tail 3·85, tarsus 1·3.

Hab. Japan.

a. Ad. st.	Nagasaki, Japan.	Purchased.
b. Ad. sk.	Nagasaki, Japan.	J. Gould, Esq. [P.].

Subsp. *a.* **Scops plumipes.**

Ephialtes plumipes, *Hume, Rough Notes*, ii. p. 397.
Ephialtes semitorques, *Jerd. Ibis*, 1871, p. 348.
Scops plumipes, *Hume, Nests and Eggs Ind. B.* i. p. 68.

Obs. Mr. Hume has kindly sent me two specimens of this Scops Owl recently described by him. It holds much the same relation to *S. semitorques* of Japan that *S. lettia* of the Himalayas does to *S. elegans* of China, being smaller and, in its brown phase, darker, while it further differs in possessing a grey phase, resembling the grey phase of *S. malabaricus*, which was called by Jerdon *S. griseus*. The measurements of Mr. Hume's two specimens are as follows:—

	Total length. in.	Wing. in.	Tail. in.	Tarsus. in.
a. ♀. Kotegurh, 1868	8·5	6·8	3·7	1·45
b. ♂. Huttoo, near Narkanda, Sept. 26, 1869	8·5	6·6	3·5	1·3

Hab. Himalaya Mountains.

9. **Scops lettia.**

Scops lettia, *Hodgs. As. Research.* xix. p. 176; *id. in Gray's Zool. Misc.* p. 82; *Gray, Hand-l. B.* i. p. 46; *Schl. Mus. P.-B. Revue Accipitr.* p. 9; *Hume, Nests and Eggs Ind. B.* i. p. 67.
Ephialtes lempiji, *Gray, Cat. Mamm. &c. Nepal Coll. Hodgs.* p. 51;

Jerd. B. Ind. i. p. 138; *Blyth, Ibis,* 1866, p. 256; *Jerd. Ibis,* 1871, p. 256.
Lempijius lettia, *Bp. Rev. et Mag. de Zool.* 1854, p. 543.
Ephialtes lettia, *Hume, Rough Notes,* ii. p. 393; *Jerd. Ibis,* 1871, p. 347.

Adult (grey phase). General colour greyish brown, more or less inclining to an ochraceous shade, all the upper surface thickly mottled with wavy cross lines of black, the feathers at the same time varied with ochraceous, forming concealed irregular bars of this colour; forehead and sides of crown greyish white, the feathers crossed with narrow wavy lines of brown, extending right to the ear-tufts, which are greyish white or buff on the inner web, the outer webs brown, slightly freckled with sandy buff; crown rather more rufous or sandy brown, the feathers closely vermiculated with blackish lines, and much varied with ochraceous buff spots; round the nape an indistinct collar of ochraceous feathers, all with more or less characterized cross vermiculations of dark brown; below this first collar the hind neck is again dark brown, exactly like the head, and this is succeeded by another ochraceous buff collar round the hind neck, the frecklings of brown on these feathers being somewhat coarser, and the blackish tips rather broader; scapulars externally ochraceous buff, some of them almost entirely of this colour on both webs, the vermiculations disappearing towards the outer webs; wing-coverts dark brown, freckled with blackish vermiculations exactly like the back, the central streaks to the feathers darker, both the median and greater series whitish at the tips, the frecklings gradually disappearing; primary coverts sandy brown, freckled with dark brown, the latter colour forming also three or four distinct bars across them; quills brown, barred with lighter brown, these light bars paler externally and mottled with dark brown vermiculations, the primaries externally whitish on the interspaces, causing a chequered appearance on the outer aspect of the wing, the inner secondaries greyish like the back and mottled with blackish vermiculations precisely in the same manner; tail ochraceous brown, regularly barred with darker brown, the dark bars being about ten on the outer feathers, less distinct at their tips, the light interspaces freckled with dark brown vermiculations; lores whitish, the long shaft-stripes blackish at tip; over the eye a patch of whitish feathers, narrowly margined at the tip with blackish; sides of face greyish white, indistinctly waved with narrow cross lines of brown, the hinder ear-coverts broadly tipped with black; behind the ear-coverts an indistinct ruff of pale ochraceous buff feathers, rather broadly tipped with blackish, a little narrower under the throat; chin whitish; rest of under surface greyish white, the blackish shaft-lines not very strongly pronounced, all the feathers narrowly waved with cross lines of brown, disappearing on the centre of the abdomen and under tail-coverts, which are pure white; leg-feathers silky white, with a few indications of narrow brown cross markings; under wing-coverts yellowish white, mottled with cross lines of dark brown near the edge of the wing, which is white, the lowest series ashy brown, yellowish white at the bases, and resembling the inner lining of the quills, which are ashy brown below, barred with slightly paler brown, inclining to yellowish white near the bases of the feathers; "bill bluish horn (more or less yellow in dried skins); feet fleshy grey; iris variable, yellow in the young, brown in the old birds" (*Hodgson*). Total length 8·5 inches, wing 6·3, tail 3·3, tarsus 0·35.

"Sexes alike" (*Hodgson*).

Obs. The brown phase of this plumage is very similar to the one above described in character and markings, but is brown on the upper surface where the other bird is greyish. The under surface, however, is not very different. Mr. Hodgson has given four figures of the species in his 'Drawings;' and in one of them he considers the grey and brown phases to be summer and winter plumage. Although amongst his four skins in the Museum we doubtless possess the types of his species, I could not make the specimens so thoroughly agree with any of the paintings as to satisfy myself that I could identify the actual individuals figured.

Hab. Himalayas.

a. Ad. st.	India.	J. R. Reeves, Esq. [P.].
b. Ad. sk.	Nepaul.	B. H. Hodgson, Esq. [P.].
c. Ad. sk.	Nepaul.	B. H. Hodgson, Esq. [P.].
d, e. Ad. sk.	Nepaul.	B. H. Hodgson, Esq. [P.].
f. Ad. sk.	Darjiling.	R. B. Sharpe, Esq. [P.].
g. Sternum.	Nepaul.	B. H. Hodgson, Esq. [P.].

Subsp. *a*. **Scops elegans** *.

Ephialtes elegans, *Cass. Pr. Philad. Acad.* 1852, p. 185.
Scops semitorques, *Swinh. Ibis*, 1863, p. 217, et 1867, p. 409; *id. P. Z. S.* 1863, p. 262; *David, N. Arch. Mus.* vol. vii. *Bull.* p. 4.
Scops lempiji, *Swinh. Ibis*, 1861, p. 265.
Scops elegans, *Gray, Hand-l. B.* i. p. 45.
Ephialtes glabripes, *Swinh. Ann. N. H.* (4) vi. p. 152.
Lempijius glabripes, *Swinh. P. Z. S.* 1871, p. 343; *id. Ibis*, 1874, p. 269.

Nestling. Shrouded in down of a pure white colour; feathers of the upper surface light ochraceous buff, regularly barred across with

* *S. glabripes* must be, I think, the bird described by Mr. Cassin as *Ephialtes elegans*, notwithstanding a slight difference in the measurements. The specimen described by Mr. Cassin bore a label "En mer, côtes du Japon, lat. 29° 47′ N., long. 126° 13′ 20″ E." It is impossible to speak with certainty of this species without examining the type; but I quote the following remarks of Mr. Cassin which accompany the original description, and which apply very well to Mr. Swinhoe's *S. glabripes*. "This is one of the handsomest of the small species of Owls, and bears some resemblance (rather remote, however, and on a small scale) to the large Horned Owls. It is somewhat like *S. semitorques* (Temm.), but wants the white on the throat and breast, and is smaller and differently coloured, and has the toes bare. It resembles also to some extent *S. lettia* (Hodgs.) and *S. lempiji* (Horsf.), but is larger than either, and is otherwise different. The brown stripes and lines on all parts of the plumage are unusually delicate." The dimensions are given as follows:—"Total length of skin from tip of bill to end of tail about 8½ to 9 inches, wing 6¾, tail 3½."

dark brown, and all broadly tipped with white, giving a hoary appearance to the upper parts; under surface of body white, tinged with ochraceous buff here and there, and obscurely barred across with dark brown: even in the nestling the tarsal feathering very dense.

Adult female. Above dull sandy brown, everywhere freckled with blackish vermiculations, much broader and more distinct in the centres of the feathers, giving in some indications of mesial streaks; head and neck rather more rufescent than the back, the feathers coarsely mottled or dotted with small spots of sandy rufous, the centres of the feathers of the crown blackish, with broad lateral cross lines of black; on the head and hind neck are three bands—one on the occiput not very distinct, composed of greyish white feathers scantily freckled with black cross lines, the shaft-streaks being black; on the nape a second and more distinct band of greyish white feathers coarsely mottled with sandy rufous at the tips and vermiculated with black, minute frecklings of the latter colour being apparent on most of the white feathers; round the hind neck a third and very distinct collar of ochraceous buff plumes less barred and mottled with blackish than the two preceding ones, and consequently more distinct; scapulars rather greyer than the rest of the back, the outermost yellowish buff externally, with scarcely any blackish frecklings; wing-coverts rather darker brown than the back, and a little more rufous, the blackish vermiculations being very coarse, the greater coverts fulvescent towards their tips, forming a uniform spot on the outer web, but mottled with blackish vermiculations near the tip of the inner one; spurious quills externally notched with fulvous; primary coverts blackish brown, crossed with about four bars of deep tawny buff, more or less obscured with blackish vermiculations; quills dark brown, barred with paler brown, these being represented on the outer web by not strictly conterminous bars of sandy buff, everywhere slightly obscured with blackish vermiculations, the outer webs of the primaries largely notched with buffy white, giving a chequered appearance to this portion of the wing, the innermost secondaries mottled and vermiculated like the scapulars; tail dark brown, varied towards the tips with vermiculations of sandy buff and crossed with seven bands of the same on the outer feathers, only about six being distinguishable on the centre ones; frontal and loral plumes ashy white, narrowly margined with blackish, these plumes forming at the same time a very conspicuous eyebrow; the adjacent sides of the crown are at the same time ashy white, extremely finely vermiculated with brown, so as to form a streak along the sides of the head, continued even to the ear-tufts, which are 1·25 inch long, and are greyish white on the other side, with fine brownish cross lines, the outer webs blackish at tip and mottled with rufous spots; above the hinder part of the eye a small patch of rufous feathers mottled with black, produced backwards to the base of the ear-tufts, the outermost small plumes of which resemble them in colour; sides of face ashy white, slightly tinged with rufous below the eye, and banded across with narrow blackish cross lines; behind the ear-coverts a tolerably distinct ruff of whitish feathers washed with sandy rufous before the tip, which is black, this ruff continued across the throat, where a few of the plumes are somewhat deeper rufous, crossed with blackish lines; chin white, the lower feathers washed with fulvous and crossed with numerous fine lines of dark brown; across the fore neck a faint indication of a white band; centre of the abdomen, vent, and under tail-coverts white, the latter with a few narrow rufous cross lines; rest of under surface ashy whitish, strongly inclining to sandy colour, especially on the chest and sides, all the feathers crossed with narrow brown, occasionally rufous, lines, and varied here and there with blackish central streaks; tibial plumes yellowish buff, inclining to buffy white on the tarsi, narrowly barred across with dark brown; under wing-coverts pale sandy buff, rather more rufescent near the edge of the wing, which is white, the marginal coverts slightly marked with dark brown bars, the lower series ashy brown, yellowish white at base, resembling the inner lining of the quills, which are ashy brown, barred with yellowish white on the inner web; iris black. Total length 11·5 inches, wing 7, tail 4, tarsus 1·5.

Adult male. Very similar to the female, but rather smaller. Total length 10 inches, wing 6·85, tail 4·3, tarsus 1·4.

Hab. China (Ningpo, Amoy, Foochow, Canton); Formosa.

a. Ad. st.	North China.	Purchased.
b. ♂ ad. sk.	Ningpo, China, April 1872.	R. Swinhoe, Esq. [C.].
c. ♀ ad. sk.	Ningpo, China, March 1872.	R. Swinhoe, Esq. [C.].
d. ♀ ad. sk.	Ningpo, China, May 1872.	R. Swinhoe, Esq. [C.].

Subsp. β. **Scops erythrocampa.**

Scops, sp., *Swinhoe, Ibis*, 1861, p. 29.
Scops lempiji, *Sclater, Ibis*, 1861, p. 29.
Lempijius erythrocampe, *Swinh. Ibis*, 1874, p. 269.

Adult (Fokien). Above sandy brown, coarsely vermiculated with blackish brown, a good many of the feathers blackish in the centre, causing a slightly streaked appearance to the upper surface, the scapulars externally ochraceous, very few of them with any blackish vermiculations on the outer web; head rather darker than the back, the feathers blackish in the centre, minutely notched with rufous buff, some of the occipital feathers spotted with white, forming a very indistinct bar, the nape with a second bar of white-mottled feathers not plainer than the former one; the hind neck ochraceous buff, slightly barred with brown, some of the feathers barred with white, and forming a very distinct collar; loral plumes very distinctly white, the shafts terminating in blackish bristles; the feathers over the eye and the forehead white, the plumes narrowly streaked and barred with blackish; sides of the crown greyish white, forming an indistinct band running along the sides of the head and continued along the depressed ear-tufts, the inner webs of which are greyish white, with tiny brown vermiculations, the outer webs dark brown at tip and barred with the same on the basal portion, which is sandy buff; feathers round the eye clear rufous, extending to the base of the ear-tufts; ear-coverts greyish, shaded with rufous and indistinctly barred across with dark brown, the hinder ear-coverts tipped with black, and resembling the ruff, which is ochraceous buff tipped with black, these tips being less distinct on the gular feathers, which are whitish, crossed with fine linear bars of dark brown; sides of neck greyish white, more or less washed with rufous and covered with wavy vermiculations of dark brown; fore neck inclining to white; rest of under surface ashy white, with broad central black markings on the chest-feathers, much narrower on the abdomen, all the plumes of the lower surface crossed with wavy bars of dull brown, wider on the abdomen and flanks, the under tail-coverts with scarcely any cross bars left; thighs bright orange tawny, the tarsal plumes whitish, with dark brown cross bars; under wing-coverts fulvous, the edge of the wing whitish, the adjacent feathers dark brown, the lowest series ashy brown, yellowish at base, resembling the inner lining of the quills, which are ashy brown below, barred with yellowish buff on the inner web; upper wing-coverts dark brown like the back, the median ones very broadly centred with black, the greater ones greyish at their tips, except on the outermost, which are fulvous at the ends, the spurious quills externally notched with buff; primary coverts sandy brown, with very minute vermiculations of brown and crossed with bars of dark brown; quills dark brown, barred with ashy brown on the inner web, these light bars represented by sandy buff ones on the outer webs, all freckled with blackish, excepting on the primaries, which are barred externally with sandy buff; tail-feathers dark brown, very distinctly banded with sandy buff, more or less vermiculated with brown, the bars seven on the centre feathers, including the terminal one; "bill pale flesh-grey, with a pale yellowish rim to the mandibles; iris golden burnt sienna, but so narrow that this colour is seldom visible, the immense pupil filling up all the space between the lids; skin round eye madder-brown; toes naked, light flesh-colour; claws light brownish grey, with blackish tips" (*Swinhoe*). Total length 8·5 inches, wing 6·9, tail 3·6, tarsus 1·5. (*Mus. R. Swinhoe.*)

Obs. But for Mr. Swinhoe's remark as to the difference in the colour of the eye, I should have been content to unite this bird to *S. elegans*. The yellow tibia is by no means a character of the species, as it exists in all the little Owls of this group; and *S. erythrocampa* further resembles *S. elegans* in the extent of feathering on the foot, as well as in the complete barring of the tail. It depends for its separation, therefore, on its rufous face and differently coloured eye; but it is by no means a very distinct bird.

Hab. China: Canton, South Fokien.

a. Ad. st.	North China.	Purchased.

10. Scops lempiji.

Strix lempiji, *Horsf. Tr. Linn. Soc.* xiii. p. 140.
Strix noctula, *Temm. Pl. Col.* ii. pl. 99.
Scops noctula, *Steph. Gen. Zool.* xiii. pt. 2, p. 52; *Schl. Mus. P.-B.* Oti, p. 24.
Scops javanicus, *Less. Traité*, p. 107; *Pucher. Rev. et Mag. de Zool.* 1849, p. 26.
Scops lempiji, *Blyth, J. A. S. B.* xv. pt. 1, p. 182; *id. Cat. B. Mus. A. S. B.* p. 36; *Bp. Consp.* i. p. 47; *Strickl. Orn. Syn.* p. 200; *Gray, Hand-l. B.* i. p. 46; *Schl. Mus. P.-B. Revue Accipitr.* p. 13; *Salvad. Ucc. Borneo*, p. 19.
Megascops lempiji, *Kaup, Contr. Orn.* 1852, p. 112; *id. Tr. Z. S.* iv. p. 229.
Lempijius indicus, *Bp. Rev. et Mag. de Zool.* 1854, p. 543.
Lempijius noctula, *Bp. Rev. et Mag. de Zool.* 1854, p. 543.
Ephialtes lempiji, *Gray, Gen. B.* i. p. 38; *Horsf. & Moore, Cat. B. Mus. E. I. Co.* i. p. 71; *Wall. Ibis*, 1868, p. 24; *Hume, Str. F.* ii. p. 469.

Young (Malacca, *Lord Ellenborough*). General colour light sandy rufous, everywhere barred with dusky blackish cross lines, more thickly pronounced on the head; quills and tail, as far as developed, resembling these parts in the adult; forehead and feathers over the eye pale fawn-colour, with very few dusky cross lines; ear-coverts and sides of face bright bay, with a few wavy cross lines of blackish on the hinder ones; under surface pale sandy buff, numerously barred with dusky brown, these cross markings nowhere very distinct; bill yellowish horn.

Adult (Malacca, *A. R. Wallace*). Above clear sandy brown, coarsely vermiculated with wavy cross lines of black, many of the feathers with broad but irregularly formed streaks of black; scapulars more or less sandy buff on the outer web, the cross lines apparent at the tips of the lower ones, but nearly absent on the upper scapulars, some of which are nearly entirely sandy buff on both webs; round the hind neck a tolerably distinct collaret of sandy buff feathers, only varied with blackish markings towards the tips of the feathers; crown of head blackish in the centre, the plumes laterally rufous buff, with very narrow black cross lines; forehead and a broad streak on each side of the crown sandy buff, the feathers narrowly streaked with black along the shaft, and varied towards the tip with a few blackish cross lines, the ear-tufts (1 inch long) which form part of the light lateral band on the sides of the crown, being pale sandy buff slightly varied with zigzag lines of black towards the tips, which are very broadly black; over the eye a patch of buffy white feathers, each terminally margined with a narrow line of black; lores sandy buff, with whitish bases; sides of face sandy buff, indistinctly barred across with narrow lines of dusky black, the hindermost ear-coverts rather broadly tipped with the same; behind the ear-coverts an indistinct ruff of sandy buff feathers extending across the centre of the throat, many of the plumes narrowly barred across with blackish lines, and all rather broadly tipped with black; rest of under surface of body entirely sandy buff, varied with narrow cross lines of dark brown, the chest with broad central spots or streaks of blackish brown, much narrower and more linear on the abdomen and flank-feathers; under tail-coverts nearly uniform sandy buff, with a few wavy lines of blackish near the tips of the feathers; leg-feathers sandy buff, the tarsal plumes crossed with distinct wavy blackish cross lines; under wing-coverts sandy buff, the inner ones nearly uniform, the outermost mottled or barred with dark brown near the edge of the wing, which is whitish, the lower series entirely ashy brown, the inner lining of the quills nearly uniform ashy brown, barred with yellowish white near the base of the inner web, and with sandy buff near the tips of the feathers; upper wing-coverts darker than the back, being blackish brown, the spurious quills externally notched with sandy buff, the innermost of the least series sandy buff, mottled with zigzag blackish lines, the greater series dark brown, widely notched with sandy buff towards the tip of the outer web; primary coverts blackish brown, with indistinct indications of sandy buff bars, thickly obscured with blackish vermiculations; quills dark brown, barred with lighter or ashy brown on the inner webs, these light bars being represented by corresponding ones of sandy buff on the outer webs of the primaries, giving a chequered appearance to the external aspect of the wing; the secondaries outwardly sandy buff, vermiculated with lines of dark brown, the innermost sandy brown, mottled and vermiculated exactly like the scapulars; rump and upper tail-coverts darker sandy brown than the back, thickly obscured with blackish vermiculations; tail dark brown, mottled with sandy buff, nowhere forming very distinct bars, and particularly broken up towards the tips of the feathers; on the outer ones the bars a little more distinct, about eight being distinguishable, all the light markings everywhere somewhat obscured by brown vermiculations. Total length 8·5 inches, wing 5·8, tail 3·1, tarsus 1·25.

Hab. Borneo, Bangka, Java, Sumatra, Malayan peninsula, ranging northwards into Tenasserim, and occurring in the neighbourhood of Calcutta.

a. Ad. st.	Sunda Islands.	Purchased.
b. Ad. st.	Sunda Islands.	Purchased.
c. Juv. st.	Borneo.	Hugh Low, Esq. [C.].
d. Ad. sk.	Sumatra.	A. R. Wallace, Esq. [C.].
e. Ad. sk.	Malacca.	A. R. Wallace, Esq. [C.].
f. Ad. sk.	Malacca.	R. B. Sharpe, Esq. [P.]
g. Ad. st.	Malacca.	W. Harvey, Esq. [P.].
h. Ad. sk.	Malacca.	W. Harvey, Esq. [P.].
i. Ad. sk.	Malacca.	Purchased.
k. Ad. sk.	Malacca.	Purchased.
l. Pull. sk.	Malacca.	Earl of Ellenborough [P.].
m. Ad. sk.	Singapore.	Earl of Ellenborough [P.].

Obs. The present species in its typical form has but two phases, a brown and a rufous one, the latter being the usual colouring, the former very rare in comparison with it. It never shows a grey phase like *S. lettia*, and is always distinguished by its completely unfeathered toes.

The amount of feathering on the lower joint of the tarsus varies slightly with individuals; but specimens of *S. lempiji* never seem to have the plumes descending so far down on the outer toe as in *S. lettia*; and in this respect it very closely approaches *S. malabaricus*. Indeed, as stated below, they are but races one of the other. Mr. Hume has sent over for my inspection two very interesting specimens, one from Thayetmyo and another from Calcutta, which form an intermediate race between the Malaccan and South-Indian birds. They differ in their very pale coloration and fulvescent tone of plumage, and in the excessively delicate character of the vermiculations. A good idea of these differences will be seen on an examination of the plates in Mr. Hume's forthcoming work on the birds of the Indian empire.

It may be noted that although closely resembling *Scops malabaricus*, the present species nevertheless presents some differential characters, especially in the absence of mesial black streaks, either on the upper or under surface, the black markings being more in the form of spots and bars.

Subsp. *a*. **Scops umbratilis.**

Ephialtes lettia, *Swinh. Ibis*, 1870, p. 88.
Ephialtes umbratilis, *Swinh. Ibis*, 1870, p. 342.
Lempijius umbratilis, *Swinh. P. Z. S.* 1871, p. 344.

Adult male (type of species). Above dull sandy brown, the feathers mesially black, and mottled with coarse blackish vermiculations, relieved here and there with a broader bar of sandy buff; head rather blacker than the back, varied only with a few lateral spots and notches of rufous sandy; forehead buffy white, the feathers narrowly margined with black at the tips, extending backwards so as to form a very distinct eyebrow; the sides of the crown fulvescent, with narrow cross lines and vermiculations of blackish, forming an indistinct band on the sides of the head, continuous with the ear-tufts, the inner webs of which are ochraceous, narrowly barred with black; on the occiput and nape a few fulvous feathers scarcely forming bands; on the hind neck a very distinct collar of ochraceous buff, some of the feathers varied with dark brown cross bars, and resembling the sides of the neck, which are ochraceous, mottled with brown; the scapulars externally clear ochraceous, tipped with dark brown, the inner web also mottled with ochraceous; upper wing-coverts dark brown, more obscure than the back, and vermiculated with rufous sandy, the greater series mottled with ochraceous on the outer webs; primary coverts dark brown, obscurely vermiculated with rufous sandy colour near the tips, and indistinctly barred across with the same; quills dark chocolate-brown on the inner webs, with faint indications of pale ashy brown bars, the outer web barred with sandy brown, obscured with vermiculations of dark brown, disappearing on the primaries, which are externally barred with clear sandy colour, none of these, however, inclining to whitish; upper tail-coverts coloured like the back, but a little more dingy; tail-feathers dark brown, profusely mottled with sandy buff near the tips, and barred most irregularly with the same colour, the bars on the centre feathers not even being conterminous; loral plumes whitish at base, washed with rufous near the tips, and indistinctly barred across with black; all the feathers round the eye and the ear-coverts sandy rufous, obscurely barred across with black, paler and more whitish on the lower margin of the ear-coverts, where they join the ruff, which is composed of buffy white feathers, broadly tipped with black, the plumes on the throat having additional narrow cross lines of black; the chin-feathers whitish; rest of under surface dull fulvous, very thickly obscured with cross vermiculations of dark brown, many of the feathers with longitudinal black streaks, narrower and less pronounced on the abdomen and flanks, where the feathers are whiter and less thickly mottled; leg-feathers dull tawny, thickly barred with dark brown; under tail-coverts white, with a few vermiculatory markings of brown near the tips; under wing-coverts sandy buff, the outermost mottled with brown, the lower series ashy brown, with concealed yellowish bases, the inner lining of the quills ashy brown, with faint indications of paler bars, and barred with sandy buff near the base of the inner web; bill yellowish. Total length 9 inches, wing 6·4, tail 3·4, tarsus 1·5. (*Mus. R. Swinhoe.*)

Hab. Hainan.

Obs. The type specimen kindly lent me by Mr. Swinhoe shows that this is a form of *S. lempiji* differing from the ordinary examples of that species by its less rufous coloration and greater amount of vermiculation. At the same time, however, it so closely resembles greyish brown birds which I have seen from Java, in Lord Walden's collection, and which we have in the Museum from the "Sunda Islands," as to seem but a large edition of that species; but, until more specimens arrive, it will be difficult clearly to settle its position.

Subsp. β. **Scops malabaricus.**

Scops javanicus, *Jerd. Madr. Journ.* x. p. 89 (*nec Less.*).
Scops malabaricus, *Jerd. Madr. Journ.* xiii. p. 119; *Hume, Str. F.* i. p. 454.
Scops griseus, *Jerd. Madr. Journ.* xiii. pt. 2, p. 119.
Scops lettoides, *Blyth, J. A. S. B.* xiv. pt. 1, p. 182 (*ex Jerd. MS.*).
Ephialtes malabaricus, *Hume, Rough Notes*, ii. p. 402; *Jerd. Ibis*, 1871, p. 348.
Ephialtes griseus, *Hume, Rough Notes*, ii. p. 398; *Jerd. Ibis*, 1871, p. 348.
Ephialtes jerdoni, *Walden, Ann. N. H.* (4) v. p. 417.
Ephialtes bakkamuna, *Holdsw. P. Z. S.* 1872, p. 417.
Scops bakkamæna, *Hume, Nests and Eggs Ind. B.* p. 69.

Adult (rufous phase). General colour above light sandy, with tolerably distinct vermiculations of black, all the dorsal feathers mesially streaked with the latter colour, many of these central markings being in the form of spear-shaped spots; the scapulars externally sandy buff, blackish at the tips of the outer webs, on many of which the vermiculations are nearly obsolete; head darker than the back, blackish brown in the centre of the crown, all the feathers centred or marginally spotted with rufous sandy, the lores buffy whitish, tipped with black; feathers over the eye sandy buff, with narrow black tips, these plumes forming a tolerably distinct band down the sides of the head continuous with the ear-tufts, which are sandy buff on the inner web and at the base, the outer webs rufous sandy colour, barred with dark brown, many of the smaller feathers rufous, with coarse vermiculations of dark brown; on the nape and occiput some tolerably distinct bars of sandy buff give an indication of a faint band; the hinder neck pale sandy buff, tipped with blackish brown, and forming a very distinct collar running round to the sides of the neck, which are also sandy buff, slightly vermiculated with brown; over the hinder part of the eye a small patch of rufous feathers running backwards to the base of the ear-coverts; cheeks and ear-coverts rufous sandy, the latter slightly inclining to silvery white on their lower margin, the hinder ones broadly tipped with black, and merging in an indistinct ruff running down the sides of the head, and composed of sandy buff plumes, narrowly tipped with black, these tips becoming narrower on the gular feathers, which are also crossed with fine wavy lines of brown; chin fulvescent; on the fore neck a faintly indicated white band; rest of under surface sandy-coloured, with very few traces of vermiculations; the chest broadly streaked with black mesial stripes, somewhat broken up near the base on some of the feathers and dilating towards the tip on others, those on the abdomen and flanks much narrower, and disappearing entirely on the under tail-coverts, which are whitish; leg-feathers buffy whitish; under wing-coverts also buffy white, with a few markings of sandy buff or brown near the outer edge of the wing, the lower series ashy brown, buffy white at base, resembling the inner lining of the quills, which are ashy brown below, with scarcely any indications of lighter bars, excepting on the inner quills, where they are yellowish white towards the base; upper wing-coverts coloured like the back, and vermiculated in the same manner, with several distinct indications of sandy buff bars, the median and greater series almost uniform sandy buff at their tips, with very few vermiculations, and giving the appearance of a buff-coloured spot; spurious quills indistinctly notched with buffy white on the outer web; primary coverts dark brown, with a few indistinct vermiculated cross bars of sandy buff; quills dark brown, barred on the inner web with ashy brown, these represented on the outer web by sandy bars, more or less vermiculated with dark brown, and inclining to buff on the primaries; the innermost secondaries resembling the back and scapulars; tail brown, vermiculated, like the quills, with sandy buff towards the tips, crossed with about seven bands of clearer sandy colour. Total length 8 inches, wing 5·5, tail 3·25, tarsus 1·15.

The grey phase is so very similar to the corresponding plumage of *Scops lettia*, that a description is unnecessary. It may be mentioned as exactly resembling that species, but to be distinguished by the less amount of feathering on the toes, as seen in the following woodcut.

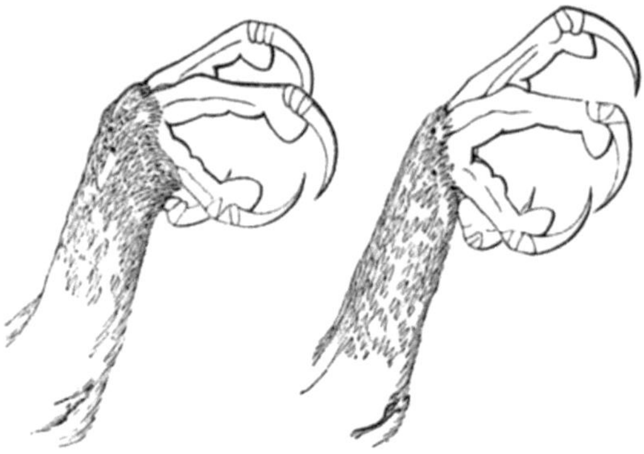

Leg of *Scops lettia*. Leg of *Scops malabaricus* (*griseus*, Jerd.).

Hab. Southern India and Ceylon.

a. Ad. sk.	India (*W. Ewer*).	J. Gould, Esq. [P.].
b. Ad. sk.	S. India.	J. Gould, Esq. [P.].
c. Ad. sk.	Madras.	Dr. Jerdon [P.].
d. Ad. sk.	Madras.	Dr. Jerdon [P.].
e. Ad. st.	Madras.	Rev. H. H. Baber [P.].
f. Ad. sk.	Madras.	Rev. H. H. Baber [P.].
g. Ad. sk.	Madras.	J. Brenchley, Esq. [P.].
h. Ad. sk.	Malabar.	Purchased.
i. Ad. sk.	Ceylon.	J. Gould, Esq. [P.].
k. Ad. sk.	Ceylon.	Purchased.
l. Ad. st.	Ceylon.	E. L. Layard, Esq. [C.].
m. Juv. st.	Ceylon.	E. L. Layard, Esq. [C.].

Obs. In this species we have a parallel case with the *Spizaeti* of India, where the hill form, *S. nipalensis*, differs only slightly in plumage, but depends for its characters on the exaggerated feathering of its tarsus for separation from *S. cirrhatus* (*cf.* Sharpe, Cat. B. i. p. 261). The Malabar Scops Owl resembles in the main the Himalayan *S. lettia*, but differs in the smaller extent of the tarsal feathering, which does not cover the toes. If I am right in considering *Scops griseus* to be the same species as *S. malabaricus*, of which it is the grey phase, then it is certain that *S. lettia* never shows such a thoroughly rufous phase as the typical *S. malabaricus*. In its rufescent form it is nothing but a large *S. lempiji*. The Ceylonese bird appears to be identical with the South-Indian, though in the Museum the majority of the specimens are rufescent. I have not, indeed, seen a really grey-plumaged bird from Ceylon, the darkest examined being of a very dark brown, blackish on the head and greyish underneath, with very coarse vermiculations.

11. Scops leucotis.

Strix leucotis, *Temm. Pl. Col.* i. pl. 16.
Otus leucotis, *Steph. Gen. Zool.* xiii. pt. 2, p. 57; *Smith, S.-Afr. Q. Journ.* ii. p. 316; *Rüpp. Syst. Uebers.* p. 12.
Asio leucotis, *Less. Man. d'Orn.* i. p. 116.
Scops leucotis, *Swains. Classif. B.* ii. p. 217, et *B. W. Afr.* ii. p. 124; *Strickl. & Sclater, Contr. Orn.* 1852, p. 142; *Strickl. Orn. Syn.* p. 203; *Hartl. Orn. W.-Afr.* p. 20; *Gray, Hand-l.* i. p. 46; *Bocage, Jorn. Sci. Lisb.* i. pp. 132, 330, ii. p. 338; *Gurney, ed. Anderss. B. Dam. Ld.* p. 40; *Sharpe, ed. Layard, B. S. Afr.* p. 74.
Ephialtes leucotis, *Gray, Gen. B.* i. p. 38; *id. Cat. Accipitr. B. M.* 1848, p. 97; *Cass. Cat. Strigidæ Philad. Mus.* p. 11; *Bp. Consp.* i. p. 45; *Layard, B. S. Afr.* p. 41.
Ptilopsis leucotis, *Kaup, Tr. Z. S.* iv. p. 225, et *Contr. Orn.* 1852, p. 111; *Bp. Rev. et Mag. de Zool.* 1854, p. 542; *Antin. & Salvad. Viagg. Bogos.* p. 39.
Ægolius leucotis, *Heugl. Syst. Uebers.* p. 14.
Bubo leucotis, *Schlegel, Mus. P.-B.* Oti, p. 17; *Finsch & Hartl. Vög. Ostafr.* p. 106; *Heugl. Orn. N.O.-Afr.* p. 115; *Schl. Revue Accipitr.* p. 6.

Adult male. General colour above ashy grey, narrowly vermiculated across with dark brown, the dorsal plumes with more or less distinct mesial streaks of black; scapulars pure white on the outer web, forming a distinct longitudinal patch down the wing; head much lighter grey than the back, the feathers in the centre of the crown blackish, mottled with white, which is distributed in duplicate spots on either web; sides of the crown paler and more whitish, delicately pencilled with transverse linear vermiculations; auricular tufts 1·8 inch long, coloured and pencilled like the crown, black on the outer web for nearly the apical half of the plume, the tip also black; wing-coverts coloured like the back, but more obscurely mottled, the dark mesial streaks being browner and somewhat less pronounced than on the back, the median series paler and having the ground-colour of the feathers whiter than the other series; edge of the wing all round pure white; primary and greater coverts greyish white, broadly barred across with dark brown, the light interspaces freckled and obscured with brown vermiculations; quills greyish brown, barred across with brown, these bars being about nine in number on the primaries, which incline to greyish white on their outer margin; the secondaries similarly coloured and having whitish tips, the interspaces of these quills decidedly grey, freckled with brown vermiculations, the innermost secondaries coloured more like the back, vermiculated with brown, but the dark cross bars nearly obsolete; tail ashy grey, crossed with about nine bars of dark brown, irregular in shape, the interspaces freckled with brown vermiculations; lores, eyebrow, and entire face pure white, the hinder margin of the ear-coverts black, forming a distinct broad bar of black down the sides of the head; over the eye a patch of greyish brown feathers; sides of neck pale greyish white, with very minute transverse frecklings of brown; under surface of body pale ashy grey, all the feathers crossed with narrow linear frecklings of brown, the black shaft-stripes very distinctly indicated; across the throat a tolerably broad band of white, separating the throat from the chest; centre of breast, abdomen, flanks, leg-feathers, and under tail-coverts pure silky white, the black shaft-streaks distinct on the flank-feathers, which, however, have little or no trace of the transverse brown frecklings; under wing-coverts pure white, the outermost of the lower series white at base, ashy brown for their terminal half, and resembling the inner lining of the quills, which are dull greyish white, barred with ashy brown; bill and feet bluish white; iris bright orange. Total length 9·5 inches, wing 7·7, tail 3·85, tarsus 1·7.

Young. Above more ochreous grey than the adults, regularly transvermiculated with brown; underneath browner and more washed with yellowish on the lower parts than the old birds.

Obs. But very little difference exists in the plumage of this Owl, which appears to constitute only one species all over Africa. Two specimens in the Museum collection (one from the White Nile, the other from the Gambia) have the hinder part of the crown nearly uniform black; and as this is accompanied by a much greater breadth of the mesial streaks to the feathers of the upper surface, and by a yellowish tint underneath, perhaps the birds may be not quite adult. In the Damara specimen described, the legs are nearly pure white, with only the faintest remains of tiny brown streaks on the lower parts; but in some examples the legs are very distinctly streaked.

Hab. The continent of Africa south of the Sahara.

a. Pull. sk.	River Gambia.	Purchased.
b, c. Ad. st.	River Gambia.	Governor Rendall [C.].
d. Ad. sk.	White Nile.	F. Galton, Esq. [P.].
e. Ad. sk.	Tete, Zambesi.	Dr. Kirk [C.].
f. Ad. sk.	Damara Land.	C. J. Andersson, Esq. [C.].
g. ♂ ad. sk.	Otjimbinque, Damara Land, Sept. 20th, 1866.	C. J. Andersson, Esq. [C.].

12. Scops sagittatus.

Ephialtes sagittatus, *Cass. Journ. Ac. Philad.* ii. p. 96, pl. 12; *id. Cat. Strigidæ Phil. Mus.* p. 11; *Gray, Gen. B.* iii. App. p. 30 *a.*
Lempijius sagittatus, *Bp. Rev. et Mag. de Zool.* 1854, p. 543.
Scops sagittatus, *Gray, Hand-l. B.* i. p. 46.

Adult. Above bay or chestnut-red, of a tolerably uniform character, the hind neck varied with tiny fulvous spots on the feathers, these spots margined with a narrow blackish bar; all these markings very indistinct on the hinder crown and occiput, and absent on the top of the crown; on the hind neck many feathers are yellowish buff, crossed with bars of dusky blackish, the fulvous markings on the interscapulary region being larger and more arrow-shaped, as well as on the scapulars, the outer webs of which are buffy whitish, with distinct blackish spots and markings, the inner web varied with tiny blackish vermiculations; wing-coverts deep bay, the sagittate markings on the least series rufous buff, larger and more yellow on the greater and median coverts, which are also very minutely vermiculated with blackish, as on the inner scapulars; the outer webs of the spurious quills notched with buffy white; primary coverts blackish on the inner web, bay on the outer, mottled with brown vermiculations, darker near the base, and having, near the tips of the shafts, longitudinal fulvous streaks; quills externally light bay, broadly barred with rufous or rufous brown, the interspaces whitish on the outer web of the primaries, giving a chequered appearance to the outer aspect of the wing, all the quills blackish on their inner webs, barred with lighter or more rufous brown, the primaries rufous at the tips of both webs; secondaries deep bay, almost uniform externally, or with a slight appearance of dusky bars, the innermost secondaries with indications of whitish spots resembling the scapulars; upper tail-coverts deep chestnut, like the back, with similar small fulvous spots; tail deep bay, tipped with fulvous, barred across with blackish, but very indistinctly, the dark bars scarcely visible on the outer webs of the feathers, the centre ones also almost uniform; forehead greyish white, and very broad, narrowly barred with blackish, the hinder feathers tipped with the same colour as the head, and resembling the crown, the grey colour of the forehead extending backwards along the side of the head to the ear-tufts, which are 1·2 inch long, the basal plumes being greyish white, narrowly barred with black, like the forehead, the longer plumes externally bay, barred and broadly tipped with blackish, and further varied with a few zigzag lines of the same, the inner webs for the most part greyish white, slightly barred and vermiculated with dark brown lines; lores and feathers round the eye chestnut, with lighter shaft-lines of buff; ear-coverts greyish white, with indistinct mottlings, the hinder ones tipped with blackish; behind the ear-coverts an indistinct ruff of whitish plumes, tipped with blackish, and narrowly crossed with blackish lines; cheeks greyish white, with dusky brown cross lines; chin-feathers chestnut; rest of under surface pale sandy buff, much deeper on the chest, which is vermiculated with blackish wavy cross lines, broader and more rufescent on the centre of the chest, which has also distinct spots of white, these spots generally containing a central irregular spot of dark brown; on the breast and flanks the brown vermiculations wider apart and less distinct, varied with white bars and spots of brown; the under tail-coverts paler, and with scarcely any markings at all; leg-feathers bay, paler and more fulvescent on the tarsus, distinctly barred with blackish; under wing-coverts rufous buff, with a patch of rufous-brown feathers near the edge of the wing, which is whitish, the lower series blackish brown, fulvous at base, resembling the inner lining of the quills, notched with buff on the inner webs, the tips of the quills rufous below, the dusky bars disappearing; bill yellow; feet apparently yellow in life. Total length 10 inches, wing 7·15, tail 4·5, tarsus 1·25.

Hab. Malacca.

a. Ad. sk.	Malacca.	Purchased.
b. Ad. st.	Malacca.	

13. Scops balli.

Ephialtes spilocephalus, *Ball, Str. F.* i. p. 53.
Ephialtes balli, *Hume, t. c.* p. 407.

Adult (type of species). General colour above rufous chocolate, with a few fine, nearly obsolete vermiculations of black, the whole upper surface with more or less concealed spots of rufous buff, very minute on the crown, but much more distinct on the back, where they form almost bars, all the eight spots having a blackish margin: scapular feathers externally barred with white, the interspaces very pale, and inclining to ochraceous; wing-coverts coloured like the back, the bases dark brown, the tips chocolate rufous, with rather more distinct vermiculations of black, the spots very plain, and forming notches on the outer web of some of the greater series; primary coverts exactly like the other, brown at base, rufescent at tips, but without the light spots; quills dark brown, nearly uniform on the inner web, and chocolate-rufous at the tips of the primaries and on the outer webs of the secondaries, which are minutely notched with fulvous, the primaries very distinctly chequered with white on the outer webs, the innermost secondaries spotted with rufous buff, resembling the scapulars; tail dark brown near the base, crossed with about six bars of dull rufous, these becoming obsolete towards the tips of the feathers, which are rufous chocolate, very finely vermiculated with black, the outer feather externally notched with fulvous, and all the feathers of the wings and tail barred with fulvous on the inner web, especially towards the base; forehead and feathers over the fore part of the eye creamy white, narrowly tipped with brown; loral plumes rufous, the shafts ending in black hair-like bristles, barred obscurely with black; ear-coverts rufous chocolate, barred and tipped with black; cheeks fulvous, all the feathers tipped, and some of them barred across, with dull black; general colour of under surface rufous sandy, obscured with grey, and finely vermiculated with blackish cross lines, the bases of the feathers, especially those of the flanks, strongly rufescent, broadly barred with fulvous, inclining to white near the tips of most of the feathers; and before this whitish tip, appears generally a slight indication of a diamond-shaped spot of black, evidently the remains of a streak; all the above markings, though distinct on the abdomen and flanks, are less plainly characterized on the chest, which is consequently more dusky; leg-feathers orange-tawny, narrowly barred with dark brown; under wing-coverts fulvous, slightly marked with brown and spotted with sandy buff near the edge of the wing, which is whitish, the lower series dark brown, fulvescent at base, resembling the inner lining of the quills, which are dark brown, notched on the outer web and barred on the inner with fulvous. Total length 7·5 inches, wing 5·45, tail 3·2, tarsus 1·15. (*Mus. Ind. Calc.*)

Obs. Another specimen, very kindly lent me by Mr. Hume, differs considerably from the one described. It is not such a deep rufous in colour, and is much more thickly spotted both above and below, recalling *Scops spilocephalus*; the under surface is much paler and greyer, the white bars very large and distinct, and extending even on to the chest. Total length 7·5 inches, wing 5·6, tail 3·2, tarsus 1·15.

Hab. Andaman Islands.

Obs. It has been suggested by Mr. Hume that the bird from the Andamans, named *Scops modestus** by Lord Walden, must be the young bird of *S. balli*; and I confess that, until I examined and compared the types, I entertained a similar impression. Lord Walden, however, having kindly lent me the original specimen of *S. modestus* for examination, I have come to the conclusion that the two species are quite distinct. Lord Walden has two specimens precisely similar; and they seem to me to represent the young of some species of the *S. malayanus* type. Immature they certainly are; but they present too many differences for me to refer them to *S. balli.* The type of the latter has been lent to me by Mr. Hume; and as the wings and tail in *S. modestus* are doubtless those of the adult bird, I draw attention to the following characters, which, as it seems to me, must separate the two; for in no other species of *Scops* is such a difference known between the young and the adult stages.

Scops balli, ad.	*Scops modestus*, juv.
Greater wing-coverts and *secondary quills* dull brown, externally rufous chocolate, with minute vermiculations of black, and a few small notches of fulvous.	*Greater wing-coverts* and *secondary quills* alternately barred with sandy rufous and dark brown, the latter bars rather broken up into vermiculations, especially on outer margin; the greater coverts with white spots near the tip of the outer web, not present in the secondaries.
Primary coverts nearly uniform blackish brown, vermiculated with rufous chocolate at the tips. Primaries dull brown, rufescent at tips, notched with white on outer web, the interspaces inclining to rufous chocolate.	*Primary coverts* and *primaries* dull brown on inner web, but regularly banded with sandy rufous and dark brown on outer web; some of the primaries with whitish notches.
Tail for the most part rufous chocolate, like back, with indications of lighter bands, the outer feather externally notched with whitish.	*Tail* regularly banded with dark brown and sandy rufous, the dark bars somewhat broken up into vermiculations on the centre feather.

Again, the feathering of the tarsus is very different, not extend-

* *Scops modestus,* Wald. Ann. N. H. (4) xiii. p. 123, et Ibis, 1874. p. 129.

ing nearly so far down the leg in *S. balli* as it does in *S. modestus*. This will be seen at a glance in the following woodcuts.

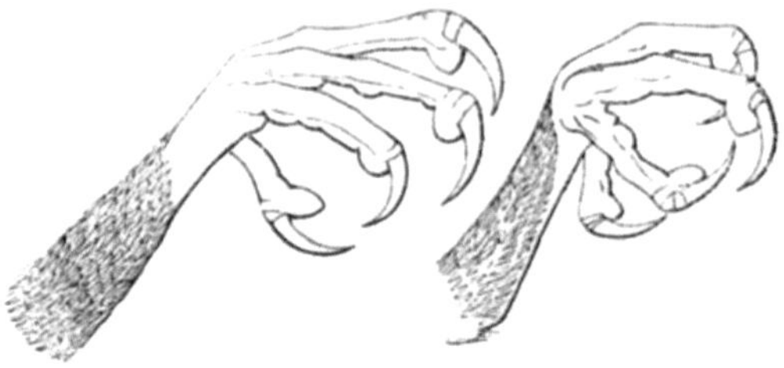

Leg of *S. balli* (type). Leg of *S. modestus* (type).

The following is a description of the type specimen of *S. modestus*:—

Immature (type of species). General colour above dull sandy brown, slightly rufescent in places, everywhere crossed with coarse zigzag lines of black; scapulars externally white, washed on the margin and at the tip with rufous, and irregularly mottled with black; wing-coverts pale sandy brown, barred with darker brown, the greater ones tipped with whitish; the bastard wing notched with buffy white; primary coverts sandy rufous, barred with brown; quills sandy rufous, crossed with vermiculated bars of dark brown, the primaries with white notches on the outer web; tail sandy rufous, strongly vermiculated with blackish forming four distinguishable bars, the light interspaces shading off into white on the margin of both webs; lores white, as well as the feathers over the eye, forming a tolerably distinct eyebrow; feathers of the side face sandy colour, barred with white, especially broad on the cheeks and hinder ear-coverts; chin white; fore neck and chest sandy brown, with irregular bars of blackish, and here and there an indication of a whitish bar; rest of under surface white, slightly washed here and there with sandy buff, and crossed with irregular bars of brown, the leg-feathers similarly coloured; under tail-coverts white, with a zigzag bar of brown near the tip of the feather; under wing-coverts white, marked with brown near the outer edge of the wing, the lower series brown, barred with sandy buff on the inner web, resembling the inner lining of the quills, which are barred on the inner web and notched on the outer with sandy buff, the whole of the quills mottled with this colour towards the tips. Total length 6·8 inches, wing 4·75, tail 2·3, tarsus 1. (*Mus. Lord Walden.*)

Hab. S. Andamans.

14. **Scops rufescens.**

Strix rufescens, *Horsf. Tr. Linn. Soc.* xiii. p. 140.
Scops rufescens, *Sw. Classif. B.* ii. p. 217; *Blyth, J. A. S. B.* xiv. p. 181; *Strickl. Orn. Syn.* p. 201; *Gray, Hand-l. B.* i. p. 46; *Schl. Mus. P.-B. Revue Accipitr.* p. 10; *Salvad. Ucc. Born.* p. 19.
Otus mantis, *Temm. & Schl. Faun. Jap. Aves*, p. 25.

Ephialtes mantis, *Gray, Gen. B.* i. p. 38; *Wall. Ibis*, 1868, p. 24.
Scops mantis, *Strickl. Orn. Syn.* p. 201; *Schl. Mus. P.-B.* Oti, p. 25; *Gray, Hand-l. B.* i. p. 46.
Ephialtes rufescens, *Horsf. & Moore, Cat. B. Mus. E. I. Co.* i. p. 71.
Lempijius mantis, *Bp. Rev. et Mag. de Zool.* 1854, p. 543.

Adult (type of *Scops mantis*). Everywhere bright tawny rufous, browner on the head and upper parts generally, these also varied with small triangular cross bars of fulvous, which become larger and more distinct on the wing-coverts; the outer margin of the scapulars paler and more inclining to tawny fulvous; ear-coverts 1·05 inch long, and coloured like the head; lores and forehead clear fulvous, with little black spots along the shaft of the feathers; sides of face rufous, inclining to dark brown round the eye; behind the ear-coverts a line of black feathers running down the sides of the neck; under surface of body tawny rufous, paler on the lower parts, and varied with little tiny spots of black along the shafts, less distinct towards the abdomen, and absent on thighs and under tail-coverts, which are uniform tawny buff; quills dark brown, barred across with fulvous, paler on the outer margin of primaries, but less distinct on the secondaries, which are almost uniform rufous like the back, but with indistinct bars of dark brown on the inner web; tail rufous, mottled with dark brown vermiculations, making it look darker than the back, with remains of five or six more or less plainly indicated fulvous bars; under wing-coverts buffy white, more tawny on outer edge, and having a line of dark brown feathers near the outer margin. Total length 9·4 inches, culmen 0·9, wing 5·2, tail 2·6, tarsus 1·05. (*Mus. Lugd.*)

Hab. Malacca, Java, Sumatra, Borneo.

a. ♀ ad. sk.	Mount Ophir, Malacca.	A. R. Wallace, Esq. [C.].
b. Ad. sk.	Malacca.	R. B. Sharpe, Esq. [P.].

15. Scops icterorhyncha.

Scops icterorhyncha, *Shelley, Ibis*, 1873, p. 138.

Adult (type of species). General colour above pale sandy, with scarcely any appearance of vermiculations at all on the back, the head, however, being scantily vermiculated; all the upper surface indistinctly stellated with buffy white spots or small bars, all these light spots relieved by a narrow line of dark brown; the hind neck more distinctly mottled with bars of buffy white, the bases to the feathers being also of this light colour, still without showing any evidences of a neck-collar; ear-tufts 0·85 inch long, vermiculated with brown zigzag lines, like the rest of the crown, but distinctly mottled with buff transverse bars; outer scapulars yellowish white externally, with irregular but distinct cross bars of black, confined to the outer web; wing-coverts rather more rufous than the back, with arrow-head spots or bars of sandy buff, inclining to whitish on the median series, the greater ones with a large spot of white at the tip of the outer web, the spurious quills also externally notched with buffy white; primary coverts sandy buff, with minute blackish vermiculations, the inner webs also inclining to blackish; quills sandy-coloured, barred with dark brown on the inner web, the primaries externally notched with white; the innermost secondaries light sandy, slightly freckled with dull blackish vermiculations, and crossed with five or six pale fulvous bars; tail pale sandy colour, with a few nearly obsolete vermiculations of blackish, and crossed with about eight fulvous bars, disappearing towards the tips of the feathers; lores sandy rufous, inclining to blackish at the tips; a broad frontal patch of feathers extending on each side over the eye, obsoletely freckled with brown, and obscurely tipped with blackish; ear-coverts dull sandy brown, with whitish shaft-lines, and obscurely barred with dusky brown; behind the ear-coverts an indistinct ruff of fulvous feathers, narrowly tipped with black, these tips disappearing on the gular feathers, which have the shaft-stripes obscurely black; rest of under surface pale ashy buff, obsoletely freckled with vermiculations of pale brown, many of the feathers inclining to dull whitish on the breast, increasing into tolerably distinct bars on the abdomen and under tail-coverts; a few of the feathers on the sides of the breast with some tiny blackish central streaks; leg-feathers sandy colour, with narrow linear markings of rufous brown; under wing-coverts fulvescent, those near the edge inclining to sandy brown, the lower series ashy brown, fulvous at base, resembling the inner lining of the quills, which are brown below, barred on both webs with fulvous in about equal proportions. Total length 7·5 inches, wing 5·1, tail 2·7, tarsus 0·95. (*Mus. G. E. Shelley.*)

Hab. Gold Coast.

B. *Species of the New World.*

a. Total length not exceeding 11 inches; ear-tufts not more than 1·5 inch long.
- *a′.* Tarsus feathered for more than half its length.
  - *a″.* Toes bare.
    - *a‴.* Size small; wing not exceeding 5½ inches.
      - *a⁴.* Tarsus thickly feathered right to the junction of the toes; sides of crown grey, like sides of neck; top of crown and sides of face washed with orange, the white spots on the scapulars also washed with orange .... *flammeolus*, p. 105.
      - *b⁴.* Tarsus bare on extreme lower portion; no orange on sides of face or on top of head, nor on the scapular spots; sides of crown not strikingly different from the crown itself .... *barbarus*, p. 107.
    - *b‴.* Size large; wing exceeding 6 inches. *brasilianus* and allies, pp. 108–112.
  - *b″.* Toes feathered .... *asio* and allies, pp. 114–121.

b'. Tarsus bare for half its length .... *nudipes*, p. 121.
b. Size very large; total length from 18 to 20 inches; ear-tufts about 2 inches long.
c'. Head rufous chocolate; feathers round the eye and ear-coverts entirely rufous .. *cristata*, p. 122.
d'. Head blackish; feathers round the eye black; ear-coverts rufous, broadly tipped with black .... *stricklandi*, p. 124.

16. Scops flammeolus.

Strix flammeola, *Licht. MS. in Mus. Berol.*, undè
Megascops flammeola, *Kaup, Tr. Z. S.* iv. p. 226.
Ephialtes flammeola, *Licht. Nomencl. Av.* p. 7.
Scops flammeola, *Scl. P. Z. S.* 1858, p. 96; *Schl. Mus. P.-B.* Oti, p. 27; *Scl. & Salv. P. Z. S.* 1868, p. 57; *iid. Ex. Orn.* pl. 50; *Gray, Hand-l. B.* i. p. 47; *Elliot, B. N. Am.* i. pl. xxviii.; *Coues, Key N. A. Birds*, p. 203; *Schl. Revue Accipitr.* p. 14; *Ridgw. B. N. A.* iii. p. 58; *Scl. & Salv. Nomencl. Av. Neotr.* p. 117; *Coues, B. N. W.* p. 304.

Adult (*grey phase*). Above grey, with the usual blackish cross vermiculations, most of the feathers also centred with a spade-shaped streak of black, the scapulars externally white, slightly spotted and tipped with black, and inclining to orange-buff on the outer web, in marked contrast to the back; head rather darker than the back, but more mottled, the black centres to the feathers broader and more spotted with white or buff; on the occiput a slight indication of a greyish white band, the feathers being barred with white, on the nape a second bar of greyish white, not quite so distinct; on the hind neck a very distinct collar of white, streaked and barred with black, and washed with orange-buff; sides of neck grey, very finely lined with blackish vermiculations; wing-coverts a little darker than the back and similarly vermiculated, the least series inclining to ochraceous, the spurious quills externally notched with ochraceous buff, the median and greater coverts grey, the outer webs pure white, very slightly vermiculated, and then only at the tip; primary coverts dark brown, with slight indications of sandy buff bars, inclining to white near the base; quills dark brown, with faint indications of sandy bars towards the tips, the secondaries with irregular sandy bars on the outer web, the primaries externally notched with white, the innermost secondaries greyish white, coarsely mottled with brown; tail-feathers dark brown, with tolerably distinct bars of sandy brown, inclining to white on the outer web of all, but especially of the external feathers, the bars on the tail about seven in number; lores buffy white, washed with rufous near the tip, the shaft-lines produced into hair-like black bristles; forehead and feathers over the eye white, with narrow blackish tips, the sides of the crown also greyish, extending to the ear-tufts, which have the inner webs greyish white, with narrow cross lines of black; sides of face greyish white, the feathers round the eye dusky, all barred across with dull brown, washed with orange on the hinder margin of the ear-coverts, which are tipped with black, and thus assimilate to the ruff, which is composed of greyish white feathers tipped with black, before which is a subterminal bar of orange, the plumes on the throat crossed with several greyish black lines; chin fulvous; rest of under surface white, with wavy cross lines of blackish, many of the feathers with broad central streaks of black, very distinct on the abdomen, where the vermiculations are fewer, being almost obsolete on the under tail-coverts, which are white washed with orange-buff; the broad central streaks on the breast-feathers in many cases edged with rufous; leg-feathers whitish, barred with brown on the outer aspect, and streaked with the same on the inner face of the tarsus; the tibia yellowish; under wing-coverts yellowish, the outer edge of the wing white, the adjacent feathers mottled with brown, the lower series ashy brown, like the inner lining of the quills, barred with fulvous, especially on the inner web. Total length 6·5 inches, wing 5·35, tail 2·7, tarsus 0·85.

Rufous phase. Orange-rufous where the other bird is grey, the markings exactly similar, and the three bands visible on the head and hind neck, the occipital one being the most difficult to trace; under surface very coarsely mottled, the cross bars being broad, and the central black streaks very distinct; all the lower parts strongly pervaded with orange-buff, forming a distinct patch on the throat. Total length 6·5 inches, wing 5·15, tail 2·8, tarsus 0·85.

Obs. Both the specimens above noticed are from Mexico. Mr. Salvin has kindly lent me two from Guatemala, which, besides being larger, differ also in the peculiar way in which the orange and grey are intermingled; the stripes on the sides of the head, as well as the three characteristic bands, are very highly developed; the eyes are surrounded by bright orange-buff, which is also the pervading colour on the ear-coverts. Measurements are as follows:—

	Total length. in.	Wing. in.	Tail. in.	Tarsus. in.
a. Ad. Dueñas, 1863	6·5	5·4	2·6	0·9
b. Ad. Dueñas, January 1863....	7·0	5·5	2·7	1·0

The present species, which is the representative in America of *Scops giu*, appears to be very variable; and although grey birds often have an admixture of orange in their plumage, sometimes this disappears, perhaps with age, as in the British Museum is a perfectly grey bird, on which scarcely a tinge of orange colouring remains, either above or below, while the whole appearance of the specimen is dingy, owing to the closeness and frequency of the vermiculations.

Hab. California; extreme southern portion of United States, ranging through Mexico to Guatemala.

a. Ad. st.	Mexico.	Purchased.
b. Ad. sk.	W. Mexico.	A. Dorman, Esq. [C.].
c. Ad. sk.	Valley of Mexico.	M. A. Bouvier.

17. Scops barbarus.

Scops flammeola, *Salvin, Ibis*, 1861, p. 355.
Scops barbarus, *Scl. & Salv. P. Z. S.* 1868, p. 57; *iid. Ex. Orn.* p. 101, pl. li.; *Gray, Hand-l. B.* i. p. 47; *Scl. & Salv. Nomencl. Av. Neotr.* p. 117.

Adult, grey phase (type of species). Above brown, barred across with sandy buff, the bars on the hinder neck paler and more white, forming a kind of indistinct collar, the head much darker brown, spotted and barred with dull rufous, the forehead spotted with white, these white spots continued on the sides of the crown to the nape, where the feathers are barred with white, so as to form a very distinct nuchal collar; scapulars white on the outer web, tipped with black and forming a distinct shoulder-patch; wing-coverts rather darker than the back, spotted and barred with sandy buff, the median and greater series marbled with fulvous markings, the outer webs with more or less perfect oval spots of white, the spurious quills notched with white externally; primary coverts almost entirely dark brown, with only a few rufous markings near the base; quills dark brown, the inner webs uniform or with only the faintest indications of lighter bars, the outer webs slightly notched or barred with rufous sandy, the primaries very distinctly notched with white externally, the innermost secondaries marbled with white or sandy buff like the greater coverts; tail tolerably uniform dark brown, with a few nearly obsolete sandy bars or spots towards the tips, the outer feathers notched with white; lores whitish, tipped with black, these feathers extending backwards over the fore part of the eye; sides of face greyish white, regularly barred with blackish, the plumes above and below the eye rufescent; all the feathers of ear-coverts, eyebrow, and lores with their shafts enormously produced into hair-like bristles, forming a fringe round the face; ear-tufts blackish, barred with white, so as to assimilate to the nuchal collar; ruff round the neck regularly barred with white and black, being broadly tipped with the latter colour, the gular plumes less distinctly barred, the cross bars being dusky; chin whitish, as also the centre of the abdomen; rest of under surface marbled with black and white, the latter forming distinct bars on the chest, but much larger on the flanks, some of the feathers here showing traces of rufous bars; under tail-coverts white, with a slight tinge of rufous and a subterminal spot of blackish; leg-feathers sandy rufous, barred with dark brown, the tarsal plumes whitish on the lateral and hinder aspects, not descending to the junction of the toes, which are consequently bare; under wing-coverts yellowish, the outer edge of the wing white, with a few bars of dark brown on the adjacent feathers; bill yellow, dark brown at base. Total length 7 inches, wing 5·35, tail 3, tarsus 1·05. (*Mus.* Salvin and Godman.)

Red phase. Totally different from the grey phase. General colour above bright bay, obsoletely barred with dull brown, the feathers fulvescent near their bases, most of the dorsal plumes mesially streaked and tipped with black, some of them with rather distinct bars of the same; the head much more distinctly streaked with black; round the nape a white collar, the frontal plumes and feathers over the eye, as well as the base of the lores, white, the elongated shafts of the latter blackish; ear-tufts bay like the head, the inner webs fulvous; sides of face rufous, with narrow whitish shaft-lines, the hinder ear-coverts tipped with black and assimilating to the ruff, which consists of buffy white feathers, distinctly tipped with black, the gular plumes whitish, barred with rufous; chin whitish; rest of under surface white, barred across with rufous, some of these rufous bars fringed with conterminous narrow lines of brown, many of them centred with black, the streaks dilating into bars on the chest-feathers; under tail-coverts white, with only a subterminal spot of rufous; leg-feathers tawny on the thighs, shading off into buffy white on the tarsus; under wing-coverts yellowish, with a few rufous markings near the edge of the wing; scapulars with the usual white spots on the outer web; wing-coverts rather deeper bay than the back, but otherwise marked in the same manner, with large oval spots of white on the outer web; quills dark brown, numerously chequered with rufous bars on the outer web, inclining to fulvous on the primaries; upper tail-coverts nearly uniform bay; tail dark brown, narrowly banded with rufous, the bars broader, and inclining to buff on the outer web; bill greenish horn-colour, yellowish at tip and on the lower mandible; feet yellowish. Total length 7 inches, wing 5·25, tail 2·9, tarsus 1·05. (*Mus.* Salvin and Godman.)

Hab. Santa Barbara, Vera Paz, Guatemala.

18. Scops brasilianus.

Le Hibou de Brésil, *Briss. Orn.* i. p. 499.
Strix brasiliana, *Gm. S. N.* i. p. 289.
Choliba, *Azara, Apunt.* ii. p. 218.
Strix choliba, *Vieill. N. Dict. d'Hist. Nat.* vii. p. 39.
Strix decussata, *Licht. Verz. Doubl.* p. 59.
Strix atricapilla, *Temm. Pl. Col.* ii. pl. 145.
Strix crucigera, *Spix, Av. Bras.* i. p. 22, pl. 9.
Strix undulata, *Spix, Av. Bras.* i. p. 23, pl. 10.
Scops atricapilla, *Steph. Gen. Zool.* xiii. pt. 2, p. 51, pl. 39; *Cuv. Règne An.* 1829, p. 347; *Bp. Consp.* i. p. 46; *Kaup, Contr. Orn.* 1852, p. 112; *Strickl. Orn. Syn.* p. 202; *Burm. Th. Bras.* ii. p. 128.
Scops lophotes, *Less. Traité,* p. 107; *Pucher. Rev. et Mag. de Zool.* 1849, p. 22; *Bp. Consp.* i. p. 46; *Strickl. Orn. Syn.* p. 204; *Scl. & Salv. Ex. Orn.* p. 102; *Gray, Hand-l. B.* i. p. 47.
Scops portoricensis, *Less. Traité,* p. 107; *Pucher. Rev. et Mag. de Zool.* 1849, p. 26; *Scl. & Salv. Ex. Orn.* p. 102; *Gray, Hand-l. B.* i. p. 47.
Athene undulata, *Gray, Gen. B.* i. p. 35.
Ephialtes choliba, *Gray, Gen. B.* i. p. 38; *Pelz. Orn. Bras.* p. 9.
Ephialtes brasiliensis, *Gray, Gen. B.* i. p. 38.
Ephialtes portoricensis, *Gray, l. c.* p. 38; *Des Murs, Iconogr. Orn.* pl. 26.
Ephialtes lophotes, *Gray, l. c.* p. 38.
Ephialtes atricapilla, *Gray, l. c.* p. 38, pl. 13. fig. 2 (head); *Pelz. Orn. Bras.* p. 9.

Scops choliba, *D'Orb. Voy. Am. Mérid. Ois.* p. 132; *Tschudi, Faun. Peruan.* p. 118; *Strickl. Orn. Syn.* p. 204.
Ephialtes watsoni, *Cass. Pr. Philad. Acad.* 1848, p. 123; *id. Journ. Philad. Acad.* ii. p. 95, pl. xii. fig. 1.
Otus brasiliensis, *Temm. & Schl. Faun. Jap.* p. 25.
Scops brasiliensis, *Bp. Consp.* i. p. 46; *Kaup, Contr. Orn.* 1852, p. 112; *Schl. Mus. P.-B.* Oti, p. 21; *id. Revue Accipitr.* p. 11.
Scops watsoni, *Bp. Consp.* i. p. 46; *Gray, Hand-l. B.* i. p. 47.
Megascops atricapilla, *Kaup, Tr. Z. S.* iv. p. 228.
Megascops brasiliensis, *Kaup, l. c.* p. 228.
Ephialtes argentina, *Licht. Nomencl. Av.* p. 7; *Schl. Mus. P.-B.* Oti, p. 21.
Asio atricapillus, *Bp. Rev. et Mag. de Zool.* 1854, p. 543.
Asio brasiliensis, *Bp. l. c.* p. 543.
Asio watsoni, *Bp. l. c.* p. 543.
Asio lophotes, *Bp. l. c.* p. 543.
Scops decussata, *Burm. Th. Bras.* ii. p. 126.
Scops argentina, *Gray, Hand-l. B.* i. p. 47.
Scops brasilianus, *Gray, Hand-l. B.* i. p. 47; *Scl. & Salv. Nomencl. Av.* p. 117.

Nestling (Bahia: *Wucherer*). General colour sandy buff, slightly tinged with rufous on the back, everywhere regularly barred with dusky brown, the head paler and having whitish downy tips to the feathers of the forehead; under surface sandy buff, rather inclining to golden buff on the belly and flanks, where also the dusky cross bars are more indistinct than they are on the chest; leg-feathers whitish, and several of the abdominal plumes with fluffy white tips.

Adult. Above brown, vermiculated with darker brown, most of the feathers with more or less distinct fulvous or buffy white bars or spots, especially on the hind neck, where they form a tolerably well-marked collar; the plumes of the upper surface mesially black, forming either streaks or bars; the crown very broadly streaked, the buff spots here very tiny, giving a minutely spotted appearance to this part, inclining to white on the frontal feathers; on the nape a more or less concealed band of light-coloured feathers, for the most part white, tinged with fulvous near the base, and subterminally crossed with two or three zigzag lines of brown; scapulars greyer than the rest of the back, being irregularly mottled or barred with greyish white on the outer web, many of the outermost entirely white externally, tipped with black and having one or two wavy lines of dark brown; quills brown, vermiculated like the back, the least coverts somewhat rufescent, the greater and median series washed with grey externally, some few feathers showing large spots or bars of greyish white towards the tips of the outer webs, the spurious quills externally notched with white; primary coverts sandy buff, barred with dark brown; quills dark brown, regularly barred with sandy rufous, inclining to buff on the inner webs, these light bars represented on the outer webs by light sandy buff, slightly inclining to white on the outer margin of the secondaries, all of which are more or less obscured by brown vermiculations, the primaries very plainly notched with buffy white on the outer web in chequered fashion, the innermost secondaries coloured and mottled like the back, but showing also a few greyish white cross bars; tail dark brown, slightly vermiculated towards the tip, which is also slightly washed with grey, the tail crossed with ten regular bars of sandy rufous, paler and inclining to white towards the tip; lores white, the shafts produced into hair-like black bristles, the plumes over the fore part of the eye white, narrowly tipped with dark brown, forming an indistinct eyebrow; the plumes on each side of the crown slightly spotted with white, forming an inconspicuous lateral streak, to which the ear-tufts are partly continuous, these being coloured like the head externally, but having the inner webs white, with a few narrow lines of dark brown; sides of face greyish white, crossed with narrow zigzag lines of dark brown, the hindermost broadly tipped with black, and merging in the feathers of the ruff, which are buffy white, with broad black tips, the gular feathers buffy white, mesially streaked with black, and narrowly lined across with the same; under surface of body white, washed here and there with golden buff, the bases being of this colour, every feather crossed with narrow zigzag lines of blackish brown, and mesially streaked along the shaft with the same, more broadly on the chest; the under tail-coverts equally barred; leg-feathers golden buff, barred with dark brown; under wing-coverts yellowish, the outer ones narrowly barred with dark brown, the lower series ashy brown, yellowish at base, resembling the inner lining of the quills, which are light brown below, barred for their whole extent with sandy buff, inclining to yellowish near the base. Total length 9·5 inches, wing 6·25, tail 3·7, tarsus 1·35.

Hab. Brazil, ranging northwards through Guiana and Venezuela into Columbia.

a. Pull. st.	South America.	Purchased.
b. Ad. st.	South America.	E. Wilson, Esq.
c. Ad. st.	South America.	Purchased.
d. Ad. st.	South America.	Purchased.
e, f, g. Ad. st.	South America.	
h, i. Ad. sk.	Brazil.	Purchased.
k, l. Ad. sk.	Brazil.	M. Claussen [P.].
m. Pull. sk.	Bahia.	Dr. Wucherer [C.].
n. Ad. sk.	Bahia.	Purchased.
o. Juv. st.	Pará.	R. Graham, Esq. [P.].
p. Ad. sk.	Island of Mexiana.	A. R. Wallace, Esq. [C.].
q. Ad. sk.	Upper Amazons.	J. La Mert, Esq. [C.].
r. Ad. sk.	Cayenne.	M. Becœur [C.].
s. Ad. st.	Trinidad.	
t. Ad. sk.	Caracas.	M. A. Sallé [C.].
u, v. Pull. st.	Caracas.	M. A. Sallé [C.].
w. ♂ juv. sk.	Antioquia, New Granada.	T. K. Salmon, Esq. [C.].
x, y. Ad. sk.	New Granada.	Purchased.

Obs. After a very careful and extended study of the American Scops Owls, in which I had the assistance of the beautiful series contained in the Salvin-Godman collection, I have arrived at the conclusion that three different and fairly recognizable forms have been confounded under the collective title of *Scops brasilianus* in general use. The Brazilian bird always has a mixture of yellowish in the plumage of the under surface, and the two phases of its plumage are of a "brown" and "ochraceous" cast; whereas the Central-American bird has a decided "rufous" phase, very like *Scops asio*, but always to be distinguished by its bare toes: the brown phase never exhibits any yellow shade below. A third race, recognizable in both its phases, is to be seen in *Scops ustus* from the Upper Amazons. Measurements are of no help; for the dimensions of the three birds are much alike, and a summary of the specimens examined gives the following results:—

	Total length. inches.	Wing. inches.	Tail. inches.	Tarsus. inches.
S. brasiliensis	8–10·5	5·85–6·5	3·4–4·1	1·2–1·45
S. ustus..................	9– 9·5	6·45–6·8	3·4–3·9	1·3
S. guatemalæ	8·5	6·3 –6·5	3·5–3·7	1·25

There is considerable variation in sizes among the South-American birds, a specimen in the Salvin-Godman cabinet, collected by Mr. Rogers in the province of Santa Catarina, measuring as follows:—Total length 10·5 inches, wing 7·7, tail 5·2, tarsus 1·55.

Subsp. *a*. **Scops ustus.**

Scops usta, *Sclater, Tr. Z. S.* iv. p. 265, pl. lxi.; *Gray, Hand-l. B.* i. p. 47.

Adult male (grey phase). General colour above dull earthy brown, so finely vermiculated as to appear almost uniform at first glance, a few fulvescent cross markings more conspicuous on the scapulars and secondaries, very slightly indicated on the hind neck, and not forming a distinct collar; crown of head rather blacker than the back, the feathers infinitesimally freckled with sandy rufous, the ear-tufts blackish, scarcely vermiculated at all; loral plumes dull rufous, barred with black, the shafts elongated into black hair-like bristles; feathers over the fore part of the eye buffy white, with black cross bars, scarcely forming a distinct eyebrow; forehead and sides of crown greyer than the crown itself, being greyish buff, minutely vermiculated with brown, and forming a lateral band which runs round the nape, being here rather lighter, several of the feathers barred with white; feathers in front of and above the eye dark brown; ear-coverts sandy brown, indistinctly barred across with blackish brown, and narrowly shaft-streaked with white, the hindermost ear-coverts tipped with black, and merging in the ruff, which is composed of sandy buff plumes, distinctly barred across with black, more narrowly on the gular feathers; chin whitish; rest of under surface ochraceous buff, thickly sprinkled with wavy lines and vermiculations of dark brown, especially on the sides of the chest, some of the breast-feathers streaked with black and barred across with white, the flanks scantily barred with dark brown, inclining to white near the tip, the markings scanty, as also on the under tail-coverts; under wing-coverts nearly uniform ochraceous buff, with a few tiny spots of brown near the outer edge of the wing, the lower series ashy brown, resembling the quill-lining; scapulars externally yellowish buff, barred with dark brown, a few of the wing-coverts also largely spotted with yellowish on the outer web, the wing-coverts otherwise tolerably uniform like the back, the vermiculations a little more distinct; primary coverts dark brown, regularly barred across with sandy buff rather obscured with brown vermiculations; quills dark brown, barred with sandy buff, inclining to yellowish on the inner webs, the secondaries so closely vermiculated on the outer web that the light bars are nearly obscured, being, however, much more distinct on the outer webs of the primaries, inclining to white on the margins of the latter, and producing a chequered appearance; tail dark brown, thickly vermiculated with sandy buff on the margins and at the tips, and crossed with bars of the same, eight or nine of which are distinguishable. Total length 9·5 inches, wing 6·6, tail 3·9, tarsus 1·3.

Adult female (rufous phase). General characteristics as in the grey phase, but rufous where the other bird is brown, and slightly more mottled on the upper surface with rufescent cross bars; below nearly uniform rufous, deeper on the chest, some of the feathers slightly streaked with black, more narrowly on the breast and abdomen; on the chest a few dull brown vermiculations, the abdomen indistinctly barred with fulvous. Total length 9 inches, wing 6·55, tail 3·4, tarsus 1·3.

Obs. The principal characteristics of this race are the uniformity of its upper surface, and the comparative absence of streaks; scapulars fulvescent, not white. These remarks apply both to the brown and rufous phases, neither of which shows any collar on the hind neck.

Hab. Upper Amazons.

a. Ad. st.	Upper Amazons.	H. W. Bates, Esq. [C.].
b. ♀ ad. st.	Sarayacu, E. Peru.	E. Bartlett, Esq. [C.].
c. ♂ ad. sk.	Chamicuros, E. Peru, June 15th, 1867.	E. Bartlett, Esq. [C.].
d. ♀ ad. sk.	Chamicuros, E. Peru, May 10th, 1867.	E. Bartlett, Esq. [C.].
e. Ad. st.	Venezuela [?].	

Subsp. *β*. **Scops guatemalæ.** (Plate IX.)

Scops brasilianus *Auct. ex Americâ Centrali* (passim).

Nestling. Above very pale sandy buff, lighter and more yellowish on the hind neck, most of the feathers with fluffy terminal bars of white, and barred across with dusky brown; wings and tail, as far as developed, much as in the adult; under surface of body whitish, with a few scanty cross lines of pale brown.

Adult (brown phase). Above sandy brown, coarsely vermiculated with black, most of the feathers of the upper surface not streaked but crossed with a succession of broad black bars, alternately banded with sandy buff, paler and more distinct on the hind neck, but not forming any mottled neck-collar; head rather darker than the back, the feathers of the crown black, minutely mottled with tiny spots and bars of rufous sandy; the ear-coverts similarly coloured, and not to be distinguished from the crown; on the nape a few feathers barred with buffy white, forming an indistinct band; scapulars externally white, tipped with black; wing-coverts coloured like the back, the mottlings of buff paler and broader on the median and greater coverts, some of which are tipped with large oval spots of white; primary-coverts black, crossed with distinct bars of sandy rufous; quills blackish brown, mottled with sandy vermiculations near the tips, and barred with pale sandy brown on the inner webs, these bars represented by sandy buff ones on the outer web of the secondaries, distinctly inclining to white on the primaries, the innermost secondaries resembling the back, but somewhat washed with grey near the tip; tail blackish brown, mottled with sandy buff on the edges and at the tips of the feathers, which are likewise crossed with about eight bars of sandy rufous; loral plumes and feathers in front of the eye whitish, slightly tinged with rufous and tipped with brown; sides of face sandy rufous, barred across with black, inclining to white on the cheeks and on the hinder ear-coverts, which are broadly tipped with black, merging in the ruff; this is composed of buff-coloured feathers, tipped and centred near the base with black, the gular feathers mesially streaked and broadly barred with black; chin-feathers whitish; rest of under surface of body white, thickly mottled with vermiculations of dark brown, forming regular zigzag lines on most of the feathers; the breast washed with sandy buff, many of the feathers broadly marked with black in the centre; leg-feathers rufous, narrowly barred with dark brown; under tail-coverts white, the cross markings less distinct than on the belly; under wing-coverts yellowish buff, inclining to fawn-colour on the lower ones, with a few dark brown bars near the outer edge of the wing, which is white, the lower series dark brown, yellowish at base, and resembling the inner lining of the quills, which are dark brown below, inclining to ashy buff on the inner web and to buffy white on the outer one. Total length 8·5 inches, wing 6·5, tail 3·7, tarsus 1·25.

Adult (rufous phase). General colour above bright bay, with a few mesial streaks of black on the head, disappearing on the dorsal feathers, and not very distinct on the wing-coverts, a few feathers in front of and over the eye white; sides of face bay, the ear-coverts streaked with white; rest of under surface of body white, thickly mottled on the belly with cross lines and vermiculations of sandy brown, the feathers mesially streaked along the shaft with black, these streaks rather more broadly developed on the chest, which is otherwise bright bay, irregularly barred with white on the centre of the chest; quills and tail brown barred with rufous, inclining to white on the outer edge of the primaries, the innermost secondaries and centre tail-feathers nearly uniform rufous. Total length 8·5 inches, wing 6·3, tail 3·5, tarsus 1·25.

Hab. Central America, from Veragua northwards to Mexico.

a. Ad. sk.	Central America.	Capt. Kellett and Lieut. Wood.
b. Pull. sk.	Guatemala.	Purchased.
c. Ad. sk.	Guatemala.	M. A. Bouvier. (Type of species.)
d, e, f, g, h. Ad. sk.	Guatemala.	M. A. Bouvier.
i. Ad. sk.	Acoyapa, Nicaragua.	M. A. Marche [C.]. (Rufous phase.)
k. Ad. sk.	Costa Rica.	Purchased.
l, m, n. Ad. sk.	Veragua.	M. E. Arcé [C.]. (Rufous phase.)
o, p. Ad. sk.	Veragua.	M. E. Arcé [C.].

Obs. Throughout Central America the Scops Owls have a very different aspect from those of South America: and although they have all been united together as *Scops brasilianus*, the former are sufficiently well characterized to deserve a separate name. *Scops guatemalæ*, as I propose to call these birds, resembles *S. ustus* in the more *mottled*, not *streaked*, character of the upper parts; but the chief point in which it differs from its South-American allies is in the entire absence of any yellow colouring on the underparts.

The rufous phase of *S. guatemalæ* is quite different from any thing that I have seen from South America, being entirely of a foxy rufous colour, with the head never darker than the back or showing any approach to a blackish patch; the back is generally rather narrowly streaked with black, as is also the head; and there are in some examples slight indications of bars.

19. Scops asio.

The Little Owl, *Catesby, N. H. Carol.* i. p. 7, pl. 7.
Le Petit Duc de la Caroline, *Briss. Orn.* i. p. 497.
Strix asio, *Linn. S. N.* i. p. 132; *Wils. Am. Orn.* vi. pl. 42. fig. 1; *Temm. Pl. Col.* ii. pl. 80; *Audub. B. Am.* pl. xcvii.; *id. Orn. Biogr.* i. p. 486.
Red Owl, *Penn. Arctic Zool.* ii. p. 231, pl. xi. fig. 1.
Mottled Owl, *Penn. t. c.* p. 231, pl. xi. fig. 2.
Strix nævia, *Gm. S. N.* i. p. 289 (ex *Penn.*); *Wils. Am. Orn.* pl. xix. fig. 1.
Bubo asio, *Vieill. Ois. d'Am. Sept.* p. 53, pl. 21.
Bubo striatus, *Vieill. t. c.* p. 54.
Otus asio, *Steph. Gen. Zool.* xiii. pt. 2, p. 57; *Schl. Faun. Jap.* p. 25.
Asio asio, *Less. Man. d'Orn.* i. p. 117.
Otus nævius, *Cuv. Règne Anim.* 1829, p. 241.
Surnia nævia, *James. ed. Wils. Am. Orn.* i. p. 96.
Ephialtes asio, *Gray, Gen. B.* i. p. 38.
Scops asio, *Bp. Consp.* i. p. 45; *Kaup, Contr. Orn.* 1852, p. 112; *Strickl. Orn. Syn.* p. 199; *Cass. B. N. Am.* p. 51; *Schl. Mus. P.-B.* Oti, p. 27; *Gray, Hand-l. B.* i. p. 46; *Schl. Revue Accipitr.* p. 9; *Ridgw. B. N. Am.* iii. p. 49; *Coues, B. N. W.* p. 303.
Megascops asio, *Kaup, Tr. Z. S.* iv. p. 228.
Asio nævia, *Bp. Rev. et Mag. de Zool.* 1854, p. 543.

Red phase. General colour bay, with no transverse vermiculations, but distinct mesial shaft-streaks of black, the feathers of the head and neck with obsolete lateral bars of blackish; ear-tufts coloured like the head, the inner webs yellowish near the base; round the hind neck an indistinct collar, caused by yellowish buff bars and bases of the feathers showing when the plumage is disturbed; the scapulars white on the outer web, tipped with black, forming a distinct longitudinal bar down each side of the back; wing-coverts coloured like the back, the greater series externally white near the tip, the spurious quills also notched with white on the outer web; primary-coverts dusky brown on the inner, and dull bay on the outer web, barred with dark brown; quills bay, dull brown on the inner web, and barred with dark brown, the representative bars on the outer web more or less shaded with rufous, the primaries externally very broadly notched with white; upper tail-coverts uniform bay; tail dull bay-colour, barred with dusky brown, these bars nearly obsolete on the centre feathers, but numbering eight on the outer ones; lores whitish, the feathers blackish at tip, the shafts produced into hair-like bristles; feathers over the eye dull white, with central spots of bay, forming a tolerably defined eyebrow; ear-coverts whitish, washed and barred with bay, the hindermost tipped with black and merging into the ruff, which is composed of whitish feathers streaked and broadly tipped with black, before which is a subterminal shade of rufous; the sides of the neck buffy white, mottled with bay; chin white; gular feathers of the ruff whitish, the centre ones very broadly streaked with black, the lateral ones narrowly barred with bay, with a few central streaks of black; fore neck and centre of the body white, with the exception of a few nearly obsolete cross bars of bay, the latter colour, however, prevailing on the sides of the body, where the feathers are washed with fulvous and barred with white, the flanks broadly barred with the latter, most of the feathers on the sides being distinctly streaked with black; leg-feathers white, streaked with bay on the tarsus, the tibial plumes yellowish, the toes plumed; under wing-coverts yellowish white, with a few brown spots near the outer edge of the wing, which is pure white, the lower series ashy brown, yellowish at base, resembling the inner lining of the quills, which are light brown, with yellowish buff bars. Total length 10 inches, wing 6·8, tail 3·6, tarsus 1·55.

GREY PHASE. *Nestling.* General colour of the adult bird, but more fluffy, the feathers of the head and hind neck barred with dull white, those of the latter also slightly barred with yellowish buff; the entire under surface dull white, with distinct cross bars of dusky brown, becoming obsolete on the abdomen.

Adult. Above greyish brown, with indistinct vermiculations of darker brown, some of the feathers with pale bars of yellowish buff, and all having distinct mesial shaft-lines of black, the feathers of the head and hind neck with distinct lateral bars of black; ear-tufts buff or white on the inner web, narrowly barred with black, more broadly on the outer webs, which are also spotted or notched with sandy buff towards the tips; on the occiput a few feathers of lighter grey, more distinct on the nape, where a faint bar is visible, the hinder neck distinctly barred with white or yellowish buff, forming a tolerably distinct collar, extending on to the sides of the neck, which are greyish buff, with narrow zigzag lines of dark brown; scapulars externally white, tipped with black, with a few cross lines of dark brown on a few of the feathers; wing-coverts rather darker than the back, the mottlings indistinct, but the feathers broadly centred with black, the median and greater coverts with a large white spot near the tip of the outer web, the greater coverts greyer than the others, and strongly resembling the scapulars; the spurious quills and primary-coverts externally notched with white; quills brown, barred across with darker brown on both webs, those on the outer web slightly mottled with buff, the interspaces decidedly fulvescent on the secondaries, and inclining to white on the outer web of the primaries, imparting a chequered appearance to the outer aspect of the wing, the innermost secondaries resembling the scapulars, but rather plainly barred across with whitish; tail pale brown, mottled with fulvous vermiculations, especially on the margins of all the feathers, which are crossed with about eight bars of pale sandy buff; loral plumes and feathers over the eye white, very narrowly tipped with brown, forming a distinct eyebrow; above the eye a small rufous patch extending to the ear-tufts; sides of face greyish white, narrowly barred with brown, the hindermost tipped with black, and resembling the ruff, the feathers of which are whitish, broadly tipped with black, the gular feathers centred and narrowly lined with black; chin and fore neck whitish; the rest of the under surface dull white, most of the feathers broadly centred with black, diminishing in breadth on the abdomen, the breast consequently having an ocellated, the rest of the lower parts a barred appearance, the flanks also narrowly barred with rufous; tibiæ yellowish, barred with brown, disappearing on the tarsi, which are whitish; under wing-coverts yellowish white, the edge of the wing pure white, mottled with dark brown near the external margin, the greater series light brown, the bases yellowish, resembling the inner lining of the quills, which are brown, barred with yellowish on inner web. Total length 9·5 inches, wing 6·3, tail 3·2, tarsus 1·4.

Hab. United States, except the southern middle province, the north-west region, and Florida. (*Ridgway.*)

Rufous phase.

a. Ad. st.	Delaware, U.S.A.	E. Doubleday, Esq. [P.].
b. Ad. st.	Delaware, U.S.A.	E. Doubleday, Esq. [P.].
d. ♀ ad. sk.	North America.	Lord Odo Russell [P.].
e. Ad. sk.	North America.	Major-Gen. Hardwicke [P.].

Grey phase.

f. Ad. sk.	Toronto, Canada.	Mr. J. Whitely [C.].
g, h. Pull. sk.	Toronto, Canada.	Mr. J. Whitely [C.].
i. Ad. st.	Delaware, U.S.A.	E. Doubleday, Esq. [P.]

Subsp. *α*. **Scops kennicotti.**

Scops kennicotti, *Elliott, Pr. Philad. Acad.* 1867, p. 69; *id. B. N. Am.* pl. xxvii.; *Baird, Tr. Chicago Acad.* i. p. 311, pl. xxvii.; *Dall & Bann. t. c.* p. 273; *Gray, Hand-l. B.* i. p. 47.
Scops asio, *Lord, Naturalist in Vanc. Isl.* ii. p. 292.
Scops asio, var. kennicotti, *Ridgw. B. N. Am.* iii. p. 53; *Coues, B. N.W.* p. 303.

Adult. Above brown, generally of an earthy tint, all the feathers centred and barred with black, with very faint dusky vermiculations, on the occiput a few feathers barred with sandy buff, forming a tolerably distinct band, another on the nape more pronounced, and on the hind neck a third very much plainer one, the feathers distinctly barred with sandy buff, extending on to the sides of the neck, where they are very broad; outer web of scapulars yellowish buff, tipped with black, the greater wing-coverts also terminally spotted with buff, the rest of the wing-coverts coloured like the back, but the black centres rather broader; primary-coverts dusky brown, barred with rufous buff; quills dark brown, barred with sandy brown on the inner webs, paler and clearer fulvous on the outer webs, giving a strongly chequered appearance to the wing; tail dark brown, with about eight narrow bars of sandy buff; lores and plumes over the fore part of the eye greyish white, slightly tipped with brown; sides of face greyish white, washed with rufous below the eye, and regularly barred with dull brown, the hindermost of the ear-coverts tipped with black, merging in the facial ruff, which is composed of sandy-buff plumes mesially streaked and tipped with black, the gular feathers very broadly streaked with black, with lateral narrow bars of black; chin whitish, as well as a few feathers on the fore neck; rest of under surface white, washed with ochre on the breast and flanks, the sides of the chest more dusky than the rest of the lower parts, most of the feathers crossed with wavy lines of dark brown, and broadly centred with black, these mesial streaks less distinct on the breast and flanks; under tail-coverts buffy white, with a few indications of rufous or brown bars near the tip; under wing-coverts ochraceous buff, thickly mottled with brown near the outer edge of the wing, which is white, the lower series ashy brown, yellowish at base, and resembling the inner lining of the wings, which are ashy brown below, barred with sandy buff on the inner web. Total length 10 inches, wing 6·85, tail 3·5, tarsus 1·5.

Obs. This is a dark brown form of *S. asio,* with no *rufous* phase at all; moreover it is larger than the typical *S. asio.* Full particulars will be found in Mr. Ridgway's work (B. N. A. iii. p. 53), where the whole subject is treated in a more exhaustive manner than the limits of the present volume will allow.

The specimen described by me is very brown in coloration, and seems to represent the rufous phase of *S. asio*; for other examples are decidedly greyer above and whiter below. Such a specimen from British Columbia (*Lord*) measures as follows:—Total length 10·5 inches, wing 7·6, tail 4·5, tarsus 1·8.

Hab. North-western coast of North America, from the Columbia river northward: Idaho (*Ridgway*).

a, b. Ad. sk.	North America.	R. B. Sharpe, Esq. [P.].
c. Ad. sk.	Vancouver's Island.	R. Brown, Esq. [C.].
d. Ad. sk.	Vancouver's Island.	J. K. Lord, Esq. [C.].
e. Ad. sk.	West side of Rocky Mountains.	J. K. Lord, Esq. [P.].

Subsp. β. **Scops floridanus.**

Scops asio, *Allen, Bull. Harv. Coll.* ii. p. 338.
Scops asio, var. floridana, *Ridgway, B. N. Am.* iii. p. 51; *Coues, B. N. W.* p. 303.

"Similar to var. *asio*, but much smaller, and the colours deeper. The grey stage very similar to that of var. *asio*, but the red phase very appreciably different, in there being a greater amount of rufous on the lower parts, the breast being nearly uniformly coloured, and the rufous broken elsewhere into transverse broad bars, connected along the shafts. Wing 5·50–6, tail 2·75–3·10." (*Ridgway.*)

Hab. Florida and Lower Georgia.

Subsp. γ. **Scops enano.**

Scops trichopsis, *Gray, Hand-l. B.* i. p. 47; *Scl. & Salv. Nomencl.* p. 117.
Scops asio, var. enano, *Ridgw. B. N. Am.* iii. p. 48; *Coues, B. N. W.* p. 304.

Adult (grey phase). Above greyish brown, all the feathers coarsely, though indistinctly, mottled with wavy cross lines of blackish brown, with which they are also distinctly streaked down the centre; the head and neck rather more brown than the back, obscurely spotted with sandy buff, the occiput barred with white, forming a slightly indicated band, which is more distinct on the nape, and much more so on the hinder neck, where all the feathers are barred with buffy white and form a distinct collar; scapulars distinctly barred with greyish white, the outermost externally pure white, tipped with black; wing-coverts browner than the back, spotted with sandy buff, the median and greater series mottled with ashy grey, and having very large oval spots of white on the outer web, the spurious quills also externally notched with white; primary-coverts dark brown, the outer ones notched with sandy buff on the external web; quills brown, mottled with greyish vermiculations on the outer web, the innermost also mesially streaked with black, and resembling the dorsal plumes, all the quills barred with lighter brown, inclining to fulvous on the inner web and to white on the outer one, especially on the primaries, which are deeply notched with pure white; forehead and lores whitish, with dark brown tips, the feathers on the sides of the crown also mottled with white, forming a tolerably distinct eyebrow, the feathers below the eye sandy buff; sides of face dusky white, barred with dark brown, the hindermost ear-coverts broadly tipped with the latter colour, and merging in the ruff, the feathers composing which are clear sandy buff, tipped with black, the gular plumes black in the centre, with narrow transverse lateral lines of the same; chin whitish; rest of under surface white, transversely lined with blackish brown, these cross lines springing from a very broad central streak of black, extremely wide on the chest, and decreasing in width on the abdomen and flanks, the under tail-coverts nearly entirely white, excepting a few sagittate markings of dark brown near the tips; leg-feathers white, clothing the tarsi and extending on to the toes; under wing-coverts whitish, the outermost slightly streaked with brown, the lowest series ashy brown, resembling the inner lining of the quills, which are tolerably uniform ashy brown below, the innermost with a few yellowish bars near the base of some of the inner ones, all externally notched with yellowish; bill yellow. Total length 7·5 inches, wing 5·65, tail 2·8, tarsus 1·05.

Young. Very similar to the adult, but much more dingy, the white spots on the scapulars and wing-coverts obscured by vermiculations, the three bands on the hind neck also very indistinct; under surface of body dull white, crossed with numerous dusky brown bars, the chest broadly streaked with dull brown down the centre of the feathers; on the breast a few feathers are appearing white, streaked and narrowly barred with black, as in the fully adult plumage.

Rufous phase (not quite adult). General colour rufous, the markings very similar in character to those of the grey phase, but not so mottled, the white spots on the scapulars and wing-coverts not so distinct, the bands on the head and neck scarcely perceptible, the hind neck barred with fulvous, but not forming a very distinct collar; feathers of crown streaked with black down the centre; sides of face rufous, barred with dull brown; throat and chest rufous, the former barred with dusky brown, the latter broadly streaked with black and slightly mottled with whitish bars; remainder of under surface buffy white, obscurely barred with sandy rufous, many of the feathers of the new plumage appearing on the breast, white washed with sandy rufous, narrowly barred with black and broadly streaked down the centre with the same. Total length 7·5 inches, wing 5·7, tail 3·2, tarsus 1·1.

Hab From Mexico to Guatemala.

a. Ad. st.	Mexico.	
b. Juv. sk.	W. Mexico.	Alex. Dorman, Esq. [C.].

Subsp. δ. **Scops trichopsis.**

Scops trichopsis, *Wagl. Isis,* 1832, p. 276; *Bp. Consp. Av.* i. p. 46; *Strickl. Orn. Syn.* p. 201; *Salvin, Ibis,* 1874, p. 314.
Ephialtes trichopsis, *Gray, Gen. B.* i. p. 38.
Megascops trichopsis, *Kaup, Tr. Z. S.* iv. p. 227.
Asio trichopsis, *Bp. Rev. et Mag. de Zool.* 1854, p. 543.
Scops maccallii, *Cass. B. Calif. & Texas,* p. 180; *id. B. N. Am.* p. 52, pl. 39; *Baird, Mex. Bound. Comm.* iv. pl. 1; *Strickl. Orn. Syn.* p. 200; *Scl. & Salv. Ibis,* 1859, p. 220; *Gray, Hand-l. B.* i. p. 47.
Scops asio, var. maccallii, *Ridgw. B. N. Am.* iii. p. 49; *Coues, B. N. W.* p. 303.

Nestling. Above brown, the nestling plumes tipped with white, the hinder neck barred across with sandy brown; the scapulars with distinct white spots on the outer webs, and the spurious quills distinctly notched with white even at this early age; wings and tail marked much as in the adult, but more tinged with sandy brown; feathers of the face dull white, very distinctly barred with dusky brown, as also the underparts, which are white, barred with brown, wider and broader on the flanks and abdomen.

Adult. General colour above grey, slightly browner on the head, the centres to the feathers very distinctly black, causing a streaked appearance, broader on the crown, the transverse vermiculations very coarse and distinct; scapulars externally white, tipped with black; forehead and lores greyish white, with black tips to the feathers, those round the eye whitish, mesially streaked with black, extending in an indistinct line on the sides of the crown to the ear-coverts, which are very plainly notched with white on the inner web, the hind neck mottled with fulvous, forming a tolerably distinct collar, the only one traceable on the upper surface; wing-coverts grey, like the back, and mottled in the same manner, the black streaks being broader and occupying the entire centre of the greater coverts, which are largely spotted with white at the tip of the outer web, the spurious quills externally notched with buffy white; primary-coverts dark brown, barred with whitish buff; quills dark brown, barred with fulvous on the inner web and also on the outer webs, which are further varied with greyish vermiculations on the exterior margin, the light bars inclining to white on the primaries, the inner secondaries very distinctly grey like the back, and mesially streaked in the same manner; tail grey, darker brown in the centre of the feathers, each of which is crossed with seven bars of pale fulvous brown, inclining to white on the outer webs of the external feathers; sides of face dull greyish white, barred with black and distinctly tipped with the same, the ruff greyish white, slightly tinged with fulvous, and very finely barred with wavy lines of black, the stiff gular feathers broadly streaked with black, and distinctly barred with narrow lines of the same; rest of under surface of body white, plainly though narrowly barred with zigzag lines of black, the chest very broadly streaked with black, these streaks becoming much narrower on the abdomen and flanks; the centre of the body white, as also the under tail-coverts, the latter having only a few black cross markings near the tip; legs densely clothed with feathers, the tibia sandy brown, barred with blackish, the tarsus white but similarly barred; under wing-coverts yellowish, the edge of the wing and the adjacent feathers washed with sandy and barred with dark brown, the greater series ashy brown, resembling the inner lining of the quills, which are ashy brown below, notched with yellowish on the outer and barred with the same on the inner web; bill greenish horn-colour, the tip yellowish; iris yellow. Total length 9 inches, wing 6·3, tail 3·6, tarsus 1·35.

Obs. This is a small race of *S. kennicotti*; but, as far as can be determined, it has only a grey phase and no brown one. Its measurements distinguish it at once; and it may also be told by its narrowly barred under surface, every feather being streaked with black and barred with the same, from the chin to the lower abdomen and flanks. It is larger than *Scops enano*, and differs from that bird also in not having a rufous phase; the cross-barring of the under surface in the latter is of the same character in *S. enano* as in *S. asio*; that is to say, the bars are often double, whereas in *S. trichopsis* they are single and very distinct.

Hab. Mexico, ranging into the southern portion of the United States and Southern California.

a. Ad. sk.	W. Mexico.	Alex. Dorman, Esq. [C.].
b. Pull. sk.	W. Mexico.	Alex. Dorman, Esq. [C.].

20. Scops nudipes.

Bubo nudipes, *Vieill. Ois. Am. Sept.* pl. 22.
Strix psilopoda, *Vieill. N. Dict. d'Hist. Nat.* vii. p. 46.
Scops nudipes, *Cuv. Règne An.* 1829, p. 347; *Strickl. Orn. Syn.* p. 203; *Scl. & Salv. Nomencl. Av.* p. 117.
Ephialtes nudipes, *Gray, Gen. B.* i. p. 38.
Acnemis nudipes, *Bp. Rev. et Mag. de Zool.* 1854, p. 542.

Adult. Above sandy rufous, vermiculated with black, much darker on the back than on the head, the dorsal feathers black in the centre, barred and spotted with sandy rufous, these bars more distinct on the scapulars, the outermost of which are silvery white, externally tipped with black; wing-coverts decidedly darker than the back, the innermost of the least series uniform blackish brown, the rest spotted and barred with sandy rufous, the bars especially broad on the greater series, some of which have a tolerably large buffy white spot near the tip of the outer web; spurious quills externally notched with sandy rufous, inclining here and there to whitish; primary-coverts nearly uniform blackish brown, with a few bars of sandy rufous near the tips of the outer webs; quills blackish brown, the inner webs of the primaries quite uniform, excepting for a few yellowish bars near the bases of the interior feathers, the secondaries indistinctly barred with ashy brown on the inner webs, all the quills externally barred with sandy rufous, paler and more fulvous on the outer web of the primaries, the innermost secondaries mottled and barred with sandy rufous, and resembling the scapulars; tail blackish brown, with seven feebly indicated narrow bars of sandy rufous; head and neck decidedly clearer than the back, and somewhat inclining to chestnut, the feathers black in the centre, and laterally barred with the same, giving a generally barred appearance to these parts; lores and sides of face bright bay, the loral plumes blackish at tip, and the ear-coverts with a few indistinct cross bars of black; over the eye a few white-barred feathers, forming a faint eyebrow; ear-tufts lighter than the crown, orange rufous, broadly barred with black at the tips; under surface of body sandy rufous, many of the feathers coarsely vermiculated with black, the breast-feathers streaked and laterally barred with black, these black markings less distinct on the flanks and abdomen, on which parts are tolerably distinct bars of white; leg-feathers bright orange rufous, with a few narrow brown bars on the tibia; under tail-coverts white, barred across with sandy rufous; under wing-coverts fulvous, thickly mottled with brown near the outer edge of the wing, which is white, the lower series dark brown, like the inner lining of the quills; bill yellowish; feet yellowish, claws horn-colour. Total length 10 inches, wing 6·8, tail 4·1, tarsus 1·55, bare part of latter 0·85. (*Mus. Salvin and Godman.*)

Obs. My description is taken from a specimen obtained at Calobre, in Veragua, by Arcé, and kindly lent to me by Mr. Salvin. He has at the same time lent me another specimen, obtained by the same collector in Costa Rica. This latter bird differs in several points from the one described, having more of the general aspect of *Scops spilocephalus* of the Himalayas; it is clear sandy in colour, the black forming regular bars across the plumage, the subterminal one very broad, and giving the appearance of large black spots to the upper surface; feathers of the crown centred with black, the sandy-coloured interspaces forming very distinct spots, the ear-tufts being still lighter, and barred across with whitish; the white spots on the scapulars, wing-coverts, and outer webs of primaries very distinct, as also are the white bars on the lower surface, many of them being apparent also on the chest-feathers; the quills barred with ashy brown on their inner webs, inclining to sandy buff on the secondaries. Total length 9·5 inches, wing 6·8, tail 4, tarsus 1·45, bare part of latter 0·65.

21. Scops cristatus.

La Chouette à aigrette blanche, *Levaill. Ois. d'Afr.* i. pl. 43.
Supercilious Owl, *Lath. Gen. Syn. Suppl.* ii p. 63.
Strix cristata, *Daud. Traité*, ii. p. 207 (ex *Levaill.*).
Strix superciliosa, *Shaw, Gen. Zool.* vii. p. 250.
Bubo griseatus, *Cuv. Règne An.* i. p. 344; *Schl. Mus. P.-B.* Oti, p. 16; *Kaup, Tr. Z. S.* iv. p. 238; *Schl. Revue Accipitr.* p. 6.
Syrnium griseatum, *Less. Traité*, p. 113.
Lophostrix griseata, *Less. Compl. Buff.* i. p. 261.
Lophostrix cristata, *Gray, Cat. Accipitr. Brit. Mus.* 1844, p. 47; *Bp. Consp.* i. p. 45; *Kaup, Contr. Orn.* 1852, p. 115; *Scl. & Salv. Nomencl. Av.* p. 117; *Salv. Ibis*, 1874, p. 325.
Ephialtes cristatus, *Gray, Gen. B.* i. p. 38; *id. Cat. Accipitr. B. M.* 1848, p. 98.
Scops cristata, *Strickl. Orn. Syn.* p. 205; *Gray, Hand-l. B.* i. p. 47.

Adult female. Above deep and slightly reddish chocolate-brown, the crown of the head decidedly tinged with rufous, all the upper surface obsoletely vermiculated with blackish; scapulars and wing-coverts strongly washed with rufous, and irregularly studded with spots of white of unequal size, those near the outer base of the scapulars and on the least wing-coverts tinged with yellow; primary-coverts thickly vermiculated with blackish, rufescent only at the tips; quills deep brown, externally dull sandy, marked with a few irregularly shaped spots of yellowish buff, inclining to rufous on the outer web of the secondaries, and often obscured by blackish vermiculations, the inner secondaries not spotted, and coloured like the back; the lower surface of the wing much paler, and broadly barred with yellowish white on the inner web; tail rufous brown, vermiculated with blackish, the inner webs barred with yellowish white, these bars about nine in number, and varying very much in breadth and in extent; forehead and a very broad eyebrow white, slightly vermiculated with brown; the lores also whitish, tinged with rufous, and terminating in black shaft-bristles; all the feathers round the eye, as well as the entire sides of the face, and extending slightly on to the sides of the neck, bright bay; auricular tufts 2·6 inches long, brown, more or less tinged with rufous, the basal feathers entirely white, as is also more or less of the upper web of the other plumes; cheeks and throat brownish ashy, with a small spot of white at the hinder end of the former; under surface of body ashy brown, everywhere vermiculated with narrow lines of darker brown; the centre of the chest and flanks more or less white, the vermiculations nearly obsolete, the lower flanks and tibial plumes inclining to orange-buff; tarsal feathers fulvous, tinged with rufous on the upper part, and coarsely vermiculated with brown bars; under tail-coverts fulvous, with slight remains of brownish vermiculations; under wing-coverts orange-buff, with slight remains of rufous brown mottlings near the outer edge of the wing, the greater series orange-buff at base, brown at tip, resembling the inner surface of the wing. Total length 20·5 inches, culmen 1·65, wing 13·4, tail 8·5, tarsus 1·8.

Young male. Differs from adult in being rather more rufescent above, and much more coarsely vermiculated with blackish; the white spots on the scapulars and wing-coverts smaller and fewer in number; on the under surface the vermiculations much coarser, and the white on the chest and flanks much less distinct. Total length 18·5 inches, wing 12·7, tail 8·5, tarsus 1·8.

Rufous phase. General colour foxy red, covered with little brown vermiculations, pretty distinct but irregular on the back, much less plainly indicated on the hind neck, which is consequently more uniform; head like the back, but a little darker, vermiculated in the same manner; forehead itself white, the feathers of the fore part of the crown creamy white, minutely vermiculated with reddish brown, as also are some of the median plumes of the ear-tufts, the innermost plumes of the latter being pure white, the outer ones reddish brown, more or less whitish on the inner web; facial aspect deep foxy rufous, the loral plumes whitish at base; under surface of body fulvescent, vermiculated with brown, the chest inclining to foxy red, with sparsely indicated vermiculations; leg-feathers creamy buff, with slight indications of rufous bars, disappearing on the under tail-coverts, which are uniform creamy buff; under wing-coverts deep fulvous, slightly spotted with brown, and washed with rufous near the outer margin of the wing, which is fulvous, the greater series fulvous, barred with brown, and resembling the inner lining of the quills, which are brown, barred with fulvous on the inner web, inclining to rufous towards the tips of the feathers; upper wing-coverts foxy red, like the back, and vermiculated in the same manner, with also a few white spots on most of the greater and median series; primary-coverts uniform foxy red; quills rufous, paler than the back, and with the vermiculations less distinct, the primaries with a few spots of buffy white, the secondaries slightly barred with fulvous; the tail a little darker rufous than the back, paler towards the tip, and with very few vermiculations. Total length 16 inches, wing 12·2, tail 8·5, tarsus 1·8. (*Cayenne*; *Mus. Paris.*)

Hab. Lower Amazonia and Guiana, extending also into Ecuador.

a. ♀ ad. st.	South America.	Purchased.
b. ♂ imm. st.	Cayenne.	Purchased.
c. ♂ imm. sk.	Para, May 1849.	A. R. Wallace, Esq. [C.].
d. Ad. sk.	Ecuador.	Purchased.

22. Scops stricklandi.

Scops cristata, var., *Strickl. Contr. Orn.* 1852, p. 60, pl. 10.
Lophostrix stricklandi, *Scl. & Salv. Ibis*, 1859, p. 229; *iid. Nomencl. Av.* p. 117.
Scops stricklandi, *Gray, Hand-l. B.* i. p. 47.

Adult. Above chocolate-brown, thickly vermiculated with blackish, the scapulars lighter and more sandy brown; entire head uniform blackish brown; a very broad white eyebrow; auricular tufts 2·2 inches long, particoloured, the upper ones white, or white on inner web, the lower one blackish brown, some few vermiculated with sandy brown; lores whitish, the shafts terminating in long black bristles; feathers round the eye black; ear-coverts rich bay, broadly black at tip; some of the outer scapulars with large spots of yellowish white, which are also apparent on some of the outer wing-coverts; quills blackish, the outer web sandy brown, the primaries spotted externally with yellowish white; the secondaries barred with sandy brown, the innermost uniform with the back, and vermiculated in the same manner; tail rufous brown, nearly uniform towards the tips of the centre feathers, which are vermiculated near the base with dark brown, barred indistinctly with sandy rufous on the outer web, and crossed with seven or eight bars of yellowish white; under surface of body ashy brown, coarsely vermiculated with darker brown, and tinged with rufons on the breast; the chest and flanks somewhat varied with white; leg-feathers orange-buff, with a few cross lines of blackish; under tail-coverts whitish, broadly barred with dusky brown; under wing-coverts orange-buff, with a few brown markings near the outer edge of the wing, the greater ones blackish, like inner lining of wing. Total length 18·5 inches, culmen 1·8, wing 12·8, tail 8, tarsus 2.

Obs. The young of this bird is much paler than the adult, with a brownish head, which shows slight traces of vermiculations. Although the presence of these markings brings the species in its young stage near to *S. cristatus*, the black feathers round the eye and the black terminal half of the ear-coverts serve to mark the species at a glance.

Hab. Central America: Guatemala, Costa Rica, and Veragua.

a. Ad. sk.	Guatemala.	Purchased.
b. Ad. sk.	Veragua.	M. E. Arcé [C.].
c. Juv. sk.	Veragua.	M. E. Arcé [C.].

5. NYCTEA.

	Type.
Nyctea, *Stephens, Gen. Zool.* xiii. pt. 2, p. 63 (1826)..	N. scandiaca.
Nyctia, *Swainson, Classif. B.* ii. p. 217 (1837)	N. scandiaca.
Leuchybris, *Sundevall, Av. Meth. Tent.* p. 105 (1872).	N. scandiaca.

Head of *Nyctea scandiaca*, to show ear-tufts.

Range. The northern portions of both the Old and New Worlds, migrating south in winter.

1. Nyctea scandiaca.

Great White Owl, *Edwards, N. H. Birds*, pl. 61.
Le Chat-huant blanc de la Baye de Hudson, *Briss. Orn.* i. p. 522.
Strix scandiaca, *Linn. S. N.* i. p. 132.
Strix nyctea, *Linn. S. N.* i. p. 132; *Daud. Traité*, ii. p. 188; *Wolf & Meyer, Vög. Deutschl.* p. 184, with plate: *Vieill. Ois. Am. Sept.* pl. 18; *Wils. Am. Orn.* iv. p. 53, pl. 32. fig. 1; *Temm. Man. d'Orn.* i. p. 82; *Naum. Vög. Deutschl.* i. p. 417, pl. 41; *Werner, Atlas Ois. Eur. Rapaces*, pl. 31; *Sw. & Rich. Faun. Bor.-Am.* p. 88; *Audub. B. Am.* pl. 121; *id. Orn. Biogr.* ii. p. 135; *Schl. & Susem. Vög. Eur.* Taf. 41; *Yarr. Brit. B.* i. p. 134; *Schl. Vog. Nederl.* pl. 44; *Sundev. Sv. Fogl.* pl. xxiv. figs. 3, 3*.
Harfang, *Buff. Pl. Enl.* i. pl. 458.
Strix arctica, *Bartr. Trav. Carol.* p. 285 (nec *Sparrm.*).
Strix nivea, *Thunb. Sv. Akad. Förh.* 1798, p. 184.
La Chouette blanche, *Levaill. Ois. d'Afr.* i. pl. 45.
Bubo albus, *Daud. Traité*, ii. p. 210.
Ermine Owl, *Lath. Gen. Syn. Suppl.* ii. p. 60.
Strix candida, *Lath. Ind. Orn. Suppl.* p. xiv.
Strix erminea, *Shaw, Gen. Zool.* vii. p. 251.

Noctua nyctea, *Boie, Isis*, 1822, p. 770; *Less. Traité*, p. 101; *Schl. Mus. P.-B.* Striges, p. 43 (1862).
Nyctea erminea, *Steph. Gen. Zool.* xiii. pt. 2, p. 63.
Syrnium nyctea, *Kaup, Natürl. Syst.* p. 59.
Surnia nyctea, *James. ed. Wils. Am. Orn.* i. p. 92; *Selby, Ill. Brit. Orn.* i. p. 95, pl. 23; *Gould, B. Eur.* i. pl. 43; *Keys. & Blas. Wirb. Eur.* p. xxxiii; *Kaup, Tr. Z. S.* iv. p. 214; *Hewits. Eggs Br. B.* i. p. 64, pl. xviii. fig. 3; *Newt. in Baring-Gould, Iceland*, p. 408; *Degl. & Gerbe, Orn. Eur.* i. p. 118.
Nyctia candida, *Swains. Classif. B.* ii. p. 217.
Nyctea candida, *Bp. Comp. List B. Eur. & N. Am.* p. 6; *id. Consp. Av.* i. p. 36.
Syrnia nyctea, *Thomps. Ann. Nat. H.* i. p. 241; *Macgill. Hist. Br. B.* iii. p. 407.
Nyctea nivea, *Gray, Gen. B.* i. p. 34, pl. 12. fig. 2; *Cass. B. Calif.* p. 190; *Strickl. Orn. Syn.* p. 194; *Fritsch, Vög. Eur.* Taf. 12. fig. 4; *Cass. B. N. Am.* p. 63; *Newt. P. Z. S.* 1861, p. 394; *id. Ooth. Wolley*. p. 178; *Gould, B. Gt. Br.* i. pl. xxxiv.; *Coues, Key N. Am. B.* p. 205.
Nyctea nyctea, *Kaup, Contr. Orn.* 1852, p. 107.
Nyctea nivea europæa, *A. Brehm, Verz. Samml. C. L. Brehm*, p. 2.
Nyctea nivea americana, *A. Brehm, l. c.* p 2.
Leuchybris nyctea, *Sundev. Av. Meth. Tent.* p. 106.
Nyctea scandiaca, *Newt. ed. Yarr. Brit. B.* i. p. 187; *Dresser, B. Eur.* part xxiii.; *Coues, B. N. W.* p. 309.
Nyctea scandiaca, var. arctica, *Ridgw. B. N. Am.* iii. p. 70.

Nestling. Covered with down of a sooty brown colour, very dense in character on the head and neck, where the downy plumelets have a whitish tip; lores and feathers in front of the eye white, dusky brown at tip, the ear-coverts more or less white at base; all the feathers, as far as developed, pure white, the least wing-coverts barred with blackish; greater wing-coverts and quills pure white, spotted on the outer web, and barred on the inner one with blackish, with here and there a faint indication of dusky freckling, increasing on the innermost secondaries, where the interspaces are entirely clouded with dusky brown, as also are the scapulars, many of which are entirely dusky or greyish brown; round the hind neck the feathers are white, barred across with black, as also are those which are appearing on the breast, though here the dark bars are narrower; tail pure white, the developing feathers mostly unspotted, but on the centre ones are remains of indistinct bars of black.

Adult male (Upland, Sweden, Nov. 18th, 1872). Absolutely snowy white above and below, with only a few remains of longitudinal brown spots on the hinder part of the crown and on the wing-coverts, the quills with also a few remains of bars; tail pure white, with a small brown spot remaining near the tip of the centre feathers; bill and claws blackish horn-colour; iris deep yellow. Total length 23 inches, wing 16·7, tail 9·6, tarsus about 2·1.

Female. Total length 26 inches, wing 18·3, tail 10, tarsus about 2·5.

Immature male in second year's plumage (Upland, Jan. 7th, 1872). General colour above pure white, the head almost immaculate, and only showing one or two dusky brown markings, the hind neck with a larger number of brown-barred feathers; back and wing-coverts transversely barred with pale dusky brown, less distinctly characterized on the primary and outermost greater coverts; quills pure white, with a few more or less distinct spots of pale brown on the outer web, the inner web of the longest primaries having also some brown bars near the tip of the inner webs, the innermost secondaries obscurely clouded with pale brown, the cross bars very indistinct on all of these; lower back, rump, and upper tail-coverts pure white, with here and there remains of pale brown cross markings, more distinct on the tail-coverts; tail itself pure white, the two outer feathers on each side unspotted, the rest having three rows of brown markings near the tips; forehead and entire face and throat pure unspotted white; rest of under surface of body white, narrowly barred with dusky brown, these bars narrowing on the lower parts of the body, and being absent on the under tail-coverts and leg-feathers, the plumes on the latter completely hiding the entire foot, and even the nails; under wing-coverts pure white, as also is the inner lining of the quills, with the exception of a few dusky brown spots near the tips. Total length 20·5 inches, wing 16·7, tail 9·1, tarsus about 2·35.

Obs. Mr. Ridgway, in the third volume of the 'North-American Birds' (p. 61), wishes to separate the Snowy Owl into two races or "varieties," peculiar to the Old and New World respectively. The Palæarctic species he calls *Nyctea scandiaca*, var. *scandiaca* (vel *nivea*); and the Nearctic bird is separated as *N. scandiaca*, var. *arctica*. The characters which he gives for separating them are the narrowness and darker brown shade of the bars in the American Snowy Owl. But this I cannot see; for on comparing a specimen from Norway with one from Labrador, there is no perceptible difference in the barring of the plumage. Adult birds are very similar; for I have before me perfectly white examples both from Sweden and from North America. I therefore do not agree with Mr. Ridgway in this separation into two races; and I believe that we have in the Snowy Owl an analogous case to the Greenland Jerfalcon (*Hierofalco candicans*), and that the differences of plumage can be accounted for by the *age* of the bird. The young birds are strongly barred; and these bars disappear as the individuals advance in age. The females seem never to get as free from transverse markings as the males; and I am entirely able to agree with Mr. Dresser on this point (B. of Eur. part xxiii.); or, at all events, if they ever become pure white, they take a longer time to lose the barred plumage; and in numerous other families of birds this is also known to be the case. At the same time I find a difference in the Snowy Owls of Europe and North America which seems to have escaped Mr. Ridgway's notice: this consists in the much greater amount of feathering of the toes in the European bird, the plumes completely hiding toe and nail, whereas in the American bird the nails always, and even the final joint of the toe occasionally, may be clearly seen. The following are the measurements of the Museum series:—

		Total length.	Wing.	Tail.	Tarsus.
		in.	in.	in.	in.
a. Ad.	Unst, Shetland Isles	22	16·5	9	circa 2·6
b. Ad.	Icy Cape, Norway	26	17	9·2	„ 2·3
c. Juv.	Icy Cape, Norway	25	18·3	9·7	„ 2·4
d. ♂ ad.	Upland, Sweden, Nov. 18, 1872.	23	16·7	9·6	„ 2·1
e. ♂ juv.	Upland, Sweden, Jan. 7, 1872.	20·5	16·7	9·1	„ 2·35
f. ♀ ad.	Upland, Sweden, Jan. 10, 1872.	26	18·3	10	„ 2·5
g. Ad.	Archangel	21	18·5	9·8	„ 2·35
h. [♂] ad.	Lichtenfels, S. Greenland, Feb. 27, 1874.	23	17·3	9·7	„ 2·4
i. [♀] ad.	Lichtenfels, S. Greenland, June 8, 1871.	20·5	15·5	9	„ 2
k. Ad.	Hudson's-Bay Territory..	24	17·8	10	„ 2·5
l. Ad.	Hudson's-Bay Territory..	21	16·6	9·5	„ 2·3
m. Ad.	Labrador..............	22	17·5	10·3	„ 2·2
n. Ad.	Fort Simpson, Jan. 15, 1862	25	18·5	10	„ 2·55
o. Ad.	N. America............	24	18·7	9·5	„ 2·6

The Greenland birds agree with those from North America in not being quite so thickly feathered on the toes. I think it likely that an error in the sex has occurred in the two Lichtenfels examples. But little reliance can be placed on the tarsal measurements, owing to the extreme density of the feathers, which renders it almost impossible to obtain the exact dimensions.

Hab. The same as that of the genus.

a. Ad. st.	Unst, Shetland Isles.	Mr. Bullock.
b. Ad. st.	Orkneys.	
c. Pull. st.	N. Europe.	Zoological Society.
d. Ad. sk.	Icy Cape, Norway.	H. D. Woodfull, Esq. [P.].
e. Juv. sk.	Icy Cape, Norway.	H. D. Woodfull, Esq. [P.].
f. ♂ ad. sk.	Upland, Sweden, Nov. 18, 1872.	Purchased.
g. ♂ juv. sk.	Upland, Sweden, Jan. 7, 1872.	Purchased.
h. ♀ ad. sk.	Upland, Sweden, Jan. 10, 1872.	Purchased.
i. Ad. sk.	Archangel.	R. B. Sharpe, Esq. [P.].
k. [♂] ad. sk.	Lichtenfels, S. Greenland, Feb. 27, 1874.	R. B. Sharpe, Esq. [P.].
l. [♀] ad. sk.	Lichtenfels, S. Greenland, June 8, 1871.	R. B. Sharpe, Esq. [P.].
m. Ad. st.	Labrador.	Rev. Mr. Reichel [P.].
n. Ad. st.	Hudson's-Bay Territory.	Hudson's-Bay Co. [P.].
o. Ad. st.	Hudson's-Bay Territory.	Hudson's-Bay Co. [P.].
p. Ad. sk.	Fort Simpson, Jan. 15, 1862.	B. R. Ross, Esq. [P.].
q. Ad. st.	Arctic America.	Admiral Sir G. Back [P.].

6. SURNIA.

	Type.
Surnia, *Dumér. Zool. Anal.* p. 34 (1806)	S. ulula.
Nycthierax, *Sundev. Av. Meth. Tent.* p. 106 (1872)	S. ulula.

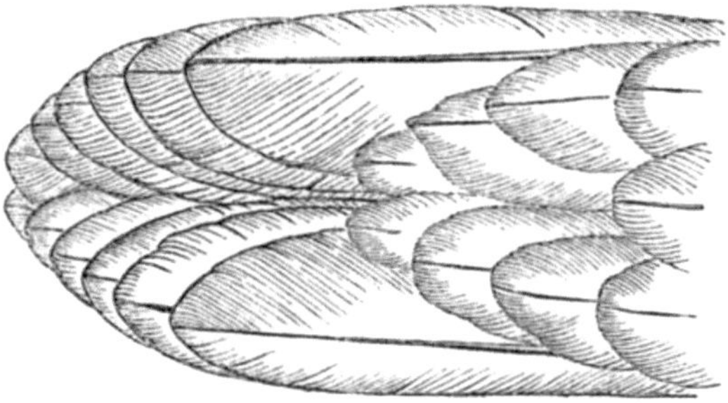

Tail of *Surnia ulula* (reduced).

Range. The northern and central portions of the Palæarctic and Nearctic regions.

1. Surnia ulula.

Strix ulula, *Linn. S. N.* i. p. 133.
Chouette à longue queue de Sibérie, *Buff. Pl. Enl.* i. pl. 463.
Strix nisoria, *Wolf & Meyer, Vög. Deutschl.* fol. ii. p. 123; *Meyer in Wolf & Meyer, Taschenb.* i. p. 84; *Naum. Vög. Deutschl.* i. p. 427, pl. 42. fig. 2; *Sund. Sv. Fogl.* p. 192, p. xxv. fig. 5.
Surnia funerea, *Duméril, Zool. Anal.* p. 34; *Bp. Comp. List B. Eur. & N. Am.* p. 6; *Gould, B. Eur.* i. pl. 45; *Keys. & Blas. Wirb. Eur.* p. xxxii; *Strickl. Orn. Syn.* p. 183; *Hewits. Eggs Br. B.* i. p. 65, pl. xviii. fig. 1; *Degl. & Gerbe, Orn. Eur.* i. p. 117; *Borggr. Vogelf. Deutschl.* p. 64; *Gould, B. Gt. Br.* i. pl. xxxv.
Strix uralensis, *Shaw, Gen. Zool.* vii. p. 277, pl. 35.
Strix doliata, *Pall. Zoogr. Rosso-As.* i. p. 316.
Surnia uralensis, *Steph. Gen. Zool.* xiii. pt. 2, p. 62.
Noctua nisoria, *Cuvier, Règne An.* i. p. 344 (1829).
Surnia ulula, *Bp. Cat. Met. Ucc. Eur.* p. 22; *Gray, Gen. B.* i. p. 33, pl. 12. fig. 1; *Bp. Consp.* i. p. 36; *Kaup, Contr. Orn.* 1852, p. 107; *id. Tr. Z. S.* iv. p. 214; *Fritsch, Vög. Eur.* tab. 11. fig. 5; *Newt. Ooth. Wolley.* p. 167; *Gray, Hand-l. B.* i. p. 39; *Sharpe & Dresser, B. Eur.* pt. xii.
Strix funerea, *Kjærb. Orn. Dan.* pl. vi. fig. 5; *Midd. Sibir. Reis.* p. 131; *Schrenck, Reis. Amurl.* p. 248; *Radde, Reis. Sibir.* p. 124.
Noctua ulula, *Schl. Mus. P.-B.* Striges, p. 42.

Adult male. Above sepia-brown, everywhere strongly mottled with white, the bases of the feathers all white, but showing less on the back than the rest of the upper parts, though here the feathers are broadly barred with white, the bases being more or less concealed; the scapulars externally pure white, forming a longitudinal patch; crown of head white, barred with dark brown, the bars more predominating and broader towards the nape, which is white varied with a few shaft-lines and margins of brown; on each side of the hind neck a large black patch; hind neck and wig white, rather broadly barred and margined with brown; wing-coverts dark sepia, externally margined with white, and showing large oval spots of the same on the median and greater series; quills rather ashy brown, with indications of lighter brown bars, on the external margin of which occurs a large white spot of irregular shape, the primaries terminally margined with white, a little broader on the secondaries; upper tail-coverts broadly barred with brown and white; tail ashy brown, crossed with nine narrow bars of dull white, all the feathers being broadly tipped with the same, the white cross bars purer in colour on the centre feathers, but obscured with brown towards the outermost ones, all the bars on the inner webs much broader and purer white; an indistinct eyebrow and sides of face white, the lores obscured by blackish bristles, the ear-coverts tipped with black, merging in a crescentic black line running down behind the ear-coverts; throat whitish, the chin obscured by black bristles like the lores; the chest white, excepting a few small specks of brown, and separated from the throat by a band of dusky brown feathers; rest of under surface of body white, narrowly banded with brown, the bars disappearing on the abdomen and under tail-coverts, becoming broader again towards the tips of the latter; leg-feathers yellowish white, with a few markings of dusky brown; under wing-coverts and axillaries white, thickly barred across with brown, the greater series with broad tips of brown, and thus resembling the inner lining of the wings, which are brown, with large oval notches of white on the inner web; bill light yellow; claws blackish brown, soles of feet yellow; iris bright yellow. Total length 14 inches, wing 9·7, tail 7·5, tarsus (about) 1.

Young female. Similar to the adults, but having the feathers of the head more fluffy, pale brown in colour, narrowly tipped with a whitish apical bar; on the back of the wing a few basal markings of white are apparent on disarranging the feathers; in other respects the upper plumage is very similar to that of the adult, but is more dingy, and none of the white spots on the scapulars or wings are so large or distinct; the bars on the centre tail-feathers are whitish brown, and not pure white; the throat and sides of face are buffy white; there is no distinct white pectoral band, and the bars on the breast are more tinged with brown, while the tarsi are not so thickly feathered. Total length 13·2 inches, wing 9·2, tail 7. (*Sharpe & Dresser*, B. of Eur. *l. c.*)

Hab. The whole of Northern Europe, across Siberia to Kamtschatka, ranging into Central Europe accidentally or in winter.

a. ♂ ad. sk.	Sodermanland, Sweden, Oct. 27, 1870.	Purchased.
b. ♂ ad. st.	North Bothnia.	Prof. Wahlberg [C.].
c. ♀ ad. st.	North Bothnia.	Prof. Wahlberg [C.].
d. Ad. st.	Archangel.	R. B. Sharpe, Esq. [P.].
e. Juv. st.	Archangel.	R. B. Sharpe, Esq. [P.].
f. Ad. sk.	Archangel.	R. B. Sharpe, Esq. [P.].
g. Ad. sk.	Southern Ural (*Strader*).	R. B. Sharpe, Esq. [P.].
h, i. Ad. st.	Siberia, Nov. 1841.	Moscow Museum [P.].
k. ♂ ad. sk.	Lake Baikal, Oct. 31, 1871.	Dr. Dybowski [C.].
l. ♀ ad. sk.	Kultuk, Oct. 4, 1870.	Dr. Dybowski [C.].

Subsp. *a*. **Surnia funerea.**

Little Hawk-Owl, *Edwards, N. H. Birds*, pl. 62.
Le Chat-huant de Canada, *Briss. Orn.* i. p. 518, pl. 37. fig. 2.
Le Chat-huant de la Baye de Hudson, *Briss. Orn.* i. p. 520.
Caparoch, *Buff. Hist. Nat. Ois.* i. p. 306.
Strix funerea, *Linn. S. N.* i. p. 133; *Werner, Atl.* Rapaces, pl. 32.
Strix caparacoch, *P. L. S. Müll. S. N. Suppl.* p. 69.
Strix hudsonia, *Gm. S. N.* i. p. 295; *Wils. Am. Orn.* vi. pl. 50. fig. 1.
Surnia canadensis, *Steph. Gen. Zool.* xiii. pt. 2, p. 62.
Surnia borealis, *Less. Traité*, p. 100.
Surnia hudsonia, *James. ed. Wils. Am. Orn.* i. p. 90; *Brehm, Naum.* 1855, p. 268.
Surnia funerea, *Rich. & Sw. Faun. Bor.-Am. Birds*, p. 92; *Audub. B. N. Am.* pl. 378; *id. Orn. Biogr.* i. p. 550; *Newt. ed. Yarr. Br. B.* i. p. 183; *Gray, B. Scotl.* p. 64; *Harting, Handb. Brit. B.* p. 91; *Sharpe & Dresser, B. Eur.* pt. xii.
Syrnia funerea, *Macgill. Brit. B.* iii. p. 404.
Noctua funerea, *Yarr. Brit. B.* i. p. 139.
Surnia ulula, *Cass. Cat. Strigidæ Philad. Mus.* p. 17; *id. B. Calif. & Tex.* p. 191; *id. B. N. Am.* p. 64; *Dall & Bann. Tr. Chicago Acad.* i. p. 274.
Surnia ulula, var. hudsonia, *Coues, Key N. Am. B.* p. 205; *Ridgw. B. N. Am.* iii. p. 75; *Coues, B. N. W.* p. 312.

Adult male. Above black, with somewhat of a purplish gloss on the nape; head blackish, thickly covered with large rounded spots of white, the bases to the feathers being also white; hind neck and wig whitish, broadly barred across with brown; all the dorsal feathers with concealed white bars, larger and more oval on the scapulars, which are white on the outer edge, forming a longitudinal patch; wing-coverts brown, the least ones edged with whitish, the greater series relieved with oval spots of white; quills brown, barred across with lighter brown, these lighter bars indicated on the outer web by a large oval spot of white, less distinct on the secondaries, some of which are barred nearly across with white; upper tail-coverts more rufous than the back, and broadly barred and tipped with white; tail dark brown, with about nine white cross bands, including the terminal one; an indistinct eyebrow and sides of face whitish or yellowish white, the lores with blackish bristles; the cheeks silvery white, the hindermost black at their extremity, merged in a broad black band running down behind the ear-coverts; a similar longitudinal black line on either side of the hind neck; throat white, as also the fore neck, the chin obscured with blackish bristles, and the centre of the throat dusky blackish; the fore neck white, separated from the throat by a band of dusky feathers; rest of under surface of body white, crossed with numerous tolerably broad bars of vinous brown or rufous, the leg-feathers narrowly marked with the same, the bars being broader on the under tail-coverts; under wing-coverts white, broadly barred with dark brown, the greater series whitish at base, ashy brown at tip, thus resembling the inner lining of the wing, which is ashy brown, broadly barred with white on the inner web; bill yellow, inclining to horn-blue on the gape and lower mandible; feet and iris as in the European species. Total length 15·5 inches, wing 9·5, tail 8.

Adult female. Similar to the male. Total length 14 inches, wing 9·2, tail 7·3, tarsus 1.

Hab. North America, to the north of 40° N. lat.

a.	Ad. st.	North America.	
b.	Ad. st.	North America.	Admiral Sir G. Back [P.].
c.	Ad. sk.	Repulse Bay.	J. Rae, Esq. [P.].
d.	♂ ad. sk.	Fort Simpson.	Bernard R. Ross, Esq. [P.].
e.	♂ ad. sk.	Western side of the Rocky Mountains.	J. K. Lord, Esq. [P.].
f.	♀ ad. sk.	Western side of the Rocky Mountains.	J. K. Lord, Esq. [P.].
g.	Ad. st.	Scotland.	Purchased.

7. CARINE.

	Type.
Noctua, *Savign. Descr. Egypte*, p. 29 (1809, *nec Fabr.*)	C. glaux.
Athene, *Boie, Isis*, 1822, p. 549 (*nec Hübn.*)	C. noctua.
Carine, *Kaup, Natürl. Syst.* p. 29 (1829)	C. noctua.

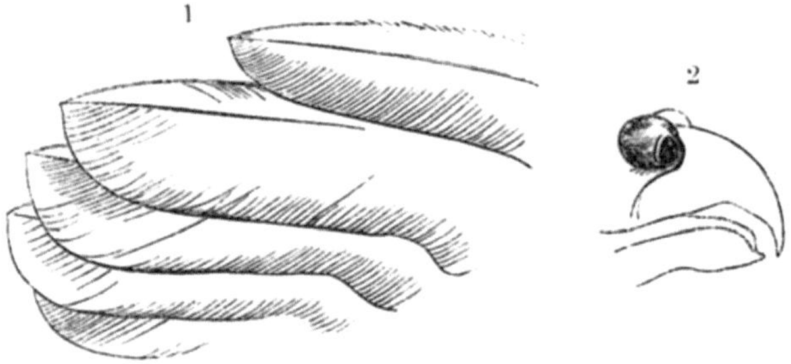

1. Wing of *Carine noctua*, to show long first primary.
2. Nostril of same (from fresh specimen).

Range. Central Europe and all the countries bordering the Mediterranean; N.E. Africa, extending across the desert subregion, through Persia and Central Asia to Northern China; the whole of India and Burmah.

Key to the Species.

a. Breast and abdomen streaked.
a'. Face brown, ear-coverts dark; general colour of breast brown, mottled with white margins to the feathers .... *noctua*, p. 133.
a''. General colour brown; below whitish, streaked with brown. *α. noctua.*
b''. General colour fawn-rufous; below pure white, streaked with rufous.
a'''. Toes scantily covered with hair. *β. glaux.*
b'''. Toes thickly feathered. *γ. plumipes.*
b'. Face white, the ear-coverts scarcely darker; entire under surface white, with a few narrow reddish streaks .... *spilogastra*, p. 138.

b. Breast and abdomen barred .... *brama*, p. 138.
c'. General colour above greyish brown, the white spots large and distinct; bands on tail *five*. *α. brama.*
d'. General colour slaty brown, the white spots very small; bands on tail six. *β. pulchra.*

1. Carine noctua.

Little Owl, *Edwards, Gleanings*, pl. 228.
La Petite Chouette ou La Chevêche, *Briss. Orn.* i. p. 514.
Strix noctua, *Scop. Ann.* i. p. 22; *Naum. Vög. Deutschl.* i. p. 493, pl. 48. fig. 1; *Schl. u. Susem. Vög. Eur.* Taf. 47. figs. 1, 2; *Nordm. in Demid. Voy. Russ. Mérid.* iii. p. 109; *Bailly, Orn. Sav.* i. p. 194; *Keulem. Onze Vogels*, pl. 1.
La Chevêche, *Buff. Pl. Enl.* pl. 439.
Strix passerina, *Gm. S. N.* i. p. 296; *Temm. Man. d'Orn.* i. p. 92; *Roux, Orn. Prov.* i. p. 78, pl. 53; *Werner, Atlas*, Rapaces, pl. 39; *Yarr. Brit. B.* i. p. 142.
Strix psilodactyla, *Nilss. Skand. Faun.* 1st ed. p. 88.
Strix nudipes, *Nilss. Orn. Suec.* i. p. 68.
Athene passerina, *Boie, Isis*, 1822, p. 549; *Brehm, Vög. Deutschl.* p. 110.
Carine noctua, *Kaup, Natürl. Syst.* p. 29; *Newt. ed. Yarr. Brit. B.* i. p. 178; *Irby, B. Gibr.* p. 58; *Sharpe, Ibis*, 1875, p. 258.
Noctua passerina, *Cuv. Règne An.* i. p. 345; *Less. Traité d'Orn.* p. 103; *Selby, Ill. Brit. Orn.* i. p. 107, pl. 26; *Stevenson, B. Norf.* i. p. 59.
Surnia noctua, *Bp. Oss. Regn. Anim. Cuv.* p. 48; *Keys. u. Blas. Wirb. Eur.* p. xxxii; *Borggr. Vogelf. Norddeutschl.* p. 64.
Scotophilus nudipes, *Jard. Brit. B.* i. p. 274.
Noctua nudipes, *Gould, B. Eur.* i. pl. 48; *Hewits. Eggs Br. B.* i. p. 67, pl. xix. fig. 3.
Athene noctua, *Bp. Comp. List B. Eur. & N. Am.* p. 6; *Gray, Gen. B.* i. p. 34; *Bp. Consp.* i. p. 39; *Kaup, Contr. Orn.* 1852, p. 105, et *Tr. Z. S.* iv. p. 210; *Schl. Vog. Nederl.* pl. 43; *Strickl. Orn. Syn.* p. 159; *Fritsch, Vög. Eur.* tab. 11. fig. 2; *Jaub. et Barth. Lapom. Rich. Orn.* p. 77; *Newt. Ooth. Wolley.* p. 151; *Gould, B. Gt. Br.* pl. xxxvii.; *Gray, Hand-l. B.* i. p. 39; *Salvad. Faun. Ital. Ucc.* p. 29; *Sharpe & Dresser, B. Eur.* part iv.
Syrnia psilodactyla, *Macgill. Brit. B.* iii. p. 417.
Athene psilodactyla, *Brehm, Naum.* 1855, p. 270.
Athene indigena, *Brehm, Naum.* 1855, p. 270.
Noctua veterum, *Schl. Mus. P.-B.* Striges, p. 28; *id. Revue Accipitr.* p. 26.
Noctua minor, *Degl. & Gerbe, Orn. Eur.* i. p. 122.

Adult. Above brown, everywhere mottled with oval white spots to the feathers, more or less concealed on the back, and more distinct on the scapular feathers; wing-coverts brown, marked in the same manner as the back with oval spots of white, those on the greater series larger and more distinct; primary-coverts brown, notched on the outer web with white, these markings slightly tinged with rufous; quills brown, barred with paler brown, shading off into whitish on the inner web, and exhibiting a whitish notch on the outer one, producing a chequered appearance; upper tail-coverts brown, barred across with yellowish white; tail brown, tipped with whitish and crossed with four other bands of whitish brown, the bands traversing the rectrices near the base, but incomplete towards the apex; head brown, thickly marked with triangular spots of white, with which all the feathers are streaked; on the nape a patch of white, the feathers being mostly of this colour, with broad terminal margins of brown; feathers of the face, as well as a distinct eyebrow, white, the lores blackish at tip, the ear-coverts also marked with brown; cheeks white, extending backwards in a triangular patch behind the ear-coverts; chin and fore neck white, separated from each other by a band of brown feathers across the throat; rest of under surface of body white, streaked with brown, some of the flank-feathers having also indications of cross markings of the same colour, the feathers of the upper breast barred and mottled with whitish; leg-feathers and under tail-coverts whitish, the latter with a few central markings of brown; under wing-coverts whitish, inclining to yellow on the inner ones and axillaries, most of the under coverts being marked with brown near the base; inner lining of wing brown, the inner webs broadly barred and the outer webs spotted with yellowish; bill yellow, slightly tinged with greenish; feet greyish yellow; iris yellow. Total length 11·5 inches, wing 6·7, tail 3·5, tarsus 1·3.

The sizes of the sexes are given as below in Sharpe and Dresser's 'Birds of Europe':—

♂ ad. Total length 8·2 inches, wing 6, tail 2·9, tarsus 1·1.
♀ ad. Total length 9 „ wing 6·1, tail 3, tarsus 1·1.

I suspect that both these birds were really males, and that the specimen described by me above represents the female sex.

Young. Very similar to the adult, but more dingy, especially on the lower surface, where the central markings to the feathers are ashy brown and very broad. All the characteristic markings of the adult are developed in the young bird; but the white markings on the upper parts are broader and more conspicuous than in the old bird, especially on the scapulars and upper wing-coverts.

Hab. The continent of Europe, occurring as far north as Denmark, but only an occasional visitant in England or Scandinavia. In N.E, Africa only near Tangier (*Irby*), thence along the northern shores of the Mediterranean to Asia Minor and Southern Russia generally. Its eastern range is as yet undefined.

a.	Juv. st.	Europe.	
b.	Ad. st.	Essex.	Mr. J. Baker [C.].
c.	Pull. st	France.	
d.	Ad. sk.	Near Paris.	M. A. Bouvier [P.].
e.	Ad. sk.	Ain, France.	M. A. Bouvier [P.].
f.	Ad. st.	Ain, France.	R. B. Sharpe, Esq. [P.].
g.	Vix ad. sk.	Seville, Spain (*H. Saunders*).	R. B. Sharpe, Esq. [P.].
h.	Ad. sk.	Geneva, Oct. 1871.	L. Fraser, Esq. [C.].
i.	Ad. sk.	Sardinia, winter of 1869.	A. B. Brooke, Esq. [C.].
k.	Ad. sk.	Greece.	J. Gould, Esq. [P.].
l.	Ad. sk.	South Ural (*Strader*).	R. B. Sharpe, Esq. [P.].
m, n.	Ad. sk.	Tangier.	Signor Olcese [C.].

Subsp. *a*. Carine glaux.

Strix noctua, *Forsk. Descr. Anim.* p. 8.
Strix passerina, *Sonn. Voy. Egypte,* i. p. 349.
Noctua glaux, *Savign. Descr. Egypte, Ois.* p. 287.
Strix persica, *Vieill. N. Dict. d'Hist. Nat.* vii. p. 26; *Pucher. R. Z.* 1849, p. 18.
Athene meridionalis, *Less. Man. d'Orn.* i. p. 110; *Gray, Gen. B.* i. p. 35; *Kaup, Contr. Orn.* 1852, p. 105, et *Tr. Z. S.* iv. p. 209; *Strickl. Orn. Syn.* p. 160; *Fritsch, Vög. Eur.* Taf. 13. fig. 7.
Noctua passerina, *Rüpp. Neue Wirb. Vög.* p. 45.
Strix noctua meridionalis, *Schl. u. Susem. Vög. Eur.* Taf. 47. fig. 3.
Noctua persica, *Keys. u. Blas. Wirb. Eur.* p. xxxii; *Degl. & Gerbe, Orn. Eur.* i. p. 123.
Athene bactriana, *Hutton, J. A. S. B.* xvi. p. 776.
Strix numida, *Levaill. Jun. Expl. Sci. Algér. Ois.* pl. 4.
Athene persica, *Bp. Rev. et Mag. de Zool.* 1854, p. 543; *Newt. Ooth. Wolley.* p. 152; *Loche, Expl. Sci. Algér. Ois.* i. p. 106; *Hume, Rough Notes,* ii. p. 407; *Salvad. Faun. Ital. Ucc.* p. 30.
Noctua veterum meridionalis, *Schl. Mus. P.-B.* Striges, p. 29; *id. Revue Accipitr.* p. 26.
Noctua veterum, *Heugl. Orn. N.O.-Afr.* i. p. 118.
Athene glaux, *Gray, Hand-l. B.* i. p. 39.
Carine meridionalis, *Shelley, B. Egypt,* p. 177.
Carine glaux, *Irby, B. Gibr.* p. 58; *Sharpe, Ibis,* 1875, p. 258.
Carine bactriana, *Sharpe, Ibis,* 1875, p. 258*.

Adult female. General colour pale fawn-brown, mottled all over with white spots, more or less concealed on the back, and plainer on the scapulars and greater wing-coverts, the latter being rather darker brown on their inner webs; primary-coverts and quills dark brown, tipped with white and broadly barred on the inner web with white, this bar not extending across the feathers, but represented on the outer web with a white notch, producing a chequered appearance; upper tail-coverts pale fawn-brown, rather more rufous than the back, largely spotted with white, which occupies the basal half of the feather; tail pale fawn-colour, tipped with whitish and crossed with four bands of buffy white, continuous near the base, the apical bands, however, not completely traversing the feather; head rather darker fawn-brown than the back, streaked with white down the centre of the feathers, dilating towards the tip into a distinct white spot; hind neck white, the feathers being mottled with brown, most of the feathers edged and tipped with the latter colour; facial aspect white, the lores with dark brown shaft-stripes, the ear-coverts also broadly streaked with brown; cheeks white, continued in a triangular patch to behind the ear-coverts; chin and fore neck pure white, separated from each other by a band of feathers of a pale rufous fawn-colour; rest of under surface of body pure white, broadly streaked with rufous fawn-colour, these streaks becoming narrower on the abdomen and disappearing on the under tail-coverts; leg-feathers entirely white; under wing-coverts white, the inner ones and the axillaries washed with yellowish, the outer ones streaked with brown,

* It is quite possible that Hutton's title will turn out to belong to the next species, in which case it will have precedence.

the greater series dark brown at base, ashy brown at tip, resembling the inner lining of the wing, which is ashy brown, broadly barred with white on the inner web; bill and irides pale yellow. Total ength 10·5 inches, wing 7, tail 3·8, tarsus 1·25.

Adult male. Similar to the female. Total length 9·5 inches, wing 6·4, tail 3·6, tarsus 1·25.

Nestling. Dusky brown, and covered with down of this same colour, the white spots on the wing-coverts very strongly developed, the latter nearly entirely white, with central streaks of brown; the greater coverts brown, the outer webs whitish; quills brown, tipped with ashy grey, the outer webs notched with rufous buff; under surface dull white, with central streaks of brown.

The above description is from a nestling collected by Mr. L. Fraser, and having the feathers only slightly developed. An older bird, obtained from the same collector, exhibits the markings of an adult bird, but is very much dingier, as is a fully grown young individual shot by Mr. Edward Bartlett, near Cairo. This bird is dark brown, mottled with whitish or buff, and very closely resembles the young of *C. noctua.* The under surface is whitish, streaked with brown, the latter colour entirely predominating, and giving a clouded appearance to the lower parts; quills and tail dull brown, crossed with five bands of fulvous or pale fawn-colour; head almost uniform, with scarcely any fulvous spots.

This species is certainly very closely allied to *C. noctua*; and the young appear to be almost indistinguishable from each other. It may be regarded as a desert form of the European bird, with a distinct range, being pale fawn-colour where *C. noctua* is dark brown: hence it has a generally pale appearance. On raising the wing the inner lining in *C. glaux* is very pale, and inclines to ashy brown, the white bars on the inner web very broad and contrasting strongly, the bars on the inner web of the first primary being *four* in number. In *C. noctua* the wing-lining is much darker brown, and only *three* bars are apparent on the inner web of the first primary, all the bars on the inner web being not quite so broad. These differences of wing-coloration are really very little more than part of the general pale tone predominant in *C. glaux*, and hardly constitute specific characters. The specimen of *C. noctua* from Greece is so thoroughly intermediate, that I am inclined to look upon the two birds as races of one form.

Hab. The southern countries of the Mediterranean, excepting the immediate vicinity of Tangier; thence extending into N.E. Africa, and eastwards through Palestine and Persia to Afghanistan and Thibet.

a. Pull. st.	Tunis.	L. Fraser, Esq. [C.].
b. Juv. st.	Tunis.	L. Fraser, Esq. [C.].
c. Ad. sk.	Tunis.	L. Fraser, Esq. [C.].
d. Ad. st.	Egypt.	W. B. D. Turnbull, Esq. [P.].
e, f. Ad. sk.	Egypt.	Sir Samuel Baker [P.].
g. Imm. st.	Thebes.	Dr. A. Leith Adams [P.].
h. Juv. sk.	Cairo, June 26, 1864.	E. Bartlett, Esq. [C.].
i. ♂ ad. sk.	Egypt, March 5, 1868.	Capt. G. E. Shelley [P.].
k. ♀ ad. sk.	Fayoom, Feb. 24, 1871.	Capt. G. E. Shelley [P.].

l. ♀ ad. sk.	Wilderness of Judea.	Canon Tristram [C.].
m. ♂ ad. sk.	Beersheba.	Canon Tristram [C.].
n. Ad. st.	Ruins of Suza.	K. Loftus, Esq. [P.].
o. ♀ ad. sk.	Shiraz, Persia.	Major St. John [C.].
p. ♂ ad. sk.	Niriz, east of Shiraz, June 2, 1872.	W. T. Blanford, Esq. [C.].

Subsp. β. **Carine plumipes.**

Athene nudipes, *G. R. Gray, Cat. Mamm. &c. Nepal Hodgs.* p. 50 (1846).
Strix noctua, *Radde, Reis. Sibir.* p. 123.
Athene noctua, var., *Dybowsk. & Parv. J. f. O.* 1868, p. 331.
Athene bactriana, *Gray, Hand-l. B.* i. p. 39.
Athene plumipes, *Swinh. P. Z. S.* 1870, p. 448, et 1871, p. 342; *Taczan. J. f. O.* 1872, p. 349.
Athene persica, *David, N. Arch. Mus.* vii. *Bull.* p. 4 (1871, *nec Vieill.*).
Athene noctua orientalis, *Severtz. Turkest. Jevotn.* p. 63.
Athene orientalis, *Sharpe, Zool. Record,* x. p. 17.
? Carine glaux, *Dresser, Ibis,* 1875, p. 110.
Carine plumipes, *Sharpe, Ibis,* 1875, p. 258.

Adult (type of species). General colour deep fawn-brown, the head streaked with buffy whitish, all the streaks dilating towards their tips, some of which have a subterminal spot; hind neck largely mottled with white, many of the feathers broadly barred across with white, the outer webs of many of them also of the latter colour; interscapulary region tolerably uniform, the white spots being concealed, but very distinct and oval in shape on the scapulars; wing-coverts coloured like the back, and distinctly spotted with white, the spots on the median and greater coverts larger and more oval in form; primary-coverts dusky brown, barred across with fawn-colour; quills brown, rather more dusky at tip, barred across with lighter and more rufous brown, the ends of the feathers greyish white, all the rufescent bars shading off into a white spot on the outer web, and giving a chequered appearance to the wing; upper tail-coverts fawn-colour, the bases white; tail pale fawn-colour, tipped with whitish, and crossed with five rather irregular bands of fulvous fawn; lores pure white, as also a small superciliary streak, the former with black shaft-lines; ear-coverts fawn-colour, with whitish shaft-streaks, the feathers under the eye whitish; cheeks and throat white, extending in a triangular patch backwards below the ear-coverts; lower part of fore neck white, separated from the throat by a band of rufescent feathers mottled with whitish; rest of under surface fawn-buff, inclining to white on the centre of the body, streaked with rufous brown, the centres to the feathers very broad and distinct on the breast, much narrower on the sides of body, and absent altogether on the lower flanks, vent, and under tail-coverts; leg-feathers whitish, light fawn on their outer aspect; under wing-coverts whitish, with a few slight markings of brown here and there, the lower series whitish at base, ashy brown at tip, and thus resembling the inner lining of the quills, which are pale ashy brown, barred with buffy white on the inner web. Total length 8 inches, wing 6·1, tail 3·4, tarsus 1·15. (*Mus. R. Swinhoe.*)

Obs. The typical specimen of *Athene plumipes*, which Mr. Swinhoe has lent me, differs, as he has himself pointed out, from *C. glaux* in having the toes more feathered. Latterly M. Severtzoff has described, from Turkestan, a small Owl, which he separates as *Athene orientalis*, on the self-same characters (*cf.* Dresser, *l. c.*). With regard to Mr. Blanford's remarks on the Thibetan bird from Hodgson's collection, of which a single example only is to be found in the Museum, he is quite right that Hodgson never published a name for the species. Mr. G. R. Gray identified it as *Athene nudipes* (Nilss.), and he indicates that the name of *Athene gymnopus* had been given by Hodgson to the bird; but no such name occurs in the list published by the latter gentleman in J. E. Gray's 'Zoological Miscellany' (*l. c.*), while his drawing of the species (no. 876) bears, in his handwriting, the name of *lagopus*, evidently much more in accordance with the feathered foot of the bird than *nudipes* or *gymnopus*. As, however, no description of this Thibetan specimen was ever published, the name given by Mr. Swinhoe will have to be used if the species should not ultimately prove to be the true *A. bactriana* of Hutton, with which Mr. G. R. Gray identified it.

Hab. Eastern Siberia, Mongolia, and Northern China, near Pekin, ranging southwards into Central Asia, Turkestan, and Thibet.

a. Ad. st. Thibet. B. H. Hodgson, Esq. [P.].

2. Carine spilogastra.

Athene spilogaster, *Heugl. J. f. O.* 1863, p. 15.
Noctua spilogastra, *Heugl. Orn. N.O.-Afr.* p. 119, pl.
Carine spilogastra, *Sharpe, Ibis*, 1875, p. 258.

Adult female. Above of a pale smoky rufous colour, varied with whitish spots and bars, the white colour washed with buff; tail-feathers whitish, with six or seven smoky rufous bands; the face, lower part of breast and abdomen white, slightly washed with buff, and with scattered obsolete rufescent streaks; upper breast varied with the same colour; tarsi, which are feathered in front, white; the toes above covered with white bristly hairs; under wing-coverts entirely whitish; bill yellowish; cere and nails, which are very short, blackish horn-colour; iris yellow; naked orbits blackish; toes fleshy grey. Total length 7·6 inches, wing 5–5·1, tail 2·9–3, tarsus 1–1·25. (*Heuglin.*)

Hab. Abyssinian coastland.

3. Carine brama.

Strix brama, *Temm. Pl. Col.* ii. pl. 68.
Noctua brama, *Stephens, Gen. Zool.* xiii. pt. 2, p. 115; *Less. Traité*, p. 103; *Schl. Mus. P.-B.* Striges, p. 29; *id. Revue Accipitr.* p. 27.
Surnia brama, *Bp. Oss. Regn. An. Cuv.* p. 58.
Noctua indica, *Frankl. P. Z. S.* 1831, p. 115.
Noctua tarayensis, *Hodgs. As. Research.* xix. p. 175.
Athene indica, *Blyth, J. A. S. B.* xi. p. 457.
Athene brama, *Blyth, Ann. N. H.* xii. p. 93; *Gray, Gen. B.* i. p. 34, pl. 12. fig. 3; *Cass. Cat. Strigidæ Philad. Mus.* p. 14; *Blyth, Cat.*

B. Mus. A. S. B. p. 39; *Bp. Consp.* i. p. 39; *Kaup, Contr. Orn.* 1852, p. 105, et *Tr. Z. S.* iv. p. 210; *Bp. Rev. et Mag. de Zool.* 1854, p. 543; *Horsf. & Moore, Cat. B. Mus. E. I. Co.* i. p. 65; *Strickl. Orn. Syn.* p. 166; *Jerd. B. Ind.* i. p. 141; *Gray, Hand-l. B.* i. p. 39; *Hume, Rough Notes,* ii. p. 404; *id. Nests &c. Ind. B.* p. 70.

Carine brama, *Sharpe, Ibis,* 1875, p. 258.

Adult. General colour above greyish brown, mottled with white in the form of more or less concealed spots on the back, the scapulars barred with white, more broadly on the outer webs; wing-coverts brown, like the back, spotted with white, more largely on the outer webs of the median and greater series; primary-coverts brown, externally notched with white, resembling the quills, which are notched on the outer web, and broadly barred on the inner one with white, the inner secondaries more narrowly barred across the feather; upper tail-coverts brown, like the back, broadly barred across with white; tail brown, paler at the tip, and crossed with five white bands, incomplete and not conterminous on the centre feathers; head brown, barred and spotted with white, the hind neck broadly barred with white, and forming a kind of indistinct collar; lores and a tolerably distinct eyebrow pure white, the former obscured with blackish shaft-lines; ear-coverts brown, like the head, barred with whitish; cheeks pure white, extending backwards in a triangular patch behind the ear-coverts; throat and fore neck also pure white, separated from each other by a narrow band of brown-tipped feathers; rest of under surface white, barred transversely with brown, these bars broader and closer on the flanks, the sides of the upper breast brown, barred with white; abdomen, under tail-coverts, and leg-feathers pure white, unspotted; under wing-coverts also pure white, very slightly streaked with brown, the lower series brown, barred with white, and thus resembling the inner lining of the wing, which is ashy brown, barred with white on the inner, and notched with white on the outer web; cere dusky, much swollen above the nares; bill horny green; feet dingy greenish, front of toes sparsely clad with whity-brown bristles, claws blackish horny; iris bright pale yellow. Total length 9·5 inches, wing 6·1, tail 3·5, tarsus 1·2.

Young. Very similar to the adult, and marked in the same manner, but of a more dingy appearance, especially below; the head not so thickly marked with bars, and preserving a more spotted appearance.

Hab. Indian peninsula, extending into Baluchistan.

a. Juv. sk.	India.	J. Gould, Esq. [P.].
b. Ad. st.	India.	J. Hillier, Esq. [P.].
c. ♀ ad. sk.	Rajkote, Kattiawar, Aug. 20, 1871.	Major J. Hayes Lloyd [P.].
d. Ad. st.	Nepal.	B. H. Hodgson, Esq. [P.].
e, f, g, h. Ad. sk.	Nepal.	B. H. Hodgson, Esq. [P.].
i. Ad. sk.	Behar.	B. H. Hodgson, Esq. [P.].
k, l. Ad. sk.	Deccan.	Purchased.
m. Juv. sk.	Deccan.	R. B. Sharpe, Esq. [P.].

n, o, p. Ad. sk.	Kamptee.	Dr. B. Hinde [P.].
q, r. Juv. sk.	Kamptee.	Dr. B. Hinde [P.].
s. Ad. sk.	Western Ghats.	J. Gould, Esq. [P.].
t, u. Ad. st.	Madras.	Rev. H. H. Baber [P.].
v. ♀ ad. sk.	Kalagan, Baluchistan, March 19, 1872.	W. T. Blanford, Esq. [C.].

Subsp. *a*. **Carine pulchra.**

Athene pulchra, *Hume, Str. F.* i. p. 469; *id. op. cit.* iii. p. 39.
Carine pulchra, *Sharpe, Ibis,* 1875, p. 258.

Adult male. General colour above slaty brown, with more or less concealed spots of white, very small on the head, where they are rather plainly defined; hind neck barred with white, forming an indistinct kind of collar; scapulars barred across with white, broader on the outer web; wing-coverts dark slaty colour, externally notched with white, the median and greater series with large rounded spots of white on the outer web; primary-coverts and quills dark slate-colour, notched on the outer web and barred on the inner one with white; tail slaty brown, crossed with six narrow bars of white, rather broader on the outer feather; forehead and eyebrow white, the lores developing into hair-like bristles; fore part of cheeks whitish; ear-coverts dark slaty, barred across with white; entire throat white, extending backwards in a triangular patch below the ear-coverts, a black mark running down the sides of the neck; rest of under surface white, barred with ashy brown, the bars getting fainter on the lower flanks; leg-feathers and under tail-coverts pure white; under wing-coverts white, the lower series ashy brown at tip, and resembling the inner lining of the quills, which are ashy brown below, barred with white; bill brownish, the tip and the culmen greenish yellow. Total length 8 inches, wing 5·5, tail 3·5, tarsus 1·05. (*Mus. R. G. Wardlaw-Ramsay.*)

Obs. The above description is taken from a Thayetmyo skin, collected by Lieut. Wardlaw-Ramsay, and kindly lent me by Lord Walden. Mr. Hume has also been so good as to lend me a bird from the same place; but it does not show the characters so well as the bird described by me. His skin measures:—Total length 7 inches, wing 5·4, tail 3·1, tarsus 1·1. I consider that *C. pulchra* may very fairly be separated from *C. brama*, as the characters seem quite constant, to judge from three other Burmese specimens examined by me in Lord Walden's collection. Mr. Hume also writes:—"I have now four good specimens before me, two of them sent by Mr. Oates; and it appears to me certain that they must be separated. I do not suppose that these birds weigh much more than half what our Indian ones do. In length they vary from 7·25 to 7·8 inches against 8 to 9·5 inches in *C. brama*, while the wings vary from 5·3 to 5·8 inches against 6 to 6·65 inches; and an expanse of 19 to 19·5 inches against 21 to 22·5 inches in the Indian bird. No doubt the general character of the plumage is the same as in *C. brama*; but the spotting of the head is smaller and neater. The general colour of the upper surface is a darker and purer brown; the dark throat-band is more strongly marked. The tail exhibits five or six narrow transverse bands, or traces of these, against four or five far broader and more distinctly marked bands in *C. brama.* The scapulars, too, seem to be generally less barred and more spotted than is usually the case in Indian specimens.

"Generally the birds have a very different appearance, and, I think, would be separated by any one who examined a series of both; but no mere description will sufficiently explain this difference, and the very marked difference in size is the character that must be first relied on for separating the species."

Hab. Burmah and Pegu.

a. Imm. sk. Tonghoo, Burmah. Purchased.

8. HETEROGLAUX.

Heteroglaux, *Hume, Str. F.* i. p. 466.

Wing and nostril of *Heteroglaux blewitti* (from a skin).

Range. Central Provinces of India.

1. Heteroglaux blewitti.

Heteroglaux blewitti, *Hume, Str. F.* i. p. 467; *Ball, op. cit.* ii. p. 382.

Adult female. Above ashy brown, nearly uniform, excepting for a few half-concealed white bars on the hind neck and scapulars, rather more distinct on the latter, some of them being entirely white on the outer web; wing-coverts a little darker than the back, with a few spots of white, larger and more distinct on the greater and median series, the bastard wing notched externally with white, and the primary-coverts also spotted with the same; quills dark brown, barred across with lighter brown, shading off into white notches on the margins of both webs, very large on the inner, and producing a chequered appearance on the outer web; upper tail-coverts a little darker than the back, and spotted with white; tail dark brown, tipped with white, and crossed with four other bands of white besides; frontal plumes, a broad eyebrow, and feathers round the eye white, the lores with blackish hair-like shafts; the ear-coverts ashy brown, barred with whitish; throat, fore neck, and cheeks white, with a tolerably broad bar of dark brown feathers across the lower throat; centre of the abdomen and leg-feathers pure white, the breast clouded with grey, with more or less concealed white bars, the sides of the body distinctly barred with dark brown, the under tail-coverts white, with remains of brown bars near the tips; under wing-coverts white, many of the feathers with brown bases, causing a slightly streaked appearance, the lower series whitish at base, ashy brown at tip, thus resembling the inner lining of the quills, which are ashy brown below, spotted on the outer and broadly barred on the inner web with yellowish white; bill light yellow; iris bright yellow. Total length 9·5 inches, wing 5·8, tail 3·2, tarsus 1·05. (*Mus. A. O. Hume.*)

Hab. Central India; "the dense forests of the eastern portion of the central provinces, on the borders of the Tributary Mehals" (*Hume*).

9. SPEOTYTO.

	Type.
Speotyto, *Gloger, Handb. Naturg.* p. 226 (1842)	S. cunicularia.
Pholeoptynx, *Kaup, Isis*, 1848, p. 769	S. cunicularia.

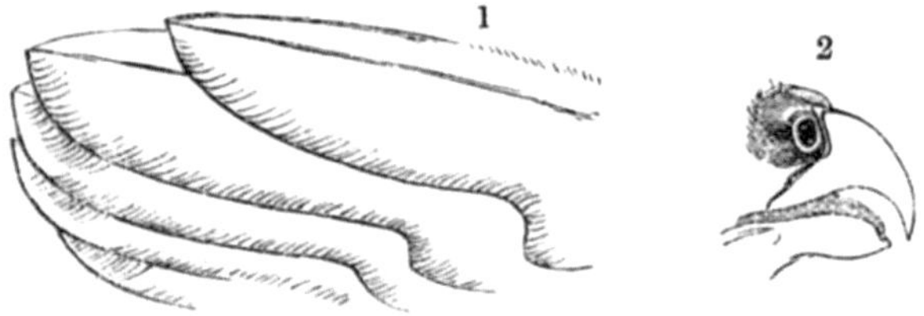

1. Wing of *Speotyto cunicularia.*
2. Nostril of same (from fresh specimen).

Range. Southern and western portions of United States, throughout the whole of Central and South America.

Key to the Species.

a. Inner lining of quills fulvescent; the first primary with five or six isabelline bars, much broader at the base of the inner web than the intervening brown interspace; primary-coverts spotted like the quills .... *cunicularia*, p. 142.

b. Inner lining of quills dark brown; the first primary with four fulvous bands on inner web, smaller and narrower than the brown interspace; primary-coverts blackish brown, with a single fulvous bar faintly indicated ...... *guadeloupensis*, p. 148.

1. Speotyto cunicularia.

La Chouette de Coquimbo, *Briss. Orn.* i. p. 525.
Coquimbo Owl, *Lath. Gen. Syn.* i. p. 145.
Strix cunicularia, *Molina, Stor. Chil.* p. 343; *Gm. S. N.* i. p. 292 (*ex Lath.*); *Audub. B. N. Am.* pl. ccccxxii.; *id. Orn. Biogr.* v. p. 264; *Bp. Am. Orn.* i. p. 68, pl. 7. fig. 2; *Max. Beitr.* iii. p. 248.
La Chouette de St. Domingue, *Sonn. ed. Buff.* iv. p. 145.
Strix dominicensis, *Gm. S. N.* i. p. 296.
Urucurea, *Azara, Apunt.* ii. p. 211; *Hartl. Ind. Azara*, p. 4.
Strix grallaria, *Spix, Av. Bras.* i. p. 21; *Temm. Pl. Col.* 146.
Noctua grallaria, *Steph. Gen. Zool.* xiii. pt. 2, p. 67.
Ulula cunicularia, *Jard. ed. Wils. Am. Orn.* iii. p. 325.

Noctua cunicularia, *Darwin, Journ. Nat. Beagle*, p. 145; *D'Orb. Voy. Am. Mér. Zool.* p. 128; *Burm. Th. Bras.* ii. p. 440; *id. La Pl. Reis.* ii. p. 138; *Schl. Mus. P.-B.* Striges, p. 30; *id. Revue Accipitr.* p. 27.

Otus cunicularia, *Cuv. Règne An.* 1829, p. 341.

Surnia cunicularia, *Bp. Oss. R. A. Cuv.* p. 50.

Noctua urucurea, *Less. Traité*, p. 103; *Tschudi, Faun. Peruan.* p. 116.

Nyctipetes cunicularia, *Swains. Classif. B.* ii. p. 218.

Athene cunicularia, *Bp. Comp. List B. Eur. et N. Am.* p. 6; *id. Consp.* i. p. 38; *Darw. Voy. Beagle, Birds*, p. 31; *Gray, Gen. B.* i. p. 35; *Cass. Cat. Strigidæ Philad. Mus.* p. 15; *Strickl. Orn. Syn.* p. 160; *Baird in Cooper B. Calif.* p. 437; *Pelz. Orn. Bras.* pp. 9, 299; *Gray, Hand-l. B.* i. p. 43.

Athene dominicensis, *Gray, Gen. B.* i. p. 35; *Bp. Consp.* i. p. 38; *Strickl. Orn. Syn.* p. 161; *Gray, Hand-l. B.* i. p. 43.

Noctua dominicensis, *Tschudi, Faun. Peruan.* p. 115.

Athene socialis, *Gamb. Pr. Phil. Acad.* iii. p. 47.

Pholeoptynx cunicularia, *Kaup, Isis*, 1848, p. 769; *id. Contr. Orn.* 1852, p. 106; *id. Tr. Z. S.* iv. p. 211; *Scl. & Salv. Nomencl. Av. Neotr.* p. 117.

Athene hypugæa, *Woodh. in Sitgr. Exp. Zuni & Color. Rivers*, p. 62; *Peale, U.S. Expl. Exp.* p. 78; *Cass. Cat. Strigidæ Philad. Mus.* p. 15; *id. B. Calif. & Texas*, p. 118; *id. B. N. Am.* p. 59; *Blakist. Ibis*, 1863, p. 50; *Gray, Hand-l. B.* i. p. 43.

Pholeoptynx hypogæa, *Salvin, Ibis*, 1860, p. 401.

Spheotyto cunicularia, var. grallaria, *Ridgw. N. Am. B.* p. 89.

Spheotyto cunicularia, var. hypogæa, *Ridgw. in Coues's Key*, p. 208; *id. N. Am. B.* p. 90; *Coues, B. N. West*, p. 323.

Spheotyto cunicularia, var. cunicularia, *Ridgw. l. c.* p. 90; *Coues, l. c.* p. 323.

Adult. General colour sandy brown, the head distinctly marked with oval streaks of buffy whitish, some of the feathers also transversely barred with the same; hind neck more mottled with whitish, the feathers having a broad subterminal bar of this colour; back mottled with white spots and bars on all the feathers, the latter being often concealed and of a sandy buff colour; scapulars broadly barred across with white; wing-coverts sandy brown, a little darker than the back, spotted with white, the greater and median coverts with a large spot of white near the tip of the outer web, all this series of coverts being also barred with sandy buff; primary-coverts brown, barred with sandy buff; quills brown, tipped with whitish and barred with sandy brown, broadly on the inner web, the outer web notched with buffy whitish, especially the secondaries; tail-feathers brown, tipped with whitish and crossed with five other bands of sandy buff, inclining to whitish on the centre feathers, the bars with a conterminous shade of brown both above and below; lores and an indistinct eyebrow whitish, the former obscured by black hair-like shaft-lines; ear-coverts brown, streaked with fulvous just behind the eye; cheeks whitish, extending backwards behind the ear-coverts; throat and fore neck white, separated from each other by a band of sandy buff feathers, barred with darker brown; sides of the neck buffy whitish, streaked with light brown; rest of under surface creamy white, transversely barred with light sandy brown, these bars decreasing in width on the lower part of the body, and being entirely absent on the under tail-coverts, the sides of the upper breast more thickly barred with brown; thighs yellowish buff, perfectly uniform, the tarsi also clothed near the top with feathers of the same colour; under wing-coverts yellowish buff, the outermost white, as also the edge of the wing, the greater series dark brown at tip; quills brown, barred across with lighter brown, inclining to sandy buff, the bases of the inner webs entirely of the latter colour; bill greyish yellow, darker at base; feet dusky grey, claws black; iris yellow. Total length 9 inches, wing 6·8, tail 3·6, tarsus 1·95.

The above is probably a female, as a sexed male measures—total length 8 inches, wing 6·5, tail 3·1, tarsus 1·65.

Young (Corrientes, D'Orbigny), More uniform than the adults both above and below; ear-coverts dark brown like head; throat and cheeks white, the latter as well as the sides of neck with a silvery gloss; rest of under surface covered with creamy buff down, the ordinary feathers on the chest commencing to appear; wings and tail apparently like the adult; leg-feathers white, and reaching nearly to the base of the toes. (*Mus. Paris.*)

Hab. The same as that of the genus.

a, b, c. Ad. sk.	Western side of the Rocky Mountains.	J. K. Lord, Esq. [P.].
d. Ad. sk.	Magdalena Bay, California.	Capt. Kellett and Lieut. Wood [P.].
e. Ad. st.	California.	Purchased.
f. Ad. sk.	California.	Purchased.
g. Juv. sk.	California.	Purchased.
h. ♂ ad. sk.	Mexico.	Purchased.
i. Ad. sk.	Puebla, Mexico.	M. A. Boucard [C.].
k. Ad. sk.	Peru.	H. Whitely, Esq. [C.].
l, m. ♂ ♀ ad. st.	Tambo valley, Peru.	H. Whitely, Esq. [C.].
n. ♀ ad. sk.	Tinta, May 21, 1868.	H. Whitely, Esq. [C.].
o, p, q, r, s. Ad. sk.	Chili.	E. C. Reed, Esq. [C.].
t. Ad. st.	Maldonado.	Charles Darwin, Esq. [C.].

Obs. The Chilian birds are much larger than those from Central America or the United States, and measure in the wing 6·95–7·4 inches (Mr. Ridgway even gives 7·5 inches; and the Tinta specimen in the collection reaches to 7·8 inches); whereas the wing of the northern bird varies from 6·45–7·25 inches. Specimens, however, from the Tambo valley are somewhat intermediate in plumage, and measure 6·6–6·7 inches in the wing; their tarsus, however, is the shortest of any, viz. 1·5 inch. Mr. Ridgway, however, has examined one which measured 1·8 inch in the tarsus; and therefore we come to the conclusion that the species is subject to great variation, and that it is impossible to define distinct races by constant characters. The bands on the tail are generally six in the North-American birds, and seven in the Chilian; but of nine specimens examined by me, two had six and two had only five bands: so this character is equally inconstant.

The following are a few notes on the series examined by me:—

British Columbia. Three specimens from the western side of the

Rocky Mountains, collected by the late Mr. J. K. Lord, have fallen under my notice. They are all very much alike in the character of their markings; but one is much more sandy-coloured than the others, the spots on the upper surface, and also the bands on the tail, being distinctly sandy buff instead of whitish; all have six bands on the tail, including the one at the tip. The throat-band, in two out of the three birds, is blackish brown, with fulvous cross bars on the feathers. The measurements are as follows:—

		Total length.	Wing.	Tail.	Tarsus.
		in.	in.	in.	in.
a. Ad.	British Columbia (*Lord*)....	9	6·8	3·6	1·95
b. Ad.	British Columbia (*Lord*)....	8	6·85	3·5	2
c. Ad.	British Columbia (*Lord*)....	9·5	6·7	3·4	1·95

California. One specimen before me from this country seems to be a younger bird; it is of a dark earthy brown, covered with little oval spots of sandy buff, the crown-spots rather more longitudinal than those of the back; the bars on the tail are sandy-coloured also; so that I suspect the tail-bands grow whiter with age. Underneath the same dark tint prevails; the throat-band and sides of chest earthy brown, mottled with spots and bars of sandy buff or fulvous white; the bands on the breast and flanks very broad and rather rufous, not perceptibly narrowing towards the flanks. Another specimen, though not adult, is rather older than the one above mentioned; the head is slightly barred, and the spots on the upper plumage whitish, as also the bars on the tail. A third specimen does not seem to differ from the British-Columbian birds, but is very bleached. The three above-mentioned specimens have the following measurements:—

		Total length.	Wing.	Tail.	Tarsus.
		in.	in.	in.	in.
a. Ad.	California	9	7·25	3·5	1·95
b. Imm.	California	9·5	6·5	3·3	1·85
c. Imm.	California	9·5	7·2	3·7	1·8
d. Ad.	Magdalena Bay (*Kellett & Wood*).	9	6·5	3·35	1·9

At the same time I must observe that the birds considered to be immature by me cannot be very young, judging from Mr. Ridgway's description of the latter plumage.

Mexico. Specimens from this country are the same as the North-American ones, and I cannot see any difference at all. In three specimens procured from Mr. Boucard, two have rather broader bands on the breast, being probably younger birds. They measure as follows:—

		Total length.	Wing.	Tail.	Tarsus.
		in.	in.	in.	in.
a. ♂ ad.	Puebla (*Boucard*)	9	6·45	3·2	1·85
b. Ad.	W. Mexico	9	6·8	3·5	1·75
c. ♂ ad.	Mexico	8	6·5	3·1	1·65

San Domingo. The British Museum contains two specimens from this island; and, after examining them carefully, I fail to find any good reason why they should have been separated from the ordinary North-American bird. The bars are narrower on the under surface when compared with examples from other countries; but as I believe these bars narrow with age, I cannot look upon them as evidencing specific distinctness. The under wing-coverts are also spotted with brown near the outer edge; but this is a character which seems to obtain in certain specimens from South America, though at present I have not seen any from North America which had any markings on the under wing-coverts.

		Total length.	Wing.	Tail.	Tarsus.
		in.	in.	in.	in.
a. Ad.	S. Domingo (*Sallé*)	8·5	6·4	3·1	1·85
b. Ad.	S. Domingo (*Sallé*)	8	6·45	3·1	1·75

Uruguay. The Museum contains a single bird from Maldonado, collected by Mr. Darwin. It is a dark-coloured bird, and seems to be identical with the true *S. cunicularia* of Chili.

		Total length.	Wing.	Tail.	Tarsus.
		in.	in.	in.	in.
a. Ad.	Maldonado (*Darwin*)	9·5	7·1	3·6	1·95

Chili. I have examined a series of nine specimens from this country, all of them collected by Mr. E. C. Reed. Placed side by side with the North-American skins, they present a very much darker aspect, and are decidedly larger. I have not as yet seen a light-coloured bird from Chili. The bars on the tail vary very much in number, being five in two specimens, six in the next two, and seven in the remaining five. This character, therefore, cannot be regarded as of specific worth; and, indeed, I think Mr, Ridgway perfectly right in looking at the Chilian bird as merely a large representative form of the North-American Burrowing-Owl. I give the measurements of the nine Chilian specimens examined by me.

		Total length.	Wing.	Tail.	Tarsus.
		in.	in.	in.	in.
a. Ad.	Chili (*Reed*)...........	11·5	7·4	4·1	2·1
b. Ad.	Chili (*Reed*)...........	11·0	7·2	3·9	2·35
c. Ad.	Chili (*Reed*)...........	10·0	7·1	3·8	1·95
d. Ad.	Chili (*Reed*)...........	11·0	7·3	3·7	1·9
e. Ad.	Chili (*Reed*)...........	10·0	7·1	3·6	2·0
f. Ad.	Chili (*Reed*)...........	11·0	7·3	4·0	2·0
g. Ad.	Chili (*Reed*)...........	10·0	6·95	3·7	1·9
h. Ad.	Chili (*Reed*)...........	10·5	7·3	3·9	1·9
i. Ad.	Chili (*Reed*)...........	10·0	7·0	3·7	1·95

In all the above the character given by Mr. Ridgway, of the

greater extent of white on the quills and tail-feathers, holds good; and although the character is variable to a great extent, these southern specimens always seem to have the tarsi more densely feathered than northern examples.

Peru. Mr. Ridgway separates the Burrowing-Owls of this country as *S. grallaria* (Spix). The difference given for this separation are the following characters:—"Brown markings of the lower parts irregularly transverse and ragged. White spots on the upper parts nearly equal in extent to the brown. Wing 6·15–6·40, tail 2·90–3·60, culmen ·58–·62, tarsus 1·50–1·80, middle toe 0·65." Three specimens collected by Mr. Whitely give the following dimensions:—

	Total length. in.	Wing. in.	Tail. in.	Tarsus. in.
a. ♀ ad. Tinta (*Whitely*)	11·5	7·8	4·5	2·0
b. ♂ ad. Tambo valley (*Whitely*).	8·5	6·7	4·3	1·5
c. ♀ ad. Tambo valley (*Whitely*)..	9·0	6·6	—	1·5

The first of these three specimens has seven bars on the tail, and is identical with the large Chilian form. The Tambo-valley examples are very small, and are much more largely spotted with white than any other birds which have come under my notice; the tails are imperfect, but appear to have five bands. They are slightly darker than the North-American specimens, and are, perhaps, a small race of the Chilian form.

2. Speotyto guadeloupensis.

Speotyto cunicularia, var. guadeloupensis, *Ridgway, B. N. Am.* iii. p. 90 (1874); *Coues, B. N. W.* p. 323.

Adult. Very dark earthy brown, spotted on the entire upper surface with sandy isabelline, these spots being much smaller on the head and interscapular region, more distinct and larger on the hind neck; wing-coverts deeper brown than the back, the least ones minutely spotted with isabelline, the median and greater coverts, as well as the scapulars, having larger and paler spots of the same colour; primary coverts nearly uniform blackish brown, with a slight indication of a single fulvous notch on the outer web; quills dark brown, with faintly indicated bars of sandy-buff on the inner web, represented by a notch of the same colour on the outer one, these markings more distinct on both webs of the secondaries; tail dark earthy brown, crossed with five rows of sandy-buff spots (counting the terminal band), the basal one very indistinct; lores dull white, obscured by blackish bristles; forehead and a narrow superciliary streak sandy buff; ear-coverts very dark brown, slightly shaded with fulvous, the feathers under the eye minutely barred with sandy buff; cheeks buffy white, extending backwards in a triangular patch behind the ear; throat and fore neck also buffy white, separated from each other by a band of dark brown feathers, mottled and barred with sandy isabelline, like the adjoining sides of the neck; rest of under surface barred with buffy white and dark brown in about equal proportions, the whitish bars on the flanks and abdomen a little the broader; the sides of the upper breast very dark brown, spotted and barred with buffy white; thigh-feathers deep fulvous, the tarsi very scantily feathered; under tail-coverts whitish, unspotted; under wing-coverts ochraceous buff, mottled with brown markings, the lower series ochraceous at base, dark brown at tip, thus resembling the inner lining of the quills, which are very dark brown, with only four bars on the inner web. Total length 8·5 inches, wing 6, tail 3, tarsus 1·45.

Hab. Guadeloupe and S. Nevis, West Indies.

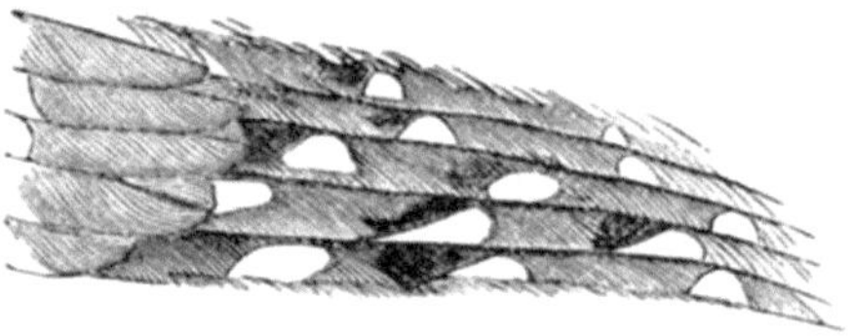

Primaries of *S. guadeloupensis*, to show pattern of coloration.

a. Ad. st. S. Nevis. Thos. Cottle, Esq. [P.].

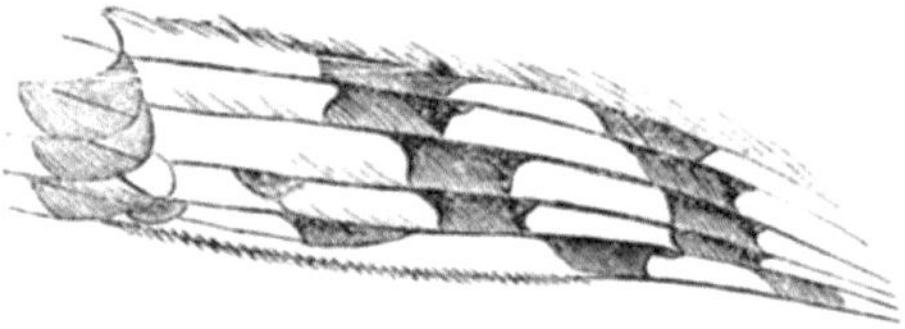

Primaries of *S. cunicularia*, to show pattern of coloration.

The specimen described agrees very well with the race called by Mr. Ridgway *Speotyto cunicularia*, var. *guadeloupensis*; and coming from neighbouring islands, these birds are very probably identical. The above account was written before I had consulted Mr. Ridgway's book, and it will be seen how well the main characters agree with his description. "Primaries without broad or regular bars of whitish on either web; primary-coverts plain brown; brown markings on the lower parts regularly transverse, and equal in extent to the white; white spots on the upper parts very small, reduced to mere specks on the dorsal region. Wing 6·40, tail 3·40, culmen ·60, tarsus 1·82, middle toe 0·85. Outer tail-feathers and inner webs of primaries with the light (ochraceous) bars only about one fourth as wide as the brown (disappearing on the inner quills)."

There is evidently some little difference between the age of Mr.

Ridgway's example and the one in the Museum, witness the greater appearance of bars on the quill-lining, and the slight evidence of a bar on the primary coverts; the character of the regular transverse barring of the underparts seems to have struck us both.

The foregoing cuts show the difference in the markings on the quills and primary-coverts in *S. cunicularia* and *S. guadeloupensis*.

10. GYMNASIO.

	Type.
Gymnasio, *Bp. Rev. et Mag. de Zool.* 1854, p. 543	G. nudipes.
Gymnoglaux, *Cab. J. f. O.* 1855, p. 466	G. lawrencii.

Range. West Indies; Cuba; Porto Rico; St. Thomas; St. Croix.

Key to the Species.

a. Above vermiculated; tarsi feathered halfway down *nudipes*, p. 149.
b. Above uniform, distinctly spotted with white; tarsi bare for nearly their entire length .... *lawrencii*, p. 150.

1. Gymnasio nudipes.

Strix nudipes, *Daud. Traité*, ii. p. 199; *Vieill. Ois. de l'Am. Sept.* i. p. 45, t. 16.
Noctua nudipes, *Steph. Gen. Zool.* xiii. p. 70; *Less. Traité*, p. 104.
Surnia nudipes, *Bp. Oss. Regn. An. Cuv.* p. 59.
Athene nudipes, *Gray, Gen. B.* i. p. 35; *Strickl. Orn. Syn.* p. 173.
Surnium nudipes, *Kaup, Contr. Orn.* 1852, p. 120; *id. Tr. Z. S.* iv. p. 250.
Gymnasio nudipes, *Bp. Rev. et Mag. de Zool.* 1854, p. 543.
Gymnoglaux nudipes, *A. & E. Newton, Ibis*, 1859, p. 64, t. 1; *Scl. & Salv. P. Z. S.* 1868, p. 328; *iid. Nomencl. Av. Neotr.* p. 116.
Gymnoglaux newtoni, *Lawr. Ann. Lyc. N. Y.* viii. p. 258.
Scops nudipes, *Gray, Hand-l. B.* i. p. 47.

Adult. Above brown, scarcely vermiculated at all on the upper surface with darker brown, but very distinctly banded across with dull fulvous, inclining to white on the hinder neck, the feathers over the eye whitish as well as the loral plumes; scapulars externally more broadly barred with buffy white, forming an irregular spot near the tips of some of the longer feathers; wing-coverts more dusky brown than the back, the least ones with a few fulvescent spots and irregular bars, the median and greater series vermiculated with buff near the tips, with indications of a tolerably distinct subterminal bar of clearer fulvous; primary-coverts uniform blackish brown, with a faint indication of buff-coloured spots or bars of fulvous near the base; quills dark brown, slightly vermiculated with sandy buff on the outer webs, especially on the innermost secondaries, which have also indications of fulvous cross bars, the inner webs of the quills with faint bars of ashy brown, a little more distinct on the secondaries, the corresponding bars on the latter quills clearer buff, and inclining to white on the primaries, giving a distinctly chequered appearance to the outer aspect of the wing; tail dull brown, vermiculated with sandy buff on the margins of both webs, and crossed with eight irregular and not conterminous bars of the same colour, inclining to buffy white on the outer web; sides of the face dull brown, barred with sandy buff, the cheek-feathers white, tipped with brown, and extending backwards in a triangular patch to behind the ear-coverts; chin-feathers dull whitish, barred across with dusky brown; rest of under surface white, the feathers slightly vermiculated with dark brown, and barred across with sandy buff, many of the feathers broadly streaked down the centre with dark brown, the white predominating more and more on the under surface towards the under tail-coverts, which are entirely white, excepting a few brown vermiculations at the tip; the chest much more dusky than the breast, and inclining to sandy buff, barred with dull brown, and here and there with white; leg-feathers sandy buff, with distinct bars of dull brown on the inner face, the outer aspect whiter; bill horn-colour, palest at the tip; legs and toes pale brown, the claws horn-colour; iris bright hazel. Total length 8 inches, wing 6·05, tail 3·4, tarsus 1·3, bare part of latter 0·9.

"The female bird seems to be more rufous than the male, with the breast and belly not so thickly mottled; but as we have only two specimens which seem to be fully adult to judge from, these distinctions may be the result of age rather than of sex." (*Newton l. c.*)

Hab. Antilles; islands of Porto Rico, St. John, and St. Croix.

a. Ad. st. Island of St. John. Edward Newton, Esq. [P.].

2. Gymnasio lawrencii.

Noctua nudipes, *Lembeye, Aves Cuba*, p. 23, pl. 4. fig. 2.
Gymnoglaux nudipes, *Cab. J. f. O.* 1855, p. 465; *Lawr. Ann. Lyc. N. Y.* vii. p. 257.
Gymnoglaux lawrencii, *Scl. & Salv. P. Z. S.* 1868, p. 328, pl. xxix.; *iid. Nomencl. Av. Neotr.* p. 117.
Scops lawrencii, *Gray, Hand-l. B.* i. p. 47.

Adult. General colour above brown, slightly tinged with rufous or lighter brown, with which the feathers are more or less distinctly margined; head nearly uniform, excepting for the dark centres to the feathers, the hind neck minutely spotted with white, a few spots of which are seen on the occiput; the white spots on the sides of the neck rather larger, those on the back fulvescent, irregular in shape, and more or less concealed; the scapulars with larger white spots externally; wing-coverts brown like the back, with very large spots, the greater series a little darker brown, the spots not so small; the primary-coverts uniform dark brown, excepting a slight wash of ochraceous on the outer web, which is also apparent on some of the other wing-coverts; quills dark brown, the secondaries lighter, all barred with paler brown (more distinct on the secondaries), these bars represented on the outer web by a white spot, larger and more distinct on the primaries; upper tail-coverts brown, slightly washed with rufous, and minutely spotted with white; tail brown, rather paler at tip, and crossed with irregular bars of lighter brown, of which about three are indicated, being more regular on the outer feathers, where they are represented by white bars on the inner web; facial aspect white, the lores overhung by hair-like brown bristles; a white eyebrow very distinct; frontal plumes also whitish; ear-coverts brown, surrounded behind by an indistinct ruff of dark brown feathers, those on the lower part whitish, margined with brown, and hence scarcely to be distinguished from the white cheeks; under surface of body white, with narrow streaks of brown, disappearing gradually on the abdomen, and absent on the under tail-coverts, which are entirely white; sides of neck rufous brown, spotted with white, descending rather on to the fore neck, and forming a demi-collar; the centre of the chest and fore neck white, narrowly streaked with brown; flanks washed with orange-buff, and more broadly streaked with brown, thus producing a somewhat mottled appearance; thighs deep orange-buff; under wing-coverts fulvous, inclining to ochre on the outer margin, which is also washed with brown, the lower series fulvous at base, ashy brown at tip, thus resembling the inner lining of the quills, which are ashy brown below, barred with fulvous on the inner web; bill greenish yellow, dusky at base; feet yellow (apparently); claws horny. Total length 7 inches, wing 5·5, tail 3·1, tarsus 1·35. (*Mus. Paris.*)

Hab. Cuba.

11. NINOX*.

	Type.
Ninox, *Hodgs. Madr. Journ.* v. p. 23 (1837) ..	N. lugubris.
Hieracoglaux, *Kaup, Isis*, 1848, p. 768........	N. connivens.
Spiloglaux, *Kaup, l. c.* p. 768................	N. novæ zealandiæ.
Cephaloptynx, *Kaup, Contr. Orn.* 1852, p. 105..	N. punctulata.
Ctenoglaux, *Kaup, l. c.* p. 109	N. hirsuta.
Rhabdoglaux, *Bp. Rev. et Mag. de Zool.* 1854, p. 544	N. strenua.

---

* I have not examined the following species:—

NINOX PODARGINA.

Noctua podargina, Hartl. & Finsch, P. Z. S. 1872, p. 90.

Adult. Pale ferruginous red, not very plainly transversely varied and marked here and there with small whitish spots; head paler fulvous, indistinctly fasciated; rump and upper tail-coverts more closely spotted with white; on the wing-coverts some larger spots of white; wings pale rufous; quills with fewer and more irregular bands, duller on the inner web; tail-feathers pale rufous, obscurely banded; breast and abdomen irregularly banded and spotted on a brighter rufous buff ground; under wing-coverts banded with whitish and dusky; bill pale; feet dark. Total length about 11 inches, culmen 0 8, wing 6·2, tail 3, tarsus 1.

Hab. Pelew Islands.

Drs. Hartlaub and Finsch (*l. c.*) give the following remarks on this species:

Wing of *Ninox scutulata*, to show proportional length of primaries.

Range. Indian peninsula and Ceylon; China and Japan: ranging southwards throughout the Malayan region to Australia and New Zealand; Madagascar.

Key to the Species.

a Breast streaked, each feather broadly centred with darker colour.
- *a'*. Hind neck not spotted.
  - *a''*. Wing-coverts uniform, with no rows of white spots.
    - *a'''*. Belly white, streaked or largely spotted with brown.
      - *a*[4]. Head grey, much lighter than the back, which is browner; tail clear grey, tipped with white, the black bands in strong contrast.
        - *a*[5]. Axillaries barred with brown and white; breast-spots brown; size large; wing 8–8·75 inches .... *lugubris*, p. 154.
        - *b*[5]. Axillaries uniform orange-chestnut; breast-spots chestnut: size small; wing 6·65 inches ...... *affinis*, p. 155.
      - *b*[4]. Head dark brown, generally uniform with the back, seldom greyer; tail dark brown, with blackish brown bars .... *scutulata*, p. 156.
      - *c*[4]. Head light brown, like the back; tail nearly uniform, with no trace of bars .... *boobook* juv., p. 169.
    - *b'''*. Belly dark ochraceous, streaked with brown .... *ochracea*, p. 167.
  - *b''*. Wing-coverts externally spotted with white; outer edge of primaries with distinct rows of white spots.......... *philippensis*, p. 167.
- *b'*. Hind neck spotted.
  - *c''*. Leg-feathers tawny or fulvous.

—"All our efforts to refer this little Owl to any known species have been without success; we must therefore consider it to be new. Its nearest ally seems to be *N. ochracea*, Schl., from Celebes. The specimen is in apparently full dress; but whether the rufous plumage may be only a phase, as in some other Owls, we are unable to decide. The light-coloured beak and the half-naked tarsi and toes, without any hairs, are especially to be noticed."

c'''. Wing-coverts uniform, or at the most slightly spotted with fulvous.
  d⁴. Interscapulium not spotted.
    c⁵. Larger; wing 8·6–10·3 inches; ear-coverts darker than crown.. *boobook*, p. 168.
      α. Darker-coloured and larger in size, the sides of the neck and markings of the under surface deeper and more rufous chestnut. *boobook.*
      β. Lighter-coloured and smaller; the markings as above more decidedly ochraceous. *a. ocellata.*
    d⁵. Smaller; wing 9·25 inches; ear-coverts grey, like the crown.... *fusca*, p. 172.
  e⁴. Interscapulium spotted.
    e⁵. General colour dark chocolate-brown; head generally uniform; the markings on the under surface deep chocolate .... *novæ zealandiæ*, p. 173.
    f⁵. General colour more of a rufous brown, the head profusely spotted with white; markings on under surface rufescent .... *maculata*, p. 174.
d'''. Wing-coverts regularly barred across with fulvous .... *dimorpha*, p. 175.
d''. Leg-feathers white or fulvous, streaked with brown exactly like the breast.
  e'''. Below streaked with brown .... *connivens*, p. 175.
  f'''. Below streaked with rufous........ *rufistrigata*, p. 177.

b. Breast uniform brown.
  c'. Secondaries uniform, with no light bars on the inner web....................... *obscura*, p. 177.
  d'. Secondaries with light bars on the inner web .... *theomacha*, p. 178.

c. Breast spotted or transversely barred.
  e'. Head spotted or barred.
    e''. Size large; wing 10·5–18·5 inches.
      g'''. Beneath whitish, with brown cross bars; ear-coverts brown, the fore part streaked with white .... *strenua*, p. 178.
      h'''. Beneath orange or yellowish rufous, barred with reddish brown.
        f⁴. Ear-coverts black; leg-feathers orange-buff, indistinctly marked with brown.................. *humeralis*, p. 180.
        g⁴. Ear-coverts dark brown; leg-feathers uniform buff .... *aruensis*, p. 181. / *franseni*, p. 181.
    f''. Size smaller; wing not exceeding 8·5 inches in length.
      i'''. Leg-feathers tawny buff.
        h⁴. Flank-feathers streaked with rufous brown, mottled with white lateral spots or markings .... *maculata*, p. 174.

i^4. Flank-feathers white, barred with light brown, like the rest of the under surface .... *superciliaris*, p. 181.
k'''. Leg-feathers white; upper surface brown, minutely spotted and barred with fulvous; sides of body brown barred with white .... *punctulata*, p. 182.
f'. Head uniform.
g''. Head dark grey; back brown .... *hypogramma*, p. 183.
h''. Head rufous; back rufous.
l'''. Belly white, barred with brown or rufous.
k^4. Ear-coverts grey; under surface of body and under wing-coverts white, narrowly barred with brown, the cross bars much narrower than the white interspaces .... *squamipila*, p. 184.
l^4. Ear-coverts rufous brown, like the head; under surface of body white, broadly barred with orange-rufous, as well as the under wing-coverts, the bars being of about the same width as the interspaces .... *variegata*, p. 185.
m'''. Belly orange-rufous, with a few rufous brown bars .... *hantu*, p. 185.
d. Breast uniform cream-colour; upper surface barred .... *tæniata*, p. 186.

1. Ninox lugubris.

Strix lugubris, *Tickell, J. A. S. B.* ii. p. 573.
Ninox nipalensis, *Hodgs. Madr. Journ.* v. p. 24, pl. 14.
Otus lugubris, *Jerd. Madr. Journ.* x. p. 87.
Ninox lugubris, *Blyth, Ann. N. H.* xii. pp. 93, 231; *Sharpe, Ibis*, 1875, p. 258.
Ninox jeridius, *Hodgs. in Gray's Zool. Misc.* p. 82.
Ninox scutellatus, *Blyth, J. A. S. B.* xvi. p. 511; *Jerd. B. Ind.* i. p. 147; *Hume, Rough Notes*, ii. p. 420.
Athene hirsuta, *Bonap. Consp.* i. p. 41.
Ninox madagascariensis, *Bonap. Rev. et Mag. de Zool.* 1854, p. 543; *Hartl. Faun. Madag.* p. 22; *Gurney, Ibis*, 1869, p. 453.
Noctua hirsuta, *Schl. Mus. P.-B.*, Striges, p. 25.
Athene lugubris, *Gray, Hand-l. B.* i. p. 41.
Athene madagascariensis, *Gray, Hand-l. B.* i. p. 42.

Adult female. Above ashy brown, the head very much greyer, the sides of the face and neck being decidedly light grey; forehead, lores, fore part of cheeks and chin very conspicuously white, the loral plumes with hair-like black shaft-lines; feathers round the eye greyish; scapulars marked externally with large bars of white, not very distinct, and often half concealed; upper wing-coverts ashy brown like the back, the outer median coverts and the primary-coverts deeper and more inclining to sepia-brown; quills dark brown, a little deeper than the back, barred across with light ashy or greyish brown, inclining to whitish on the outer web of some of the primaries and inner secondaries, most of which are narrowly tipped with white; tail very pale greyish brown, inclining to white at the base, and also tipped with white, crossed with six blackish brown bands; under surface of body white, the throat yellowish buff, streaked longitudinally with greyish brown, the chest broadly streaked with rufescent brown, the abdomen spotted, and each feather subterminally barred with the same; under tail-coverts pure white, the basal ones with a few brownish bands; leg-feathers rufous brown, very slightly mottled with whitish cross markings; under wing-coverts white, barred across with dark brown, the outermost almost entirely brown, the edge of the wing white; the greater series dusky greyish brown, barred with buffy white, thus resembling the inner lining of the wing, which is greyish brown, barred with buffy white on the inner web, these bars inclining to fulvous near the base; cere and bill green, the tip of the latter dusky; feet greyish or reddish yellow; iris bright golden yellow. Total length 13·5 inches, wing 9·15, tail 5·7, tarsus 1·35.

Adult male. Smaller than the female, and having only five blackish bands on the tail. Total length 11·5 inches, wing 8·5, tail 5·6, tarsus 1·15.

Hab. Indian peninsula and Himalayas.

a. ♀ ad. sk.	Darjiling.	R. B. Sharpe, Esq. [P.].
b. Ad. sk.	Darjiling.	Purchased.
c. ♂ (?) ad. sk.	Futtegurh, Aug. 2, 1873.	A. Anderson, Esq. [C.].
d. Ad. st.	Nepal.	B. H. Hodgson, Esq. [P.].
e. Juv. st.	Nepal.	B. H. Hodgson, Esq. [P.].
f, g. Ad. sk.	Nepal.	B. H. Hodgson, Esq. [P.].
h. Ad. sk.	Malabar.	Purchased.

Obs. Having examined the type of *Ninox madagascariensis* in the Paris Museum, I failed to see the slightest difference between this supposed Madagascar species and ordinary *Ninox lugubris* of the Himalayas; and therefore I unite them, believing that some mistake as to locality has arisen. At the same time it must be remembered that the Cuckoo (*Cuculus rochi*) from Madagascar is almost inseparable from that of the Himalayas; and therefore it is by no means improbable that a *Ninox* may appear in that island. But at present the evidence is against such a genus being found there; for none of the recent travellers have found an example. The more "guttate" spots of which Mr. Gurney (Ibis, 1869, p. 453) speaks, are, in my opinion, only a sign of maturity; for in all the species of *Ninox* the markings on the belly appear to be at first streaked and afterwards to break up into spots, perhaps without an actual moult of the feather.

2. Ninox affinis.

Ninox affinis, *Beavan, Ibis,* 1867, pp. 316, 334; *Hume, Rough Notes,* ii. p. 421; *Walden, Ibis,* 1874, p. 129, pl. v.; *Hume, Str. F.* 1874, p. 152; *Sharpe, Ibis,* 1875, p. 259.

Athene affinis, *Gray, Hand-l. B.* i. p. 42.
Ninox hirsuta, *Ball, Str. F.* 1873, p. 54.

Adult male. Above rather light chocolate-brown, the head and hind neck more ashy grey than the back, the entire upper surface being uniform, excepting a few half-concealed fulvous markings on the scapulars; lores and a few of the frontal feathers white, with long black shaft-streaks, terminating in elongated bristles; fore part of cheeks also whitish, but the ear-coverts greyish chocolate, like the head; chin whitish, streaked with brown; fore neck rufous brown, with scarcely any appearance of darker streaks; rest of under surface whitish, broadly streaked with chestnut-red, some of the feathers on the breast and flanks having these streaks somewhat inclining to dark brown; leg-feathers orange-buff; under tail-coverts white; under wing-coverts uniform deep tawny or chestnut, the outermost of the lower series dark sepia-brown, resembling the inner lining of the quills, which are sepia-brown, barred with fulvous on the inner web; upper wing-coverts uniform brown like the back, but a little more washed with rufous; primary-coverts uniform sepia-brown; quills brown, numerously barred with darker brown, and washed with rufous on the outer web, especially on the outer edge of the lighter interspaces; upper tail-coverts dark brown, barred with fulvous; tail-feathers rather light brown, paler at tip, and crossed with five bands of blackish brown, none of them so broad as the interspace; bill yellowish; feet pale yellow; iris pale yellow. Total length 10 inches, wing 6·65, tail 4·3, tarsus 5·15. (*Mus. Walden.*)

3. Ninox scutulata.

Strix scutulata, *Raffl. Tr. Linn. Soc.* xiii. p. 280 (1822).
Strix hirsuta, *Temm. Pl. Col.* i. pl. 289 (1824).
Noctua hirsuta, *Steph. Gen. Zool.* xiii. pt. 2, p. 66.
Surnia hirsuta, *Less. Man. d'Orn.* i. p. 110.
Athene malaccensis, *Eyton, Ann. N. H.* xvi. p. 228.
Athene scutulata, *Gray, Gen. B.* i. p. 35; *Kelaart, Prodr. Faun. Zeyl.* p. 28; *Horsf. & Moore, Cat. B. Mus. E. I. Co.* i. p. 68; *Strickl. Orn. Syn.* p. 170; *Wall. Ibis,* 1868, p. 22.
Strix hirsuta japonica, *Schl. Faun. Jap. Aves,* p. 29, pl. 9 B.
Athene hirsuta, *Bp. Consp.* i. p. 41; *Gray, Hand-l. B.* i. p. 41; *Pelz. Voy. Novara, Aves,* p. 26.
Athene japonica, *Bp. Consp.* i. p. 41; *Gray, Hand-l. B.* i. p. 42.
Athene borneensis, *Bp. Consp.* i. p. 41; *Gray, Hand-l. B.* i. p. 42; *Pelz. Voy. Novara, Vög.* p. 26; *Wall. Ibis,* 1868, p. 22.
Ctenoglaux scutellatus, *Kaup, Contr. Orn.* 1852, p. 109, et *Tr. Z. S.* iv. p. 219.
Ninox japonicus, *Bp. Rev. et Mag. de Zool.* 1854, p. 543; *Walden, Tr. Z. S.* viii. p. 40; *Swinh. P. Z. S.* 1871, p. 343; *Sharpe, Ibis,* 1875, p. 259.
Ninox hirsutus, *Bp. Rev. et Mag. de Zool.* 1854, p. 543; *Holdsw. P. Z. S.* 1872, p. 418; *Hume, Str. F.* 1874, pp. 151, 469, et 1875, p. 441.

Ninox borneensis, *Bp. Rev. et Mag. de Zool.* 1854, p. 543; *Walden, Ibis*, 1872, p. 18; *Salvad. Ucc. Born.* p. 18.
Noctua hirsuta japonica, *Schl. Mus. P.-B.* Striges, p. 24.
Noctua hirsuta borneensis, *Schl. t. c.* p. 24.
Athene florensis, *Wall. P. Z. S.* 1863, p. 488; *id. Ibis*, 1868, p. 23; *Gray, Hand-l. B.* i. p. 41.
Noctua hirsuta minor, *Schl. Mus. P.-B. Revue Accipitr.* p. 25.

Adult (type of *Athene malaccensis**). Above deep chocolate-brown, the head slightly darker, the scapulars with concealed white bars; wing-coverts exactly like the back and equally uniform, the primary-coverts blackish, quills dark brown, slightly washed externally with ochre, the primaries inclining to rufous ochre towards their tips, all the wing-feathers nearly uniform, with no lighter cross bands, the inner secondaries barred with white, but these bars entirely concealed; upper tail-coverts uniform chocolate-brown like the back; tail-feathers ashy brown, rather paler at tips, and crossed with five blackish bands (one basal and concealed); sides of face chocolate-brown, as also the sides of the neck, the latter slightly washed with rufous; forehead and lores whitish, the latter obscured by blackish shaft-lines; chin whitish, slightly streaked with brown; rest of under surface rufous chocolate, the throat washed with buff, the breast varied with white, generally as if streaked, the margins to the feathers being whitish; on the abdomen the white predominating and cutting across the feather, so as to form an oval spot of brown at the tips of the abdominal plumes, leg-feathers brown; under tail-coverts whitish; under wing-coverts rufous chocolate, the inner feathers washed and barred with ochraceous buff, the edge of the wing whitish; the greater series sepia-brown, nearly uniform, excepting for a few ochraceous bars near the base, thus resembling the inner lining of the wing, which is sepia-brown, with a few ochraceous bars near the base of the quills. Total length 12·5 inches, wing 8·4, tail 4·8, tarsus 1·1.

Obs. In the Transactions of the Zoological Society (viii. p. 40) Lord Walden has already clearly shown the confusion which reigns with regard to the various species of *Ninox*, thereby inducing me to pay particular attention to this subject. I have examined the series of specimens in the Norwich Museum (to the authorities of which I return my best acknowledgments) and in the collection of Mr. Swinhoe; and I think that I cannot do better than place before my readers the results of my studies of the birds from different localities.

India. The Museum contains a tolerable series from the Himalayas, the measurements being as follows:—

* This bird is described as being most probably the same as the typical Sumatran *N. scutulata* (Raffl.).

			Total length. in.	Wing. in.	Tail. in.	Tarsus. in.
a.	Ad.	Futtegurh (*Anderson*)	10·5	8·0*	4·6	1·1
b.	♀ ad.	Futtegurh (*Anderson*)	10	8	5	1·05
c.	Ad.	Darjiling (*Mus. Brit.*)	12	8·6	5·4	1·15
d.	Ad.	Darjiling (*Mus. Brit.*)	11	8·2	5·6	1·15
e.	Ad.	Nepal (*Hodgson*)	12	8·3	5·5	1·05
f.	Ad.	Nepal (*Hodgson*)	12·5	8·55	5·5	1·15
g.	Ad.	Nepal (*Hodgson*)	11·5	8·5	5·6	1·1
h.	Ad.	Nepal (*Hodgson*)	11·5	8·55	5·5	1·15
i.	Ad.	India (*Mus. Swinhoe*)	12·2	8·75	5	1·15
k.	Ad.	India (*Mus. Norv.*)	11	—	5·1	1·05

It will be seen below that in point of size the Chinese and Himalayan species are very closely allied; but the latter is quite a distinct bird. Mr. Andrew Anderson marks a specimen as "young" which appears to me to be quite old; and I mention this because it is a question to be decided by field observers in India. The Indian species is light-coloured, and when fully adult has a grey head, with a good deal of white about the face and cheeks. I do not know whether it ever has a striped abdomen; but, from the variations in the form of the central markings on the breast-feathers, I should expect that the young birds are streaked in their first plumage, and that these broad streaks break up, leaving subterminal spots. The uniformity of the first primary and the lower series of wing-coverts is also a character which we might expect to find in the young stages, these feathers being barred in the older birds. I leave these points to the consideration of some of my Indian fellow workers to determine. The tail-bands vary in the series under examination from 5 to 6.

Malabar. The only specimen which I have seen is one recently brought to the Museum by Mr. Whitely. Mr. Jerdon has united all the Indian birds under one title; and in this he is generally followed. Lord Walden seems to hint (*vide* p. 157) that the Himalayan bird, *Ninox nipalensis* of Hodgson, is distinct from the species of the peninsula of India, *N. lugubris*, Tickell; and I incline to the belief that the two are distinct: with only one specimen, however, it is impossible to decide; but the Malabar skin is more nearly allied to the Ceylon bird than to the Himalayan. Total length 11 inches, wing 8, tail 4·6, tarsus 1·1.

Ceylon. Three specimens are before me, all apparently adult. The Ceylon birds present a very marked *facies*, being very dark-coloured, and having the abdominal spots large and distinct but very thickly distributed; this gives a dark appearance to the underparts. First primary uniform in all three specimens. The bars on the tail vary in each bird, being respectively 5, 6, and 7 in number.

* Marked by Mr. Anderson as 7·7 in the flesh: perhaps we measure in a different way; and I had better state, therefore, that in the present work all the wings of the small Owls are measured with compasses, those of the larger species with a tape.

	Total length.	Wing.	Tail.	Tarsus.
	in.	in.	in.	in.
a. Ad. Ceylon (*Cuming*)	10·5	7·55	4·8	1·05
b. Ad. Ceylon (*Mus. Norv.*)..........	10·5	7·4	4·5	1·1
c. Ad. Columbo (*Swinhoe*)	11·5	7·55	4·7	1·05

I subjoin a detailed description made from a Ceylonese specimen; and if proved to be distinct, the bird will bear the title of *Ninox hirsuta* (Temm.).

Adult. Above uniform deep chocolate-brown, the head slightly darker and more blackish; sides of face uniform with crown, the lores dull whitish at base, but the frontal feathers more distinctly whitish; scapulars marked with large white spots, extending on to both webs; upper tail-coverts with a few scarcely distinguishable whitish spots; wing-coverts coloured like the back, and perfectly uniform, the primary-coverts blackish; quills dark chocolate-brown, the primaries with a few nearly obsolete bands of light brown, shading off into greyish on the outer web, the inner secondaries barred with white on each web, some of them being concealed; tail-feathers clear ashy brown, plainly tipped with white, and crossed with five black bands on the centre feathers, six on the outer ones; throat dusky chocolate, streaked with ochraceous, with which colour some of the feathers are margined; remainder of under surface dark chocolate-brown, the breast-feathers being margined with white, producing a streaked appearance; the abdomen broadly barred with white, the brown spots slightly tinged with rufous; leg-feathers brown, mottled with fulvous bars and streaks; under tail-coverts white, with a few blackish spots; under wing-coverts deep chocolate-brown, slightly barred with ochre, the edge of the wing white, the greater series sepia-brown, thus resembling the inner lining of the wing, which is almost uniform brown, only the secondaries having a few fulvescent bars near the base; bill green; feet dingy yellow; iris yellow. Total length 12 inches, wing 8·1, tail 5·1, tarsus 1·05.

Nicobar Islands. Lord Walden has kindly lent me a female collected on Trinkut by Captain Wimberley; and I am unable to separate it from Burmese specimens of *N. scutulata*. It has remains of fulvous bars near the base of the first primary, and has six bars on the tail. Total length 12 inches, wing 7·4, tail 5, tarsus 1·05. It seems to be rather smaller than the Burmese birds; and that is all the difference I can trace.

China. I have examined a good series of *Ninox japonicus*, which I will treat of in the order of their localities. I have not yet seen a typical Japanese example.

Peking. The Norwich Museum contains a specimen obtained at Peking by Mr. Whitely. It is a bird of large size; and the head varies very little in colour from the rest of the upper surface. Dark bars on tail five. Tarsus tolerably well-feathered to within about 0·3 inch of the base of middle toe. The following is a description of this specimen, which I take to be the true *N. japonica* (Schl.):—

Adult. Above chocolate-brown, the head slightly darker than the back, the scapulars with half-concealed bars and spots of white, very large, and in some instances nearly traversing the feathers; wing-coverts uniform chocolate-brown, as also the primary-coverts; quills dark brown, barred with paler brown, some of the lighter bars inclining on the outer web to fulvous, the bars on the secondaries much less distinct and narrower, except on the inner web, where they are broader and fulvous white; upper tail-coverts uniform like the back; tail light brown, tipped narrowly with whitish, and crossed with five bands of darker brown, the subterminal one rather broader than the others; lores and frontal plumes whitish, with hair-like blackish bristles directed forwards over the nostrils; cheeks and chin whitish, with dark brown streaks, the rest of the under surface pure white, thickly streaked with dark brown, the centres to the feathers being of this colour and very broad, but narrowing towards the base of the plume, and often severing the central streak into two spots, the subterminal one being generally oval in shape, and forming on the flank-feathers lateral bars of white; leg-feathers brown, mottled with fulvous; under tail-coverts pure white, with a few brown markings down the centre of some of the feathers; under wing-coverts brown, streaked with fulvous, the inner ones especially fulvescent, the median series and the axillars white, barred or spotted with brown, the lower series dark brown, spotted with fulvous bars on the inner web, resembling the inner lining of the quills, which are dark brown below, barred with fulvous on the inner web. Total length 11 inches, wing 8·6, tail 5, tarsus 1·05.

Chefoo. From this place Mr. Swinhoe has three specimens, collected in May 1873 and on the 15th of October 1874. Two of them have five, but the third six bands on the tail; the chief peculiarity lies in the reddish shade on the wing-coverts, which are thereby decidedly lighter than the back. The under surface of the body is very thickly streaked with rufous chocolate, the lower flanks barred with this colour; and from the markings on the abdomen it is evident that this change from broad central streaks to terminal spots is gained not by a moult, but by a gradual change in the pattern of the feather. The immature birds seem to have at the same time remains of fulvous on the breast, the margins to the feathers being of this colour. One of the male specimens killed at Chefoo in May is not so rufous on the wings as the other, and has the first primary perfectly uniform.

Foochow. Three of Mr. Swinhoe's specimens obtained here are in the Norwich Museum. Two of them resemble the Chefoo birds almost exactly, having the first primary unbarred. The third specimen, however, is different from any yet examined by me, for the general colour of the back is rufous brown, the head more greyish, and contrasting with the rest of the upper surface; the first primary is plainly barred below. The bands on the tail in all three specimens are five in number; but the tail is shaded with rufous in the reddish brown bird, grey in the others.

Fokien. In the British Museum is a specimen, presented by Mr. Gould, which is thoroughly intermediate between the rufous and grey plumages mentioned in the foregoing paragraph. The general colour of the bird is the usual chocolate-brown; but it is shaded with rufous on the back, and has the hind neck decidedly rufous. The bands on the tail are five, and the abdominal streaks are well broken up and the spots distinct. The first primary is nearly uniform, but has indications of apparently developing bars of rufous on the inner web.

Formosa. A specimen is in the Norwich Museum procured from the Maison Verreaux, but bearing no evidence of Mr. Swinhoe's collecting. It seems to be not quite adult, and has the scapulars rather browner than the back, the concealed white spots being for the most part tinged with fulvous; the wings decidedly rufescent; tail with five bars; bars on first primary pretty plain; abdominal streaks of irregular shape, with the subterminal spot nearly separate in some feathers.

Amoy. I have examined two specimens from Amoy, collected by Mr. Swinhoe. They are both in the ordinary chocolate plumage, with no perceptible difference in the colour of the head; the wings also are not particularly rufescent. Bands on tail five. The stripes on the abdomen in both specimens are in process of dissolution; but one (killed in November 1866) seems to be a little younger than the other, for it has a decided fulvous tinge on the under parts, and the first primary has no bars; whereas the other specimen is not so fulvescent below, and has indications of bars on the first primary.

Tingchow. The single bird in Mr. Swinhoe's collection from this locality was killed in August 1867, and is smaller than any of the other Chinese specimens yet examined (*cf.* table of measurements given below). It seems to be a fully adult bird, with the spots on the abdomen thoroughly separate and the chest-streaks narrowed. The colour is uniform chocolate, the head resembling the back, the wings with a scarcely perceptible rufous tinge. Bars on tail six, the basal one completely hidden by the tail-coverts. First primary with distinct fulvous bars on the inner web.

Hainan. Mr. Swinhoe collected a male and female in this island on the 29th of March, 1868. The former is the smaller bird; but otherwise they resemble each other, both being nearly uniform chocolate, with very little rufous on the wings. They are both somewhat washed with fulvous below, and have each five bars on the tail, the streaks on the abdomen being irregular in shape and breaking up into spots, and the bars on the first primary are plainly to be seen.

The following are the general measurements of the series examined:—

		Total length. in.	Wing. in.	Tail. in.	Tarsus. in.
a. Ad.	Peking (*Whitely*)	11	8·6	5	1·05
b. ♂ ad.	Chefoo (*Swinhoe*)	12·5	9·1	5·4	1·15
c. ♂ ad.	Chefoo (*Swinhoe*)	12	8·9	5·3	1·1
d. Ad.	Chefoo (*Swinhoe*)	12·5	8·8	5·4	1·1
e. Ad.	Foochow (*Swinhoe*) ..	11	7·9	4·9	1·05*
f. Ad.	Foochow (*Swinhoe*) ..	11	8·3	5·2	1·05
g. Imm.	Foochow (*Swinhoe*) ..	11·5	8·4	5·1	1
h. Imm.	Fokien (*Mus. Brit.*) ..	11	8 35	4·9	1
i. Ad.	Amoy (*Swinhoe*)	11	8·5	4·9	1·15
k. Juv.	Amoy (*Swinhoe*)	11	8·4	5	1
l. ♀ vix ad.	Formosa (*Mus. Norv.*)	10·5	8·3	5	1
m. Ad.	Tingchow (*Swinhoe*) ..	10	8	4·85	1
n. ♂ ad.	Hainan (*Swinhoe*)	10	8·05	4·8	1
o. ♀ ad.	Hainan (*Swinhoe*)	10	8 5	4·8	1·05

From a comparison of the Chinese birds it appears that those from the southern portion of the empire are rather smaller, and appear to me to have the feathering of the tarsus extended a little more to the base of the toes; but this character, even if constant, is very slightly developed, and by no means easy to affirm with certainty. The younger birds appear to me to be more rufescent, and to have, as a rule, the bars on the first primary less plainly distinguishable.

Eastern Kiangsi. A male procured by Père David in this district, and now in the Paris Museum, recalls exactly Mr. Swinhoe's Hainan and Tingchow specimens: breast streaked and tinged with fulvous; abdomen spotted, the spots commencing to occur also on the lower breast; first primary with a few bars; tail with five bars. Total length 10·5 inches, wing 8·3, tail 4·7, tarsus 1·05.

Yun-nan. Dr. Anderson obtained a male bird at Quaylom, on the banks of the Tah-peng river, on the 3rd of February, 1875. It has no bars on the first primary, but has five bands on the tail. Total length 11 inches, wing 8·5, tail 5·1, tarsus 1·2.

Burmah. Only three specimens have come under my notice from this locality, two of them being lent to me by Dr. Anderson. A bird from Bhamo is apparently adult, and has the first primary uniform, with four bands on the tail. Total length 12 inches, wing 8·4, tail 5·25, tarsus 1·15. A female from Sawaddy, Upper Burmah, has distinct fulvous bars on the first primary, and six bars on the tail. Total length 12·5 inches, wing 8·2, tail 5·4, tarsus 1·2. Another specimen from Rangoon has the first primary uniform, except for indications of fulvous bars near the extreme base; tail with five bands, but one of the middle feathers having six bands and no white tip. Total length 11·5 inches, wing 8·1, tail 5·5, tarsus 1·15.

Cochin China. Two specimens are in the Paris Museum from this

* Although labelled with a ticket of Mr. Swinhoe's, this specimen bears every trace of being a Malaccan skin, and, as will be seen on comparison, it agrees exactly with the Malayan bird in measurements. I think it very probable that the ticket has been accidentally changed on this bird.

country. The first, obtained by Diard, resembles Mr. Swinhoe's South-China birds: the spots on the under surface everywhere very distinct and regular, forming bars across the feathers; bands on tail four, with no perceptible trace of a fifth; first primary with a distinct fulvous bar near the base. Total length 12 inches, wing 8·25, tail 5·4, tarsus 1·1. Another specimen, collected by Germain, is rather younger than the foregoing, the chest being washed with fulvous and broadly streaked with brown, the belly having a more spotted appearance and being more broken up into spots and bars; first primary to all intents quite uniform; bars on tail five, with indications of a sixth. Total length 11·5 inches, wing 8, tail 5·5, tarsus 1·1.

Siam. A bird obtained at Bangkok by Bocourt was examined by me in the Paris Museum. It is evidently rather a young bird, the breast being still broadly streaked, the abdomen spotted; first primary with two bars; bands on tail four, with distinct indications of a fifth. Total length 11·5 inches, wing 7·7, tail 5, tarsus 1·1. The wing is perhaps not quite perfect, as the bird has been in confinement.

Malacca. Birds from this locality are generally known as *Ninox malaccensis* (Eyton), which is most probably identical with the true *N. scutulata* of Sumatra. I have carefully compared the Malaccan examples to ascertain whether they are distinct from the Chinese and Japanese forms, or whether, like many other kinds of birds, they are only the northern form occurring in Malacca during migration. In the seven specimens before me five bands are apparent on the tail, one or two bearing distinct traces of a sixth basal bar; and although the general crushed-up formation of a Malacca skin gives them a smaller appearance, they will be found to be really only a little less in size. I am therefore unable to separate *N. malaccensis* from *N. japonica.*

			Total length.	Wing.	Tail.	Tarsus.
			in.	in.	in.	in.
a.	Ad.	Malacca (*Wallace*)	11·5	7·8	5	1·1
b.	Ad.	Malacca (*Charlton*)	11·5	7·9	4·9	1
c.	Ad.	Malacca (*Mus. Brit.*)	11	7·4	4·5	1·05
d.	Ad.	Malacca (*Mus. Brit.*)	11	7·8	4·7	1·05
e.	Ad.	Singapore (*Mus. Norv.*)	10·5	7·5	4·6	1
f.	Ad.	Singapore (*Mus. Norv.*)	10·5	7·3	4·85	1
g.	Juv.	Penang (*Cantor*)	10	6·9	4·6	1·05

In nearly all the above specimens the first primary is uniform; only on one does there appear to be a trace of fulvous bars; and as this specimen has a distinctly spotted abdomen, these two characters seem to be dependent on maturity.

Camboja. The single skin that I have seen from this country was collected by the late M. Mouhot, and now belongs to the Norwich Museum. It measures as follows—total length 10·5 inches, wing 7·8, tail 5·1, tarsus 1. In general appearance it is like the South-China examples, and is adult, with a distinctly spotted abdomen; first primary mottled with fulvous near the base, but not distinctly barred.

Borneo. A skin obtained by Mr. Wallace at Sarawak is in the

Museum, and it agrees best, at first sight, with the examples obtained in Hainan by Mr. Swinhoe. The bird is adult, having several distinct subterminal spots on the abdomen and bars on the first primary. Although the general colour is rather light brown, there is very little of a rufous shade even on the wings. Bars on tail four; but I can discover traces of a fifth basal one, now apparently nearly dissolved. Total length 11·5 inches, wing 8·5, tail 5, tarsus 1·1.

The Norwich Museum contains a specimen from Borneo which is rather more rufous than the foregoing Sarawak bird. Bars on tail five, with faint traces of a sixth basal bar; first primary uniform; streaks on under surface rather narrow, and several abdominal spots distinctly developed.

In the Paris Museum I also examined a female specimen of a Bornean *Ninox*, purchased from the Maison Verreaux. The abdominal spots were very large, and in general resembling the Labuan skins in the British Museum; first primary with two or three faintly indicated bars; bands on tail four, with no trace of a fifth. Total length 10 inches, wing 8·5, tail 4·1, tarsus 0·95.

The following is a description of Mr. Wallace's specimen:—

Adult. Above uniform deep chocolate-brown, with scarcely any variations at all on the upper surface, beyond a few nearly concealed white spots on the scapulars; wing-coverts uniform with the back, the small ones on the edge of the outer wing white, forming a spot; primary-coverts blackish; quills dark brown, paler and more rufous on the outer edge, the innermost secondaries with a few half-concealed white spots and bars observable on disarranging the feathers; upper tail-coverts uniform with the back; tail ashy brown, narrowly tipped with whitish and crossed with four broad blackish bands, the subterminal interspaces slightly broader than the others; facial aspect dusky and uniform with the head, scarcely any shade of buff or whitish being apparent on the forehead or lores; throat and chest equally obscure, being dark brown like the ear-coverts, the chin very scantily streaked with white, and the chest having a few white streaks where some of the plumes are margined with dull white; rest of under surface rufous chocolate, varied with white, the bases and margins to the feathers being of the latter colour and the centres chocolate, varying in width and producing a more or less streaked appearance; under tail-coverts white, with brownish marks at the tips of a few of the feathers; leg-feathers uniform deep chocolate; under wing-coverts nearly uniform deep chocolate, some of the inner plumes varied with deep ochraceous bars, the lowest series nearly uniform sepia-brown, with a few ochraceous spots near the base, resembling thus the inner lining of the wing, which is sepia-brown barred on the inner web with pale brown, inclining gradually to ochraceous towards the base of the inner web. Total length 10·8 inches, wing 7·5, tail 4·4, tarsus 1·1.

Labuan. Four specimens collected by Mr. Low are in the Museum, adults and slightly younger birds. They are all remarkable for their very dark coloration, the scapular spots being very few and in one specimen entirely wanting. The head, sides of face, sides of neck and chest are almost uniformly brown together, so that the appearance of the fore part of the body is very dark. Two specimens I take to be adult and the other two rather younger, because of the more streaked character of the abdomen. All of them have the first primary perfectly plain; and in the one I consider the youngest of the four there are no bars either on the first or second primary; in the others there are faint traces of fulvous bars on the inner web of the second.

	Total length. in.	Wing. in.	Tail. in.	Tarsus. in.	Bars on tail.
a. Ad. Labuan (*Low*)	10	7·1	4	1	5
b. Ad. Labuan (*Low*)	10·5	7·3	4	1	5
c. Ad. Labuan (*Low*)	11	7·1	4·2	1	4
d. Juv. Labuan (*Low*)	10	7·1	4	0·95	5

In the last three specimens there is a trace of even another basal bar to the tail. The *Ninox* of Labuan is quite as distinct as any of the other races, and if the dark coloration is constant will perhaps be found worthy of the title of *N. labuanensis.*

Flores. Mr. Wallace obtained a single male bird in this island in 1862. It is the type of his *Athene florensis,* but seems not different from Chinese examples. It is adult, with a plainly barred primary; bars on tail five; abdominal spots distinct; on the wing-coverts a slight wash of rufous, one of them having a concealed white spot. Total length 12·5 inches, wing 9·1, tail 5·5, tarsus 1·2.

Mr. Swinhoe's Chefoo specimens match this bird, large as are its dimensions. The accompanying description has been prepared from the type:—

Adult male. Above uniform chocolate-brown, the head not darker than the back, the uniformity only broken by a few broad white bars on the scapulars, these also being generally half-concealed; wing-coverts brown like the back, but very distinctly washed with clay-colour or ochraceous; primary-coverts blackish; quills dark brown, slightly washed with rufous on the outer edges and barred across with lighter brown, these light bars rather inclining to ochraceous on their outer edges, the inner secondaries barred with white on both webs, the bars showing when the plumage is disarranged; upper tail-coverts uniform chocolate-brown like the back; tail paler and more ashy brown, tipped with whitish and crossed with five blackish bands; ear-coverts uniform with the head, but exhibiting slight remains of ochraceous shaft-streaks; forehead slightly white; lores whitish, obscured by blackish shaft-lines; fore part of cheeks and throat buffy white, narrowly streaked with brown; rest of under surface whitish, becoming more fulvescent on the abdomen, the whole broadly streaked with chocolate-brown in the centre, inclining to rufous on the abdomen; leg-feathers fulvous, mottled with small brown markings; under tail-coverts buffy white, the lower ones marked with oval spots of dark brown; under wing-coverts dark chocolate-brown barred with ochraceous buff, the edge of the wing whitish, the lower series sepia-brown like the wing-lining, and, like the inner lining of the quills, barred with fulvous. Total length 12·5 inches, wing 9·7, tail 5·8, tarsus 1·3.

Having thus carefully examined a large number of these Owls, I believe that there are only three forms to be distinguished specifically—two light and one dark. The light grey birds are peculiar to the Himalayas and the Andamans, the dark one ranges all over India (*vide suprà*) and Ceylon, Malaisia, through China to Japan, and southwards into the islands of the Malayan subregion. The birds vary infinitely in size, even from the same locality, as will be seen by the following

Synopsis of measurements.

			Total length. inches.	Wing. inches.	Tail. inches.	Tarsus. inches.
a.	*Ninox lugubris.*	Himalayas ..	10·0–12·5	8·0–8·75	4·6 –5·6	1·05–1·15
b.	„ „	Malabar	11·0	8·0	4·7	1·1
c.	„ *hirsuta.*	Ceylon	10·5–11·5	7·4–7·5	4·5 –4·8	1·05–1·1
d.	„ „	Nicobars	12·0	7·4	5·0	1·05
e.	„ *scutulata.*	Malacca	10·0–11·5	6·9–7·9	4·5 –5·0	1·0 –1·1
f.	„ „	Burmah....	11·5–12·5	8·1–8·4	5·25–5·5	1·15–1·2
g.	„ „	Siam	11·5	7·7	5·0	1·1
h.	„ „	Cochin China	11·5–12·0	8·0–8·25	5·4 –5·5	1·1
i.	„ *japonica.*	N. China, Formosa, Japan	10·5–12·5	8·3–9·1	4·9 –5·4	1·0 –1·05
k.	„ „	S. China, Hainan	10·5–11·5	8·0–8·5	4·8 –4·85	1·0 –1·05
l.	„ *borneensis.*	Borneo	10·0–11·5	8·5	4·1 –5·0	0·95–1·1
m.	„ *labuanensis.*	Labuan ..	10·0–11·0	7·1–7·3	4·0 –4·2	0·95–1·0
n.	„ *florensis.*	Flores	12·5	9·1	5·5	1·2

Neither does the number of tail-bands help us at all, nor the barring of the quills; for these characters are very variable, and perhaps depend on age. The result seems to be that, taking the Malaccan form (probably identical with the Sumatran) as the most typical, we find a great variation in size, the species gradually increasing to the northward, and reaching its maximum in Chinese and Japanese specimens. This larger bird seems to be migratory, if we may judge from the exact identity of *N. japonica* with *N. florensis*. The South-China bird is smaller than the North-China one; and the Nicobar bird is intermediate between these and *N. hirsuta* of Ceylon, which is the smallest and darkest of all.

a. Ad. st.	Ceylon.	Hugh Cuming, Esq.
b. Ad. sk.	Ceylon.	Purchased.
c. Ad. st.	North China.	Purchased.
d. Ad. sk.	Fokien, China.	J. Gould, Esq. [P.].
e. Ad. sk.	Rangoon, Burmah.	Purchased.
f. Ad. st.	Malacca.	Major Charlton [C.]. Type of *A. malaccensis.*
g. Juv. sk.	Malacca.	Purchased.
h, i. Juv. sk.	Malacca.	
k. Ad. sk.	Malacca.	A. R. Wallace, Esq. [C.].
l. ♂ ad. sk.	Flores.	A. R. Wallace, Esq. [C.]. Type of *A. florensis.*

m. ♀ ad. sk.	Sarawak, Borneo.	A. R. Wallace, Esq. [C.].
n, o. Ad. sk.	Labuan (*H. Low*).	R. B. Sharpe, Esq. [P.].
p, q. Juv. sk.	Labuan (*H. Low*).	R. B. Sharpe, Esq. [P.].

4. **Ninox ochracea.** (Plate XI. fig. 2.)

Noctua ochracea, *Schl. N. T. D.* i. p. 183 (1866); *id. Revue Accipitr.* p. 22 (1873).
Athene ochracea, *Gray, Hand-l. B.* i. p. 41 (1869); *Walden, Tr. Z. S.* viii. p. 38 (1872).
Ninox ochracea, *Sharpe, Ibis,* 1875, p. 258.

Adult. Above uniform chocolate-brown, the head rather duller than the back, which is slightly washed with rufous, the scapulars distinctly spotted with white on the outer web; wing-coverts dull rufous chocolate, the least series entirely uniform, the greater and median coverts externally spotted with white, these spots not numerous, and very often concealed; primary-coverts uniform sepia-brown; quills dull brown, washed on the outer web with rufous ochre, the primaries and outer secondaries minutely spotted with fulvous; upper tail-coverts uniform like the back, but a little duller and more sepia-brown; tail-feathers dull brown, obsoletely banded with lighter brown, remains of seven being apparent on the centre feathers, the outer feathers washed with rufous and banded on the extreme inner web with whitish; lores and frontal plumes whitish, the shafts blackish and terminating in bristles; sides of the face dull brown, the shaft-streaks blackish; over the eye a few white feathers, not sufficient to form an eyebrow; cheeks whitish; rest of under surface rich ochraceous orange, the chest chestnut-brown, the breast and flanks also having a few chestnut-brown streaks; under tail-coverts ochraceous buff, paler than the breast; under wing-coverts light ochraceous, slightly marked with brown near the outer margin of the wing, the greater series dark brown, notched on the inner web with yellowish, thus resembling the inner lining of the wing, which is dark brown, with yellowish bars near the base of the inner webs. Total length 11·5 inches, wing 7·9, tail 5, tarsus 1.

Hab. Celebes.

a. Ad. sk.	Celebes.	Leiden Museum [E.].

5. **Ninox philippensis.**

Ninox philippensis, *Bp. C. R.* xli. p. 654; *Walden, Tr. Z. S.* ix. p. 144, pl. xxv. fig. 1.
Noctua hirsuta philippinensis, *Schl. Mus. P.-B.* Striges, p. 26.
Athene philippensis, *Wall. Ibis,* 1868, p. 22; *Gray, Hand-l. B.* i. p. 42.

Young. Above rufous chocolate, the upper surface almost entirely uniform, with the exception of a few buffy white bars on the outer scapulars; wing-coverts uniform like the back, and only a little darker, the greater series distinctly spotted on the outer web with ochraceous or white; primary-coverts uniform dark brown; quills dark brown, notched on the outer web with ochraceous or buffy white; tail darker brown than the back, with a fulvous tip and crossed with six narrow bars of fulvous brown; forehead whitish, with long hair-like black bristles over the lores; side of face uniform rufous chocolate like the head; under surface of body fawn-colour, the feathers of the chest margined narrowly with whitish, causing a slightly streaked appearance, the abdominal plumes white, with broad fawn-coloured centres, causing this part to appear very broadly streaked; thighs fawn-colour; under tail-coverts white; under wing-coverts fawn-colour excepting the lower series, which are brown, spotted with fulvous on the inner web, exactly resembling the inner lining of the quills. Total length 8 inches, wing 6·45, tail 3·4, tarsus 1·2. (*Mus. Norv.*)

Adult. Above brown, the head slightly darker and more chocolate-brown, the scapulars with large oval marks on the outer web, some of the inner ones barred with ochraceous; wing-coverts dark brown, slightly washed with ochraceous, all of them distinctly spotted with white or ochraceous white, those on the median and greater series large and oval in shape; quills brown, margined narrowly with ochraceous, and barred across with paler brown, almost obsolete on the inner web, but indicated on the outer one by a distinct white spot, very plain, and producing a chequered appearance; upper tail-coverts rather more ochraceous brown than the back; tail-feathers sepia-brown, crossed with six narrow bands of pale ochraceous, the last one subterminal; frontal plumes whitish at base, but this colour not visible; sides of face brown like the head, the ear-coverts rather more dusky; under surface of body ochraceous brown, the chin whitish, and the throat marked with a few blackish streaks; the abdomen whiter, the centres to the feathers being ochraceous brown, producing a broadly streaked appearance; under tail-coverts white; leg-feathers ochraceous brown; under wing-coverts ochraceous, the edge of the wing white, those coverts nearest the margin being marked with dark brown; the lowest series sepia-brown, barred with pale ochre on the inner web, thus resembling the inner lining of the quills. Total length 8·5 inches, wing 6·95, tail 4, tarsus 2.

Hab. Philippine Islands.

a. Ad. st. Philippine Islands. Purchased.

6. Ninox boobook.

Boobook Owl, *Lath. Gen. Syn. Suppl.* ii. p. 64.
Strix boobook, *Lath. Ind. Orn. Suppl.* ii. p. xv.
Noctua boobook, *Vigors, Tr. Linn. Soc.* xv. p. 188; *Less. Man. Orn.* i. p. 113; *Schl. Mus. P.-B.* Striges, p. 38.
Athene boobook, *Gould, Syn. B. Austr.* part iii.; *Gray, Gen. B.* i. p. 35; *Gould, B. Austr.* i. pl. 32; *Cass. Cat. Strigidæ Philad. Mus.* p. 16; *Sturt, Exp. Int. Austr. App.* p. 17; *Bp. Consp.* i. p. 42; *Strickl. Orn. Syn.* p. 165; *Diggles, Orn. Austr.* pl.; *Gray, Hand-l. B.* i. p. 41.
Athene marmorata, *Gould, P. Z. S.* 1846, p. 18; *Cass. Cat. Strigidæ Philad. Mus.* p. 16; *Bp. Consp.* i. p. 42; *Strickl. Orn. Syn.* p. 166; *Gray, Hand-l. B.* i. p. 41.
Spiloglaux marmoratus, *Kaup, Isis,* 1848, p. 768; *id. Contr. Orn.* 1852,

p. 108, et *Tr. Z. S.* iv. p. 218; *Bp. Rev. et Mag. de Zool.* 1854, p. 544; *Gould, Handb. B. Austr.* i. p. 72.
Spiloglaux boobook, *Kaup, Isis*, 1848, p. 768; *id. Contr. Orn.* 1852, p. 108, et *Tr. Z. S.* iv. p. 217; *Bp. Rev. et Mag. de Zool.* 1854, p. 544; *Gould, Handb. B. Austr.* i. p. 74.
Ninox boobook, *Sharpe, Ibis*, 1875, p. 258.

Young (*N. boobook* auct.). Above uniform earthy brown, the fore part of the head streaked slightly with fulvous, and the hind neck varied with a few fulvous spots; scapulars with large rounded spots of white, more or less concealed; wing-coverts brown like the back, the least ones slightly washed with rufous and unspotted, the median and greater series with a few nearly concealed rounded spots of white; primary-coverts uniform dark brown; quills brown, barred across with paler brown, the primaries externally washed with rufous, the light bars fading into a whitish notch towards the ends of the latter quills; upper tail-coverts chocolate-brown, the outer ones with a few buff-coloured spots for the most part concealed; tail dark brown, paler at tip, the inner web only barred with more fulvous brown, the centre feathers with six or seven indistinct bars of lighter brown; loral plumes and a small eyebrow whitish, narrowly streaked with brown, the lores obscured by thin hair-like black bristles; ear-coverts uniform chocolate-brown; fore part of cheeks white as well as the chin; feathers under the ear-coverts on the sides of the neck ochraceous buff, with chocolate-brown streaks, the throat marked in the same manner, but the streaks broader; remainder of under surface rufous brown, mottled with white, all the feathers laterally marked with white on the breast, more largely on the abdomen and flanks, where the white spots are more oval in shape, producing an ocellated appearance; leg-feathers bright ochre, slightly streaked with brown on the outer aspect; under tail-coverts whitish, streaked and tipped with plain or ochraceous brown; under wing-coverts bright ochraceous, all streaked or barred with brown, the edge of the wing white, the lower series dark brown, barred with whitish on the inner web, exactly like the inner lining of the quills, which are dark brown below, with large rounded spots or bars of buffy white on the inner webs. Total length 14·5 inches, wing 9·8, tail 6·3, tarsus 1·75.

Adult male (*N. marmorata* auct.). Above light brown, the head spotted all over with fulvous, the hind neck very thickly mottled with buffy white spots; back uniform brown, the scapulars largely mottled with white, both webs of the latter, but especially the outer one, being largely barred with white; wing-coverts brown like the back, the bend of the wing slightly washed with rufous, the greater and median series largely spotted with white on the outer web; primary-coverts dark brown, barred with lighter brown; wing-coverts dark brown, slightly tipped with whitish, barred across with lighter and slightly rufous brown, some of the outer bars on the primaries fading off into a whitish notch externally; tail brown, barred with fulvous on the inner web, the centre feathers indistinctly marked with about six bars, confined, as on the other feathers, to the inner web; forehead and sides of head fulvous, with central brown streaks to the feathers; lores whitish, with hair-like black shaft-lines; ear-coverts brown; cheeks fulvous, narrowly streaked with brown; chin dull white; rest of under surface fulvous, the centres of the feathers pale brown, giving a broadly streaked appearance, the lateral margins to the feathers of the abdomen whiter, the central streaks indented with large white spots, causing the flanks to appear somewhat ocellated; leg-feathers ochraceous; under wing-coverts whitish, mesially streaked with brown, widening out towards the tips of the feathers; under wing-coverts deep ochraceous, streaked with brown, especially near the outer edge of the wing, the greater series dark brown, barred with white like the inner lining of the quills, which are dark brown below, with broad buffy white bars on the inner web; cere bluish grey; bill horn-colour; feet yellow; iris light brown in some, greenish brown (inclining to yellow) in other examples. Total length 16 inches, wing 10·3, tail 6·7, tarsus 1·75.

Obs. After comparing our large series of specimens in the Museum, I have come to the conclusion that the Owl called *Athene marmorata* by Mr. Gould is only the adult of the large form of *N. boobook*, which is more uniform when young, and becomes more spotted as it gets adult. This large *N. boobook* appears to be an inhabitant of South Australia, and is replaced in the north and west by a smaller race or subspecies, which is generally more rufous in colour, having the head streaked with fulvous; the hind neck also appears never to become so spotted as in *N. boobook*.

Hab. South Australia.

a. Vix ad. sk.	South Australia.	C. D. E. Fortnum, Esq. [P.].
b. Vix ad. sk.	South Australia.	Sir George Grey [P.].
c. Ad. st.	South Australia.	Sir George Grey [P.].
d. Juv. sk.	South Australia.	Sir George Grey [P.].
e. Ad. sk.	River Darling, Sept. 1844.	Captain Sturt [P.]. "Iris yellow."
f. ♂ vix ad. st.	Percy Island.	
g. Ad. st.	Australia.	A. Cunningham, Esq. [P.].
h. Pull. st.	Australia.	
i. Ad. sk.	Australia.	Purchased.
k. Ad. sk.	Queensland.	Purchased.

Subsp. *a.* **Ninox ocellata***.

Athene ocellata, *Hombr. & Jacq. Voy. Pôle Sud, Zool.* iii. p. 51, pl. 3. fig. 2; *Bp. Consp.* i. p. 42; *Strickl. Orn. Syn.* p. 174; *Gray, Hand-l. B.* i. p. 42.
Spiloglaux ocellata, *Bp. Rev. et Mag. de Zool.* 1854, p. 544.
Ninox ocellata, *Sharpe, Ibis*, 1875, p. 258.

Nestling. Clothed in whitish down, washed with golden buff on the head, neck, and breast; the feathers, as far as developed, resembling the plumage of the adult, but the white spots on the scapulars extremely large and of an oval shape on both webs; head

* I have examined the type in the Paris Museum, and am satisfied that it is Australian, and never came from "Chili."

ochraceous, all the feathers centred and tipped with dark brown; lores and forehead whitish; ear-coverts entirely dark brown, contrasting strongly with the light-coloured head and neck; under surface of body whitish, with a few narrow central streaks of brown, rather broader on the chest, which is washed with golden buff.

Adult. General colour above rather pale or earthy brown, the hind neck washed with fawn-colour; the head rather darker than the back, and slightly streaked with fulvous; scapulars brown like the back, broadly barred with white, occupying the greater part of the outer web, and showing very conspicuously; wing-coverts brown, the greater and median series with some distinct white oval spots; primary-coverts dark brown, with faint indications of lighter brown bars; quills dark brown, barred across with lighter brown, the primaries externally washed with pale rufous, the light bars fading into a buffy white notch on the outer web of the outermost primaries; lower back and rump more rufous than the rest of the upper surface, the feathers having fawn-coloured bases; tail brown, the centre feathers uniform, the rest barred with fulvous on the inner web, these bars becoming gradually more distinct towards the outer feather, the last one being barred with fulvous on the inner web, and spotted with the same on the outer one; lores and frontal plumes whitish, with black shaft-lines; sides of the head fulvous, narrowly streaked with brown, and forming a tolerably distinct eyebrow; ear-coverts dark brown, contrasting with the rest of the head; cheeks whitish, as also the feathers behind the ear-coverts, these latter streaked with brown; chin whitish; rest of under surface white, the breast yellowish, broadly streaked with rufous brown, the abdominal feathers almost entirely white, mesially streaked with brown and broadly tipped with pale fawn-colour; the under tail-coverts whitish, narrowly margined with pale fawn-colour; leg-feathers fawn-colour, whitish on the inner aspect; under wing-coverts ochraceous fawn, with scarcely any indications of brown markings beyond an occasional streak near the outside margin of the wing, the greater series brown, irregularly barred with fulvous, resembling the inner lining of the wing, which is brown, with large bars of fulvous on the inner web; "bill slate-coloured, darker at tip; iris dark brown" (*Elsey, MS.*). Total length 13·5 inches, wing 8·6, tail 5·5, tarsus 1·65.

Obs. Although closely allied to *N. boobook*, this is a much smaller and more rufous species, and differs especially in its fawn-coloured under parts and under wing-coverts, the latter being uniform. In some of the browner-looking examples, there are some markings or stripes of brown upon these under wing-coverts, but never so thickly distributed as in the true *N. boobook*. The difference in size may best be estimated by a comparison of two males, the specimens described not having the sexes determined.

		Total length. in.	Wing. in.	Tail. in.	Tarsus. in.
a. ♂ ad.	Percy Island	16	9·3	6·4	1·65
b. ♂ ad.	N.W. Australia (*Elsey*)	13	8·8	5·4	1·45

Hab. Northern and Western Australia.

a. Pull. sk.	Queensland.	Purchased.
b, c. Ad. st.	Port Essington.	Capt. W. Chambers, R.N.
d. Ad. st.	N.W. Australia.	J. R. Elsey, Esq. [P.].
e, f. ♂ ad. sk.	N.W. Australia.	J. R. Elsey, Esq. [P.].
g. Ad. sk.	"Scrub, lat. 18'S., long. 141° E., Sept. 13th, 1856."	J. R. Elsey, Esq. [P.].
h. Ad. st.	Western Australia.	Sir George Grey [P.].
i, k. Ad. sk.	Western Australia.	Sir George Grey [P.].

7. Ninox fusca. (Plate XII. fig. 1.)

Strix fusca, *Vieill. N. Dict.* vii. p. 22.
Strix maugei, *Temm. Pl. Col.* i. pl. 46.
Noctua fusca, *Steph. Gen. Zool.* xiii. pt. 2, p. 67; *Less. Traité*, p. 106.
Athene guteruhi, *Müll. Verh. Nat. Gesch.* p. 279; *Bp. Consp.* i. p. 41; *Strickl. Orn. Syn.* p. 172; *Wall. Ibis*, 1868, p. 22; *Gray, Hand-l. B.* i. p. 41.
Athene maugei, *Gray, Gen. B.* i. p. 35; *Cass. Cat. Strigidæ Philad. Mus.* p. 17.
Athene fusca, *Pucher. Rev. Zool.* 1842, p. 20; *Bp. Consp.* i. p. 38; *Strickl. Orn. Syn.* p. 170; *Gray, Hand-l. B.* i. 43.
Spiloglaux fuscus, *Kaup, Contr. Orn.* 1852, p. 108, et *Tr. Z. S.* iv. p. 218; *Bp. Rev. et Mag. de Zool.* 1854, p. 544.
Noctua guteruhi, *Schl. Mus. P.-B.* Striges, p. 26; *id. Revue Accipitres*, p. 23.
Ninox fusca, *Sharpe, Ibis*, 1875, pp. 256, 258.

Adult female. Above pale brown, mostly uniform, excepting on the hind neck, which is very plainly mottled with buffy white spots and small bars, the upper tail-coverts also varied with fulvous bars; scapulars like the back, spotted with white, these spots mostly concealed, excepting on the outer margin, where there are large white markings forming a tolerably distinct shoulder-patch; wing-coverts coloured like the back, the least series uniform, but the median and greater ones distinctly relieved with rows of white spots on their outer webs; primary-coverts uniform brown; quills ashy brown, barred with lighter brown, paler on the outer margin, turning to a distinct white spot on the secondaries; tail ashy brown, barred with eight bands of paler brown, shading into white on the outer webs, the external feathers slightly inclining to sandy brown, with distinct white spots on the outer web; forehead streaked with white, the sides of the face brown like the crown of the head; fore part of cheeks and chin whitish; throat light brown, streaked with whitish, most of the feathers being margined with this colour; rest of under surface white, broadly streaked with light (rather reddish) brown, these central streaks somewhat dilating towards the apex of the feather, but narrowing perceptibly on the under tail-coverts; leg-feathers fulvous; under wing-coverts orange-buff, the outer ones whiter with brown central streaks, the greater series ashy blackish, barred with whitish on the inner web, resembling the inner lining of the wing, which is blackish barred with dull whitish; cere dull yellow; bill bluish lead-colour, the upper mandible black at the gape; feet yellow; iris yellow. Total length 14 inches, wing 9·25, tail 5·5, tarsus 1·35.

Hab. Timor.

a.	♀ ad. st.	E. Timor.	A. R. Wallace, Esq. [C.].
b.	♀ ad. sk.	E. Timor.	A. R. Wallace, Esq. [C.].

8. **Ninox novæ zealandiæ.**

New-Zealand Owl, *Lath. Gen. Syn.* i. p. 149.
Strix novæ zealandiæ, *Gm. S. N.* i. p. 296.
Strix fulva, *Lath. Ind. Orn.* i. p. 65.
Noctua zelandica, *Quoy & Gaim. Voy. Astrol.* vol. i. p. 168, pl. 2. fig. 1.
Athene novæ zealandiæ, *Gray in Dieff. Trav.* ii. App. p. 186; *id. Cat. Accipitr. B. M.* 1844, p. 52; *id. Zool. Voy. Erebus & Terror*, p. 2; *id. Gen. B.* i. p. 35; *Bp. Consp.* i. p. 42; *Strickl. Orn. Syn.* p. 166; *Gray, Hand-l.* i. p. 41; *Hutton, Cat. B. N. Zeal.* p. 2; *Finsch, J. f. O.* 1872, p. 94, et 1874, p. 177.
Spiloglaux novæ zealandiæ, *Kaup, Isis*, 1848, p. 768; *id. Contr. Orn.* 1852, p. 108, et *Tr. Z. S.* iv. p. 218; *Bp. Rev. et Mag. de Zool.* 1854, p. 544; *Buller, B. N. Z.* p. 17, pl. 2. fig. 1; *Sharpe, Voy. Erebus & Terror*, 2nd ed. *Birds*, p. 22.
Noctua venatica, *Peale, U. S. Expl. Exp.* p. 75.
Noctua novæ zealandiæ, *Schl. Mus. P.-B.* Striges, p. 38.
Ninox novæ zealandiæ, *Sharpe, Ibis*, 1875, p. 258.

Adult female. Above deep chocolate-brown, the head darker but quite uniform brown, the forehead with a few ochraceous streaks; round the hind neck an indistinct collar of bright ochre, more or less concealed spots of which are seen on nearly all the feathers; scapulars distinctly spotted with white, these spots being larger on the outer web; wing-coverts brown like the back, darker towards the edge of the wing, the greater series slightly washed with rufous, both these and the median coverts relieved by large rounded spots of white, smaller and less thickly distributed on the least ones; primary-coverts uniform dark brown; quills dark brown, banded with lighter brown, much paler near the base of the primaries, some of the lighter cross bands of these quills ending in a white notch; upper tail-coverts dark brown, many of them with a concealed ochraceous bar near the base; tail dark brown, crossed with eight bands of lighter brown; lores and a narrow eyebrow white, the former obscured by long hair-like bristles; sides of face brown, the ear-coverts slightly streaked with whitish, the fore part of the cheeks more broadly; sides of the neck deep chocolate-brown, distinctly streaked with bright ochre; under surface of body bright ochraceous, all the feathers dark brown in the centre, producing a broadly streaked appearance, sometimes widening out and causing the ochraceous colouring of the breast-feathers to assume the form of rounded spots, these markings being white on the lower breast and abdomen, producing a very distinctly mottled appearance, the flanks barred with brown and white; leg-feathers deep ochraceous or rust-colour, almost chestnut, slightly shaded with brown on the outer aspect; under tail-coverts ochraceous, with brown centres to the feathers; under wing-coverts also deep ochraceous, thickly mottled with brown near the outer margin, the greater series ashy brown, barred with whitish on the inner web, thus exactly resembling the inner lining of the quills, which are similarly coloured; bill black, white on the ridge; toes yellow, with dark hairs; iris golden yellow. Total length 12·5 inches, wing 7·5, tail 5·3, tarsus 1·35.

Dr. Buller gives the female as slightly smaller than the male. The measurements for the latter are stated to be as follows—total length 12·5 inches, wing 8, tail 5·75, tarsus 1·5. The only specimen in the British Museum sexed as "male" has the wings and tail too imperfect to admit of comparison.

Nestling. Covered with whitish down, plumbeous beneath. (*W. L. Buller.*)

Hab. New Zealand.

a. Ad. st.	New Zealand.	Purchased.
b. Ad. st.	New Zealand.	Miss R. Stone [P.].
c. Ad. sk.	New Zealand.	Miss R. Stone [P.].
d. Ad. sk.	New Zealand.	Sir G. Grey [P.].
e. Ad. sk.	New Zealand.	Antarctic Expedition [C.].
f, g. ♂ ♀ ad. sk.	New Zealand.	Antarctic Expedition [C.].
h. Ad. sk.	Bay of Islands.	Antarctic Expedition [C.].
i. Ad. sk.	New Zealand.	Capt. Stokes [P.].
k. ♀ ad. sk.	New Zealand.	Capt. Stokes [P.].

9. Ninox maculata.

Noctua maculata, *Vig. & Horsf. Tr. Linn. Soc.* xv. p. 189; *Less. Man. d'Orn.* i. p. 113; *Schl. Mus. P.-B.* Striges, p. 39.

Athene maculata, *Gould, Syn. B. Austr.* pt. 3; *id. B. Austr.* i. pl. 33; *Gray, Gen. B.* i. p. 35; *Cass. Cat. Strigidæ Phil. Mus.* p. 16; *Bp. Consp.* i. p. 42; *Strickl. Orn. Syn.* p. 166; *Gray, Hand-l. B.* i. p. 41.

Spiloglaux maculatus, *Kaup, Contr. Orn.* 1852, p. 108; *id. Tr. Z. S.* iv. p. 217; *Bp. Rev. et Mag. de Zool.* 1854, p. 544; *Gould, Handb. B. Austr.* i. p. 76.

Ninox maculata, *Sharpe, Ibis,* 1875, p. 258.

Adult. Above chocolate-brown, the whole upper surface mottled with white spots, disposed in pairs on the feathers, being much smaller on the head and hind neck, larger and more in the form of bars on the scapulars, some of the white bars completely traversing these feathers near the base; upper wing-coverts chocolate-brown, a little more rufous than the back, the least ones with a few fulvous spots, larger and whiter on the median and greater series; primary-coverts uniform sooty brown; quills brown, barred with paler brown fading off much lighter on the outer web of the primaries, some of which have a white notch or two towards the tip of the feather; upper tail-coverts uniform chocolate-brown; tail brown, indistinctly barred with paler brown, of which about six or seven are faintly indicated on the centre feathers; forehead white; lores and a small eyebrow whitish, the former with long black hair-like bristles; sides of face chocolate-brown, the ear-coverts with a few whitish streaks; cheeks whitish, some few feathers with rufous centres; chin whitish; rest of under surface of body rufous, everywhere mottled with white, the breast-feathers streaked on each side, the belly and flank-feathers with large rounded spots of white, causing a somewhat ocellated appearance; leg-feathers orange-tawny; under tail-coverts whitish, centred and mottled with tawny brown; under wing-coverts and axillaries dull tawny, waved with small markings of brown near the outer edge of the wing, the greater series ashy brown, spotted or barred with white, and thus resembling the inner lining of the quills, which are ashy brown, with large white spots or bars on the inner web; bill dark horn-colour; feet yellowish; iris yellow. Total length 13·5 inches, wing 8·9, tail 5·2, tarsus 1·5.

Obs. The Museum contains no specimens with their proper sexes attached; but there is considerable difference in size among the specimens in the collection. A bird which I presume to be the male measures as follows—total length 12·3 inches, wing 8·1, tail 5, tarsus 1·4.

Young birds appear to be less spotted with white above and to be rather darker in their general appearance.

Hab. Van Diemen's Land.

a. Ad. sk.	[Australia.]	Linnean Society. Type of species.
b. Ad. st.	[Australia.]	
c. Imm. st.	Van Diemen's Land.	Ronald Gunn, Esq. [P.].
d. Vix ad. sk.	Van Diemen's Land.	Purchased.
e. Ad. st.	Van Diemen's Land.	J. Gould, Esq. [C.].
f. Ad. st.	Van Diemen's Land.	J. Gould, Esq. [C.].
g. Ad. sk.	Van Diemen's Land.	The Admiralty [P.].

10. Ninox dimorpha.

Athene dimorpha, *Salvad. Ann. Mus. Civ. Genov.* vi. p. 308.
Ninox dimorpha, *Sharpe, Ibis,* 1875, p. 258.

Adult female. Upper surface dusky brown, the feathers of the head and neck margined with buff; back and wings irregularly barred across with greyish buff; forehead, face, and belly pale buff, the latter relieved by longitudinal streaks down the middle of the feathers; quills and tail dusky brown, closely banded with dusky grey; tarsi densely plumed, buff, obsoletely streaked with dusky brown, the toes more scantily plumed and of a yellow colour; bill leaden grey; iris yellow. Total length 13·4 inches, culmen 1, wing 7·9, tail 6·2, tarsus 1·3. (*Salvadori.*)

Hab. Sorong, New Guinea.

Obs. The above description is a translation of Count Salvadori's original diagnosis; and from a coloured sketch of the bird which he was so good as to send me, I have been able to place it in the synopsis of species; but I am not at all sure that it is a *Ninox*, the wing looking too short: the general appearance, in fact (judging from the sketch), is that of a gigantic *Glaucidium*; and it is very probably the type of a new genus.

11. Ninox connivens.

Winking Falcon, *Lath. Gen. Syn. Suppl.* ii. p. 53.
Falco connivens, *Lath. Ind. Orn. Suppl.* ii. p. 12.
Buteo connivens, *Vieill. N. Dict.* iv. p. 481.

Noctua frontata, *Less. Traité*, p. 106; *Pucher. Rev. et Mag. de Zool.* 1849, p. 21.
Athene fortis, *Gould, P. Z. S.* 1837, p. 141; *id. Syn. B. Austr.* pt. iii.
Athene connivens, *Gray, Ann. N. H.* xi. p. 189; *id. Gen. B.* i. p. 35; *Gould, B. Austr.* i. pl. 34; *Cass. Cat. Strigidæ Philad. Mus.* p. 16; *Bp. Consp.* i. p. 43; *Strickl. Orn. Syn.* p. 164; *Gray, Hand-l. B.* i. p. 40.
Athene frontata, *Gray, Gen. B.* i. p. 35.
Ieraglaux connivens, *Kaup, Contr. Orn.* 1852, p. 109, et *Tr. Z. S.* iv. p. 220.
Noctua connivens, *Schl. Mus. P.-B.* Striges, p. 39; *id. Revue Accipitr.* p. 31.
Hieracoglaux connivens, *Gould, Handb. B. Austr.* i. p. 71.
Ninox connivens, *Sharpe, Ibis*, 1875, p. 258.

Adult. Above nearly uniform brown, the uniformity broken only by a few half-concealed white spots on the hind neck, the scapulars also barred with white, the markings on the outer wings much larger and diagonal in shape; wing-coverts brown, a little paler than the back, the least series with a few ovate spots of white externally, larger and more in the form of a notch on the median and greater series; primary-coverts dark brown, barred with lighter brown; quills brown, barred across with paler brown, inclining to ochraceous and shading into a white notch on the outer web of the primaries, some of the secondaries also externally spotted with white; upper tail-coverts brown like the back, broadly barred with white, especially on the outer feathers, most of the spots concealed; tail brown, tipped with whitish and crossed with six other rather narrow bands of whity brown; forehead and lores whitish, the latter with hair-like brown shafts; ear-coverts light brown, slightly paler than the crown; cheeks whitish, the hinder part streaked with brown and merging into the sides of the neck, which is light brown, slightly streaked with white; under surface of body dull white, everywhere broadly streaked with brown, these streaks broader and more dilated near the base on the breast-feathers; leg-feathers whitish, narrowly streaked with brown, the tibial plumes yellowish; under wing-coverts white, streaked with brown exactly like the breast, the greater series ashy brown barred with white exactly in the same manner as the inner lining of the quills; cere yellowish olive; bill light yellowish horn-colour; feet yellow, covered with fine hairs; iris bright yellow. Total length 21 inches, wing 12·5, tail 7·5, tarsus 2·1.

The specimen above described, though not sexed, is doubtless a female, as a male bird obtained by Macgillivray during the 'Rattlesnake' voyage measures as follows—total length 19 inches, wing 11·5, tail 7, tarsus 1·85. Mr. Elsey's two specimens, also sexed males, agree in the main with the above, measuring 11·3 and 11·7 inches in the wing. They are much more rufous below than the other examples.

Hab. Australia.

a.	Ad. st.	South Australia.	Sir G. Grey [P.].
b.	Ad. sk.	Interior of South Australia.	Capt. Sturt [P.]

c. Ad. st.	N.E. Australia.	Sir T. Mitchell [P.].
d. ♂ ad. st.	Cape York, Oct. 10th, 1848.	Voyage of the 'Rattlesnake' [C.].
e. Ad. sk.	Cape York.	Mr. Cockerell [C.].
f. ♂ ad. sk.	Wentworth River, long. 137°, Aug. 18th, 1856.	J. R. Elsey, Esq. [P.].
g. ♂ ad. sk.	Victoria River, Nov. 1855.	J. R. Elsey, Esq. [P.].
h. [♀] ad. sk.	Australia.	

12. Ninox rufistrigata.

Athene rufistrigata, *Gray, P. Z. S.* 1860, p. 344; *Wall. Ibis*, 1868, p. 22; *Gray, Hand-l. B.* i. p. 41.

Noctua rufistrigata, *Giebel, Thes. Orn.* p. 448; *Schl. Mus. P.-B. Revue Accipitr.* p. 31.

Ninox rufistrigata, *Sharpe, Ibis*, 1875, p. 258.

Adult (probably a female). Above chocolate-brown, the head inclining to cinereous brown, the nape and outer web of the scapulars varied with large rounded spots of white, especially on the latter; wing-coverts uniform with the back, the greater series also with large white spots on the outer web; quills brown, the primaries rufous at the base of the outer web, shading into fulvous further up the feather; the inner secondaries notched with white on the inner web, these markings much larger and more in the form of bars underneath; the primaries ashy brown below with indications of buffy white bars on the inner web, much obscured by brown mottlings; tail brown, tipped with whitish, shaded with rufous near the base of the outer web and crossed with seven bars of darker brown; forehead and lores white; sides of face brown; throat white; rest of under surface white, inclining to buff on the breast and sides of the body; the lower throat narrowly, but the breast more broadly steaked with rufous brown, inclining to clearer rufous on the abdomen, but much less pronounced on the under tail-coverts; under wing-coverts buffy white, the outermost streaked with rufous brown. Total length 18 inches, culmen 1·7, wing 11·8, tail 7·8, tarsus 1·7. (*Mus. Lugd.*)

A smaller bird, probably a male, measures—total length 15·5 inches, wing 11, tarsus 1·65. (*Mus. Lugd.*)

The type specimen, according to Mr. Wallace (*l. c.*), measured in the flesh—total length 16·5 inches, wing 11·25, tail 6·5, tarsus 2. The difference in the length of tarsus probably arises from the greater facility of measuring this part when the bird is fresh; I make it in the skin 1·85 inch.

Hab. Gilolo.

a. Ad. sk.	East Gilolo.	A. R. Wallace, Esq. [C.].	Type of species.

13. Ninox obscura.

Ninox obscura, *Hume, Str. F.* i. pp. 11, 421; *Ball, Str. F.* i. p. 55; *Walden, Ibis*, 1874, p. 129, pl. iv.; *Hume, Str. F.* ii. p. 153; *Sharpe, Ibis*, 1875. p. 258.

Adult female (S. Andaman, *Capt. Wimberley*). Above uniform chocolate-brown, the head a little darker than the back and slightly glossed with greyish; wing-coverts uniform chocolate-brown, like the back, the primary-coverts blackish; quills uniform chocolate-brown, the primaries with a very narrow external margin of whity brown, none of the feathers barred; tail dark brown, narrowly tipped with greyish white and crossed with three bands of lighter and more ashy brown, not counting the light band at the end of the tail shading off into the whitish tip; a few of the frontal feathers whitish at base, as well as the lores, the latter obscured by blackish bristles; cheeks, ear-coverts, and sides of neck deep chocolate-brown like the head, and uniform like the latter; chin whitish: rest of under surface reddish chocolate, darker and more earthy brown on the fore neck and chest, and shading off rather paler on the flanks and abdomen, on the former of which are some half-concealed bars of fulvous; under tail-coverts buffy white, barred across with dark brown; under wing-coverts uniform chocolate-brown, the lower series rather more ashy brown like the inner lining of the quills, the edge of the wing scarcely whitish; cere and ridge of upper mandible and tip of lower mandible green; bill blackish; feet yellow; claws black; iris yellow. Total length 11 inches, wing 7·55, tail 5, tarsus 1·1. (*Mus.* Lord Walden.)

Hab. Andaman Islands.

14. **Ninox theomacha.**

Athene theomacha, *Bp. C. R.* xli. p. 654; *Wall. Ibis,* 1868, p. 24; *Gray, Hand-l. B.* i. p. 41.
Spiloglaux theomacha, *Bp. Rev. et Mag de Zool.* 1854, p. 544.
Noctua hoedtii, *Schl. N. T. D.* iv. p. 3; *id. Mus. P.-B. Rev. Accipitr.* p. 24.
Ninox hoedti, *Sharpe, Ibis,* 1875, p. 258.
Ninox theomacha, *Sharpe, Ibis,* 1875, p. 258.

Adult (type of *N. hoedti*). Above very deep chocolate-brown, the entire head deep greyish brown, the forehead and cheeks slightly varied with whitish; wing-coverts chocolate like the back, the greater series with faint rufous edgings; quills blackish brown, primaries slightly margined with fulvous, the secondaries spotted with white on the inner web; tail uniform ashy brown above and below; underside of body deep chestnut-brown, slightly streaked with whitish on the throat and varied with fulvous spots and margins on the feathers of the lower abdomen; thighs and under tail-coverts almost entirely buff; under wing-coverts coloured like the breast, the greater series blackish brown, with whitish cross bars. Total length 10 inches, culmen 1·05, wing 7, tail 3·9, tarsus 1·1. (*Mus. Lugd.*)

Hab. Mysol and the N.W. peninsula of New Guinea.

15. **Ninox strenua.**

Athene strenua, *Gould, P. Z. S.* 1837, p. 142; *id. Syn. B. Austr.* part 3; *id. B. Austr.* i. pl. 35; *Gray, Gen. B.* i. p. 35; *Cass. Cat. Strigidæ Philad. Mus.* p. 16; *Bp. Consp.* i. p. 35; *Strickl. Orn.*

Syn. p. 165; *Gray, Hand-l. B.* i. p. 40; *Diggles, Orn. Austr* pl.

Athene rufa, *Gould, P. Z. S.* 1846, p. 18; *id. B. Austr.* i. pl. 36; *Cass. Cat. Strigidæ Philad. Mus.* p. 16; *Bp. Consp.* i. p. 43; *Strickl. Orn. Syn.* p. 165.

Ieraglaux strenua, *Kaup, Contr. Orn.* 1852, p. 109; *id. Tr. Z. S.* iv. p. 220.

Ieraglaux rufus, *Kaup, Contr. Orn.* 1852, p. 109; *id. Tr. Z. S.* iv. p. 220.

Rhabdoglaux rufus, *Bp. Rev. et Mag. de Zool.* 1854, p. 544.

Noctua strenua, *Schl. Mus. P.-B.* Striges, p. 40.

Hieracoglaux strenuus, *Gould, Handb. B. Austr.* i. p 68.

Hieracoglaux rufus, *Gould, l. c.* p. 69.

Ninox strenua, *Sharpe, Ibis,* 1875, p. 258.

Young. Head, neck, and under surface of body pure white, all the feathers fluffy, apparently the remains of nestling-plumage; on the crown some brown feathers barred across with white; the forehead streaked with brown, the lores obscured by blackish bristles, as also the eyebrows and checks; ear-coverts dark brown, streaked with white: the under surface with a few narrow shaft-streaks of dark brown, broader on the flanks and sides of the body, the plumes on the sides of the upper breast barred with brown; the back and all the rest of the upper surface ashy brown, transversely barred with white, some of the bars washed with rufous; quills dark brown, barred with lighter brown, these lighter bars inclining to white on either margin; tail dark brown, tipped with white and crossed besides by five narrower whitish bars.

Adult. Above dark brown, numerously banded on the whole of the upper surface with pale brown, inclining to whitish on the scapulars and feathers of the lower back; wing-coverts dark brown like the back, barred with rufous brown, more broadly on the greater series, which are likewise externally barred and tipped with white; primary-coverts blackish brown, tipped and crossed with two bars of ashy brown; quills brown, tipped with whitish, and crossed with regular transverse bands of lighter brown, whitish on the inner and inclining more or less to rufous on the outer webs; upper tail-coverts dark brown, barred with white, very few of the bars washed with rufous; tail dark brown, tipped with ashy whitish, and crossed with six bands of whity brown, inclining to pure white on the inner webs; head dark brown like the back, minutely barred and spotted on either web with paler brown; sides of face ashy brown, streaked with whitish, the lores and fore part of cheeks white with blackish shaft-streaks; behind the ears a partial ruff of whitish feathers, streaked with dark brown; sides of neck dark brown, streaked with fulvous, some of the plumes barred with whitish or ochre; throat ochraceous buff, broadly streaked with dark brown; the rest of the under surface white, barred with brown, the chest washed with ochraceous, and more closely barred than on the breast; under tail-coverts white, with broad arrow-shaped bars of dark brown; leg-feathers ochraceous, with indistinct bars of brown; under wing-coverts whitish, washed with ochre, and very thickly barred with dark brown, the greater series ashy brown, barred with white, and resembling the inner lining of the quills, which are ashy brown below, broadly barred with dull white, especially on the inner web; cere greenish olive; bill light blue at base, passing into black at tip; feet pale gamboge-yellow; iris yellow. Total length 24·5 inches, wing 15·9, tail 11, tarsus 2·1.

Obs. Another specimen in the collection, probably a female, is very similar in colour to the one described, but is larger, and has not so much of the yellow colouring on the throat. Total length 26 inches, wing 16·5, tail 11, tarsus 2·15.

Hab. Australia.

a. Ad. st.	Australia.	Lady Carington [P.].
b. Ad. sk.	N. Australia.	Purchased.
c. Juv. sk.	N. Australia.	Purchased.
d. Ad. sk.	Queensland.	H. Whitely, Esq. [P.].

16. **Ninox humeralis.**

Athene humeralis, *Hombr. & Jacq. Voy. Pôle Sud, Zool.* iii. p. 51, pl. 3. fig. 2; *Bp. Consp.* i. p. 40; *Strickl. Orn. Syn.* p. 174; *Wall. Ibis,* 1868, p. 24; *Gray, Hand-l. B.* i. p. 41; *Sharpe, Ibis,* 1875, p. 258.

Ieraglaux humeralis, *Kaup, Contr. Orn.* 1852, p. 109; *id. Tr. Z. S.* iv. p. 221.

Rhabdoglaux humeralis, *Bp. Rev. et Mag. de Zool.* 1854, p. 544.

Adult (type of species). General colour above dark chocolate-brown, everywhere on the upper surface crossed with indistinct bars of paler brown, rather more distinct on the scapulars and on the sides of the hind neck; head darker than the back, and nearly uniform, the cross bars very indistinct excepting on the sides of the crown, above the ear-coverts, and on the sides of the neck, which are brown, distinctly rayed with fulvous cross bars; ear-coverts black; cheeks rayed like the sides of the neck, the fore part whitish like the lores, the latter having very distinct blackish shaft-stripes forming overhanging bristles: under surface of body orange-buff, barred across with light brown, more dingy on the chest; the abdomen white, barred across with pale orange-brown, these bars disappearing on the under tail-coverts, which are pure white, with only a slight indication of pale bars at the tip of the long feathers; leg-feathers orange-buff, with a few indistinct brown markings; under wing-coverts orange-buff or fulvous, barred across with light brown, with a brown patch near the outside edge of the wing, this being pure white, the lower series brown, barred with white so as to resemble the inner lining of the quills, which are dark brown below, with pale greyish bars on the inner webs; the wings above chocolate-brown, with the same indistinct cross bars as on the back, less distinct on the external coverts and on the least series, which are almost uniform; primary-coverts uniform blackish brown: quills dark brown, barred across with lighter brown, the secondaries paler and the cross bars rather lighter on the extreme edge both of the primaries and secondaries; upper tail-coverts like the back, but the cross bars a little paler; tail dark brown, narrowly tipped with whitish, and barred across with lighter brown, about eight bars being distinguishable. Total length 18 inches, wing 13·4, tail 9, tarsus 1·8. (*Mus. Paris.*)

Hab. New Guinea.

17. **Ninox aruensis.**

Noctua aruensis, *Schl. N. T. D.* iii. p. 329; *id. Revue Accipitr.* p. 31 (1873).
Athene aruensis, *Gray, Hand-l B.* i p. 40.
Ninox aruensis, *Sharpe, Ibis,* 1875, p. 258.

Adult female (type of species). Above brown, of a chocolate tinge, the head a little darker, the whole upper surface crossed with indistinct bars of dull buff; quills chocolate-brown, barred with lighter brown, the secondaries much paler and edged with buffy whitish towards the tips, and exhibiting externally five bars of pale ashy brown; tail chocolate-brown, tipped with buffy white, and crossed with twelve bars of ashy brown; forehead very conspicuously white, as well as the feathers under the eye, the fore part of cheeks, and throat; ear-coverts uniform dark brown; sides of neck and underparts dull tawny rufous, barred across with buff, the bars inclining to white on the flanks and abdomen; under tail-coverts and tarsal feathers uniform fulvous; under wing-coverts also uniform buff, the outer ones crossed with rufous bars, the lower ones greyish brown, with buffy white cross bars; a large patch of white on the outer edge of the bend of the wing; under surface of wings and tail lighter than the upper, the cross bars appearing whiter. Total length 17 inches, culmen 1·65, wing 10·5, tail 7·3, tarsus 1·25. (*Mus. Lugd.*)

Hab. Aru Islands.

18. **Ninox franseni.**

Noctua fransenii, *Schl. N. T. D.* iii. p. 256; *id. Revue Accipitr.* p. 31 (1873).
Athene franseni, *Gray, Hand-l. B.* i. p. 40.
Ninox franseni, *Sharpe, Ibis,* 1875, p. 258.

Adult female (type of species). Exactly similar to *N. aruensis,* but is very much larger. Total length 20·5 inches, culmen 1·8, wing 18·5, tail 9·2, tarsus 1·9. (*Mus. Lugd.*)

Hab. Waigiou.

19. **Ninox superciliaris.**

Strix superciliaris, *Vieill. N. Dict. d'Hist. Nat.* vii. p. 33.
Strix sonnerati, *Temm. Pl. Col.* i. pl. 21.
Noctua sonnerati, *Steph. Gen. Zool.* xiii. pt. 2, p. 64; *Less. Man. d'Orn.* i. p. 111.
Surnia sonnerati, *Bp. Oss. Regne Anim. Cuv.* p. 58.
Athene sonnerati, *Gray, Gen. B.* i. p. 35; *Blyth, J. A. S. B.* xiv. p. 185.

Athene superciliaris, *Pucher. Rev. et Mag. de Zool.* 1849, p. 19; *Bp. Consp.* i. p. 39; *Strickl. Orn. Syn.* p. 168; *Gray, Hand-l. B.* i. p. 40; *Gurney, Ibis,* 1869, p. 453.
Scops superciliaris, *Blyth, J. A. S. B.* xix. p. 509.
Ieraglaux superciliaris, *Kaup, Contr. Orn.* 1852, p. 107.
Noctua polleni, *Schl. N. T. D.* i. p. 81; *id. P. Z. S.* 1866, p. 421; *id. & Poll. F. M. Ois.* p. 49, pl. 17.
Noctua superciliaris, *Schl. Mus. P.-B. Revue Accipitr.* p. 25.
Ninox superciliaris, *Bp. Rev. et Mag. de Zool.* 1854, p. 543; *Sharpe, Ibis,* 1875, p. 258.

Adult male. Above brown, uniform in the centre of the back, the scapulars with small white bars on the outer web, the whole head and neck numerously spotted with minute dots of white; wing-coverts brown like the back, the median and greater wing-coverts with large spots of white, bigger and more distinct on the latter series; quills brown, externally washed with slightly rufous brown and spotted on the outer web with white like the wing-coverts, the inner web barred with white; upper tail-coverts uniform brown like the back; tail-feathers brown, with about four indistinct rows of light brown spots, represented by notches of brown on the inner web, the spots only distinct on the outer web of the outermost feather; sides of the face brown like the back, the lores also with blackish shaft-lines; the feathers just under the eye buffy white, as well as a tolerably well-defined eyebrow; cheeks pure white, the hinder feathers tipped with brown and merging into the brown of the neck; chin whitish; throat crossed with a bar of brown feathers; fore neck whitish; rest of under surface white, crossed with regular bands of rufous brown; leg-feathers fulvous; under tail-coverts white; under wing-coverts white, marked with brown near the bend and on the edge of the wing, the greater series ashy brown, whitish at base, thus resembling the inner lining of the wing, which is ashy brown, barred with whitish on the inner web. Total length 11·8 inches, wing 7·6, tail 4·7, tarsus 1·45.

Obs. Another specimen, also a male, is apparently a younger bird. It is much more uniform above, with scarcely any white markings on the scapulars and the white dots on the crown nearly absent altogether; the under surface of the body is very much whiter, the brown bars darker and much further apart; the leg-feathers bright ochraceous. Total length 12 inches, wing 7·5, tail 4·9, tarsus 1·5.

Hab. Madagascar.

a.	♂ ad. sk.	Morondava, Madagascar.	Hr. J. C. van Dam [C.].
b.	♂ juv. sk.	East Madagascar.	Hr. J. C. van Dam [C.].

20. **Ninox punctulata.**

Noctua punctulata, *Quoy et Gaim. Voy. de l'Astrol. Zool.* i. p. 165, pl. 1. fig. 1; *Schl. Mus. P.-B.* Striges, p. 29; *id. Revue Accipitr.* p. 29.
Athene punctulata, *Gray, Gen. B.* i. p. 35; *Bp. Consp.* i. p. 41; *Strickl. Orn. Syn.* p. 173; *Wall. Ibis,* 1868, p. 22; *Gray, Hand-l. B.* i. p. 41; *Walden, Tr. Z. S.* viii. p. 38.
Cephaloptynx punctulata, *Kaup, Tr. Z. S.* iv. p. 209; *id. Contr. Orn.* 1852, p. 105.

Spiloglaux punctulata, *Bp. Rev. et Mag. de Zool.* 1854, p. 544.
Ninox punctulata, *Sharpe, Ibis,* 1875, p. 259.

Adult female. Above deep chocolate-brown, thickly covered with minute spots of dull whitish buff, a little larger on the hind neck, but nowhere inclining to bars except on the secondaries, and here very minute; quills uniform with back, the secondaries marked as above noted, the primaries with four scarcely indicated bars of paler brown, appearing whitish on the under surface, which is rather lighter brown; lores brown; over the eye a superciliary circlet of white; ear-coverts brown like the back and spotted in the same manner; entire throat and cheeks pure white; rest of under surface white down the centre of the body and a little inclining on to the flanks; a narrow bar across the upper part of the chest and the sides of the body chocolate-brown, paler and more fulvous on the flanks, spotted on the chest and barred on the breast with whitish; under wing-coverts entirely white; tail chocolate-brown, with five faintly indicated bars of buffy white. Total length 10·5 inches, culmen 1·1, wing 6·8, tail 3, tarsus 1·3.

Adult male. Very little smaller than the female. Total length 10 inches, wing 6·6.

Hab. Celebes.

a. ♀ ad. sk.	Makassar.	A. R. Wallace, Esq. [C.].

21. **Ninox hypogramma.** (Plate X.)

Athene hypogramma, *Gray, P. Z. S.* 1860, p. 344; *Wall. Ibis,* 1868, p. 23; *Gray, Hand-l. B.* i. p. 41.
Noctua hypogramma, *Schl. Mus. P.-B. Revue Accipitr.* p. 25.
Ninox hypogramma, *Sharpe, Ibis,* 1875, p. 259.

Adult male. Above uniform chocolate-brown, the head duskier and more sepia-brown; scapulars distinctly barred with white on the outer web; upper wing-coverts dusky brown, the feathers margined with dull rufous; primary-coverts uniform sepia-brown; quills sepia-brown, slightly washed externally with light brown and obsoletely banded on the inner web with paler brown; tail sepia-brown, the feathers slightly washed externally with light brown and crossed with about ten bands of pale brown and tipped with whity brown; sides of face and sides of neck uniform dark brown with the head, the fore part of the cheeks washed with ashy white; lores white, with black hair-like shaft-lines; frontal plumes also whitish; chin fulvous; rest of under surface bright tawny rufous, barred on the breast with brown, some of the interspaces on the breast whitish, the abdomen barred with tawny and white; leg-feathers entirely tawny; under wing-coverts deep tawny with fulvous cross bars, the lower series ashy brown, barred with yellowish white, the inner surface of the wing ashy brown, faintly barred on the inner web with lighter brown; cere dull yellow; bill black, bluish at base of lower mandible; feet bright yellow. Total length 14·5 inches, wing 9·5, tail 6·2, tarsus 1·4.

Adult female. Similar in colouring to the male. Total length 12·5 inches, wing 9, tail 5·5, tarsus 1·4.

Obs. The above sexes are determined by Mr. Wallace, in whose collection the specimens remained until they passed into the possession of the Museum. Another bird, also obtained by Mr. Wallace in Gilolo, differs in having the white bars on the scapulars very distinct and in having white bars on the greater wing-coverts. The hind neck shows slight indications of fulvous bars, the quills externally rufescent, with indications of fulvous spots on the outer web. Underneath the bird is much lighter than the ones described, being white, barred with rufous. As far as I can judge, this bird is an old one and probably represents the fully adult stage. Another specimen from Batchian approaches it in having the under surface lighter and also a few indications of fulvous bars on the wing-coverts.

Hab. Batchian and Gilolo.

a, b. Ad. st.	Batchian.	A. R. Wallace, Esq. [C.].
c. ♂ ad. sk.	Batchian.	A. R. Wallace, Esq. [C.].
d. Ad. st.	Gilolo.	A. R. Wallace, Esq. [C.].
e. ♀ ad. sk.	Gilolo.	A. R. Wallace, Esq. [C.].
f. ♂ sen. sk.	Gilolo.	A. R. Wallace, Esq. [C.].

22. **Ninox squamipila.** (Plate XII. fig. 2.)

Athene squamipila, *Bp. Consp.* i. p. 41; *Wall. Ibis,* 1868, p. 23; *Gray, Hand-l. B.* i. p. 41.

Ninox squamipila, *Bp. Rev. et Mag. de Zool.* 1854, p. 544; *Sharpe, Ibis,* 1875, p. 259.

Noctua squamipila, *Schl. Mus. P.-B.* Striges, p. 28; *id. Revue Accipitr.* p. 22.

Adult (type of species). Upper surface deep rufous brown, more decidedly brown on the head and wing-coverts, the greater series of the latter and the scapulars externally washed with rich ochre and distinctly barred with white; primaries brown, externally rufous barred with darker brown, the lower surface of the wing rather more ashy brown, the inner secondaries with bars of fulvous on the inner web; lower back, rump, and upper tail-coverts rufous, barred with dark brown and fulvous; tail dull rufous, more fulvous on the inner web and having about nine cross bars of darker brown; sides of face dark brown; forehead, fore part of cheeks, and eyebrow white, the latter washed with ochre; throat whitish; sides of neck and chest clear tawny rufous, with rather indistinct bars of rufous brown, the breast paler and more fulvous, the bars darker and more distinct; the belly white, with distinct bars of dark brown; under tail-coverts more indistinctly barred; thighs uniform rufous; under wing-coverts rufous, with a bar of rufous brown across the middle. Total length 13·5 inches, culmen 1·25, wing 8·25, tail 5·3, tarsus 1·25. (*Mus. Lugd.*)

Hab. Ceram, Mysol.

a. Ad. st.	Ceram.	A. R. Wallace, Esq. [C.].
b. ♂ ad. sk.	N. Ceram.	A. R. Wallace, Esq. [C.].

23. **Ninox variegata.**

Noctua variegata, *Quoy & Gaim. Voy. de l'Astr. Zool.* i. p. 166, pl. 1. fig. 2.
Athene variegata, *Gray, Gen. B.* i. p. 35; *Bp. Consp.* i. p. 41; *Strickl. Orn. Syn.* p. 173; *Wall. Ibis,* 1868, p. 24; *Gray, Hand-l. B.* i. p. 41.
Cephaloglaux variegatus, *Kaup, Contr. Orn.* 1852, p. 108; *id. Tr. Z. S.* iv. p. 216.
Rhabdoglaux variegata, *Bp. Rev. et Mag. de Zool.* 1854, p. 544.
Ninox variegata, *Sharpe, Ibis,* 1875, pp. 257, 259.

Adult (type of species). Upper surface rufous brown, with narrow concealed white bars on all the dorsal feathers, more distinct on the scapulars, where they are broader on the outer web; head rather browner than the back, washed with rufous on the sides, and slightly spotted with white on the forehead; the hind neck more rufous than the head, and approaching the colour of the back, much varied with fulvous and white spots and bars, somewhat concealed on the centre, but very plain on the sides of the neck; wing-coverts browner than the back, the least series nearly uniform, with very few spots of fulvous, the greater and median series barred with rufous buff, inclining to white bars or spots on the outer web, and producing a spotted appearance; primary-coverts dark brown, with dull fulvescent cross bars; quills brown, barred with lighter brown, shading into a white spot on the outer web, and producing a spotted appearance; upper tail-coverts rufous brown like the back, with the same half-concealed bars and spots of white; tail brown, crossed with ten bands of pale fulvous brown; lores and frontal feathers whitish, the former with distinct black shaft-lines, the latter produced a little backwards, and forming a small eyebrow; ear-coverts dark brown like the crown, slightly washed with rufous; cheeks ochraceous, the fore part whitish, the chin also whitish; fore neck ochraceous buff, streaked with brown; rest of under surface of body white, barred across with orange-rufous in about equal proportions on the breast, the abdomen, however, much whiter, the orange bars being narrower; leg-feathers pale fulvous; under tail-coverts white, with a few streaks and bars of orange-brown; under wing-coverts orange, barred with brown, but the greater series ashy brown, with buffy white bars, more resembling the inner lining of the quills, these being ashy brown, barred with pale ochraceous brown, broader on the inner web; bill yellow; toes yellow, claws horn-brown. Total length 10 inches, wing 7·9, tail 4·6, tarsus 1·3. (*Mus. Paris.*)

Hab. New Ireland.

24. **Ninox hantu.** (Plate XI. fig. 1.)

Athene hantu, *Wall. P. Z. S.* 1863, p. 22; *id. Ibis,* 1868, p. 23; *Gray, Hand-l. B.* i. p. 41.
Ninox hantu, *Sharpe, Ibis,* 1875, p. 259.

Adult (type of species). Above dull chestnut-rufous, the head darker and decidedly browner, the hind neck with a few indistinct fulvous bars; the scapulars barred with buff on the outer web; wing-coverts coloured like the back, but a little darker, only a few with obsolete tiny markings of fulvous on the outer webs; primary coverts uniform dark brown; quills brown, externally washed with the same rufous colour as the back; upper tail-coverts like the back, with a few indistinct bars of fulvous; tail dark brown, inclining to rufous brown towards the tip, and crossed with ten or eleven bars of lighter or rufous brown; frontal plumes and lores whitish, the latter with blackish shaft-stripes; sides of face brown, rather darker than the head, the cheeks and chin buffy white; rest of under surface of body orange-tawny, with indistinct bars of rufous brown, the interspaces on the lower breast inclining to whitish, the cross bars diminishing in clearness on the abdomen, and nearly absent altogether on the under tail-coverts; sides of the upper breast inclining to chocolate-brown; under wing-coverts orange-tawny, the greater series ashy brown, with two or three fulvous bars near the base; inner lining of wing slightly inclining to rufous on the outer webs of the quills, the base of the inner webs of the secondaries barred with fulvous; bill whitish horn-colour; feet white, claws pale horn-colour; iris yellow. Total length 13 inches, wing 8·3, tail 5·8, tarsus 1·3.

Obs. Another specimen obtained by Mr. Wallace is very similar to the type described, but is more distinctly barred with fulvous on the shoulders and inner secondaries, the upper wing-coverts also distinctly spotted with fulvous.

Hab. Bouru.

a. ♂ ad. sk.	Bouru.	A. R. Wallace, Esq. [C.].	Type of species.
b. Ad. st.	Bouru.	A. R. Wallace, Esq. [C.].	

25. Ninox tæniata.

Athene tæniata, *Hombr. & Jacq. Voy. Pôle Sud,* iii. p. 50, pl. 3; *Strickl. Orn. Syn.* p. 173.
Athene jacquinoti, *Bp. Consp.* i. p. 42; *Wall. Ibis,* 1868, p. 24; *Gray, Hand-l. B.* i. p. 41.
Cephaloglaux jacquinoti, *Kaup, Contr. Orn.* 1852, p. 108; *id. Tr. Z. S.* iv. p. 216.
Rhabdoglaux jacquinoti, *Bp. Rev. et Mag. de Zool.* 1854, p. 544.
Ninox tæniata, *Sharpe, Ibis,* 1875, p. 259.

Adult (type of species). General colour above dark chocolate-brown, barred over the whole upper surface with dull fulvous, or rather pale rufescent brown, the head a little darker and more thickly barred with the lighter bands, especially on the sides of the neck, the bars on the scapulars a little paler, but not perceptibly larger than on the rest of the back; wing-coverts dark brown, sparsely spotted with fulvous, the primary-coverts uniform brown; quills brown, the primaries almost uniform, with only a few indications of fulvescent spots on the outer web, these spots being much more numerous on the secondaries, which are also barred with yellowish on the inner web; upper tail-coverts coloured exactly like the back, with very faintly indicated bars, of which about four can be distinguished, the bars rather more distinct on the inner web; loral and frontal plumes whitish, as well as a very narrow eyebrow, the former very long and stiff with blackish shaft-lines; ear-coverts brown in the centre, rufescent on both webs, inclining to whitish on the upper part of the ear-coverts; cheeks and entire throat pure white; rest of under surface whitish, strongly washed with fulvous, with slight indications of dusky bars near the top of the inner face of the thighs, the colour of the sides of the neck descending somewhat on to the sides of the breast, the feathers brown, barred with pale rufous like the back, which they resemble; under wing-coverts yellowish, the edge of the wing brown, the lower series yellowish at base, brown at tip, thus resembling the inner lining of the wing, which is brown, with yellow bars on the inner webs, the bases of the inner webs entirely yellowish; bill yellow, dusky at the base of upper mandible; feet plumed; toes apparently yellow in life, claws black. Total length 9·5 inches, wing 7·65, tail 4·4, tarsus 1·7. (*Mus. Paris.*)

Hab. Solomon Islands.

12. SCELOGLAUX.

	Type.
Sceloglaux, *Kaup, Isis*, 1848, p. 768	S. albifacies.

Range. Confined to New Zealand.

1. Sceloglaux albifacies.

Athene albifacies, *Gray, Voy. Erebus and Terror, Birds*, p. 2; *id. Gen. B.* i. p. 35; *Bp. Consp.* i. p. 42; *Strickl. Orn. Syn.* p. 171; *Gray, Hand-l. B.* i. p. 41; *Hutton, Cat. B. N. Zeal.* p. 2; *Finsch, J. f. O.* 1872, p. 95, et 1874, p. 177.

Sceloglaux albifacies, *Kaup, Isis*, 1848, p. 768, et *Contr. Orn.* 1852, p. 109, et *Tr. Z. S.* iv. p. 219; *Bp. Rev. et Mag. de Zool.* 1854, p. 544; *Gould, B. Austr. Suppl.* part ii.; *Buller, B. N. Z.* p. 21, pl. 2. fig. 2; *Sharpe, Voy. Erebus and Terror*, 2nd ed. *Birds*, p. 23, pl. 1; *Rowley, Orn. Miscell.* part ii.

Strix haasti, *Buller, Essay N. Z. Orn.* p. 7.

Athene ejulans, *Potts, Tr. N. Z. Inst.* iii. p. 63.

Adult. General colour ochraceous, the whole head and neck with broad centres of dark brown, producing a streaked appearance, the feathers of the back with fulvous margins instead of ochraceous; scapulars dark brown, with large ovate spots of fulvous or white on the outer web, and with ochraceous bars across the middle of the feather; wing-coverts dark brown, washed with rufous ochre and marked with fulvous spots on both webs, those on the outer web of the greater series larger and more ovate; primary-coverts uniform dark brown; quills brown, barred across with ochraceous, and slightly washed with the same on the outer web, the light cross bars paling into whitish notches on the outer webs of the primaries, which are also tipped with whitish; lower back and rump rather more rufous ochre on their margins than the rest of the upper surface, the centres dark brown, some of the upper tail-coverts margined with fulvous;

tail dark brown, rather broadly tipped with whitish, and crossed with six other bands of buffy white, of which two are concealed by the upper tail-coverts; eyebrow and sides of face dull whitish, all the feathers mesially streaked with dark brown, the ear-coverts washed with fulvous; cheeks whitish, extending backwards in a large patch behind the ear-coverts, all the feathers with concealed bases and narrow shaft-stripes of brown; throat dusky brown; rest of under surface deep ochraceous, with broad dark brown centres to the feathers, much broader and more rufous on the flanks; under tail-feathers uniform ochraceous; leg-feathers dull whitish; under wing-coverts deep rufous ochre, with brown centres to the feathers as on the breast, the greater series ashy brown, barred with white on the inner web, exactly in the same manner as the inner lining of the wing, which is ashy brown with lighter brown bars on the inner web, inclining to white towards the base of the quills; bill black, horn-colour towards the tip; toes fleshy brown, covered with coarse yellow hairs; claws black; iris dark reddish brown. Total length 18·5 inches, wing 10·4, tail 6·5, tarsus 2·65.

Hab. New Zealand.

a. Juv. st.	New Zealand.	Percy Earl, Esq. [C.].
b. Ad. st.	New Zealand.	E. Wilson, Esq. [P.].
c. Ad. st.	New Zealand.	Capt. Stokes [P.].

13. GLAUCIDIUM.

	Type.
Glaucidium, *Boie, Isis*, 1826, p. 976	G. passerinum.
Nyctipetes, *Swains. Classif. B.* ii. p. 218	G. perlatum.
Tænioglaux, *Kaup, Isis*, 1848, p. 769	G. radiatum.
Tænioptynx, *Kaup, l. c.* p. 768	G. brodiei.
Microglaux, *Kaup, l. c.* p. 769	G. perlatum.
Microptynx, *Kaup, Contr. Orn.* 1852, p. 106	G. passerinum.
Smithiglaux, *Bp. Rev. et Mag. de Zool.* 1854, p. 544	G. capense.
Phalænopsis, *Bp. l. c.* p. 544	G. nanum.

Wing of *Glaucidium passerinum.*

Range. The greater part of the Old World, including the Palæarctic, Ethiopian, and Indian regions; but not found beyond the Indo-Malayan subregion, and absent in the Austro-Malayan subregion and the Australian region proper. In the New World distributed over the entire Neotropical region, and extending into the Southern States of North America.

Key to the Species.

a. Head spotted or streaked, not regularly barred across, sometimes nearly uniform.
- *a'*. Crown of head spotted with white or fulvous white, generally in the form of circular dots; round the hind neck a less pronounced rufescent collar.
  - *a''*. Interscapulary region barred or spotted with fulvous.
    - *a'''*. Tail regularly barred across, the transverse bars continuous.
      - *a*⁴. Tail blackish, crossed with ten or eleven bars of white; toes scantily haired .. .. *nanum*, p. 190.
      - *b*⁴. Tail brown, with six or seven bars of white.
        - *a*⁵. Toes thickly feathered; the head dotted all over with rounded white spots; breast with longitudinal dark brown marks; sides of face white, barred with dark brown .. *passerinum*, p. 191.
        - *b*⁵. Toes scantily haired; the head sparsely dotted with tiny longitudinal streaks of buff; breast with heart-shaped rufous spots; sides of face dull whitish, barred with brown, and narrow fulvous streaks on the ear-coverts .... *siju*, p. 193.
    - *b'''*. Tail with white spots or bars, not continuous across the feather.
      - *c*⁴. Tail with seven or eight bands; back and scapulars thickly varied with zigzag fulvous spots.............. *gnoma*, p. 194.
      - *d*⁴. Tail-bands not exceeding six.
        - *c*⁵. Tail deep black, darker than the wings; general aspect of upper surface blackish brown, with ochraceous bars and spots .... *jardinii*, p. 207.
        - *d*⁵. Tail pale brown; upper surface light brown, with white spots........ *perlatum*, p. 209.
  - *b''*. Interscapulary region uniform.
    - *c'''*. Head brown like the back; spots on tail four or five .... *pumilum*, p. 198.
    - *d'''*. Head greyish, contrasting with the back; spots on tail five or six ...... *griseiceps*, p. 196.
    - *e'''*. Head and back rufous, with a few lighter rufous spots; spots on tail seven .. . *cobanense*, p. 199.
- *b'*. Crown of head streaked with whitish or fulvous, with a few spots occasionally intermingled............................. *ferox* and allies, pp. 200–205.
- *c'*. Crown of head uniform grey like the back; tail with five white spots on inner web of middle feather; sides of body rufous, with browner streaks on the abdomen .... *tephronotum*, p. 211.

b. Head regularly barred across.

d'. Hind neck fulvescent, with two lateral black patches forming a collar .... { *brodiei*, p. 212. *pardalotum*, p. 214. *sylvaticum*, p. 215.

e'. Hind neck regularly barred like the head and the rest of the upper surface.

f'''. Back rufous, with very slight indications of dark cross bars; no longitudinal white shoulder-streak along the scapulars or white bar on the wing-coverts.

*e*⁴. Belly white, narrowly streaked with black .... *castanonotum*, p. 215.

*f*⁴. Belly white, broadly streaked with chestnut .... *castanopterum*, p. 216.

g'''. Back greyish or fulvous, barred across with black.

*g*⁴. Size small: bands on tail nine or ten { *radiatum*, p. 217. *malabaricum*, p. 218.

*h*⁴. Size larger: bands on tail not more than seven .... { *cuculoides*, p. 219. *whitelyi*, p. 222.

*i*⁴. Size larger: bands on tail fourteen .. *capense*, p. 223.

1. Glaucidium nanum.

Strix nana, *King, Zool. Journ.* iii. p. 427.
Glaucidium nanum, *Boie, Isis,* 1826, p. 970; *Kaup, Contr. Orn.* 1852, p. 103, et *Tr. Z. S.* iv. p. 202; *Hartl. Naum.* 1855, p. 209; *Burm. Th. Bras.* ii. p. 146; *id. Reis. La Plata St.* ii. p. 441; *Scl. P. Z. S.* 1867, p. 338; *Scl. & Salv. Ibis,* 1868, p. 188; *Hudson, P. Z. S.* 1872, p. 549; *Scl. & Salv. Nomencl. Av. Neotr.* p. 117; *Ridgw. Pr. Bost. Soc. N. H.* xvi. p. 104; *Sharpe, Ibis,* 1875, pp. 41, 259.
Athene ferox, *Fraser, P. Z. S.* 1843, p. 109 (*nec Vieill.*)
Athene nana, *Gray & Mitch. Gen. B.* i. p. 35, pl. 12; *Bp. Consp.* i. p. 37; *Hombr. & Jacq. Voy. Pôle Sud,* iii. p. 54; *Strickl. Orn. Syn.* p. 173; *Gray, Hand-l. B.* i. p. 42.
Noctua nana, *Gay, Hist. Chile,* i. p. 246; *Schl. Mus. P.-B.* Striges, p. 31.
Athene leucolaima, *Bp. Consp.* i. p. 40 (ex *Hombr. & Jacq. Voy. Atl.* pl. 4. figs. 2, 3).
Strix chilensis, *Licht. MS. in Mus. Berol. et Nomencl. Av.* p. 8 (*descr. nullâ*).
Phalænopsis nana, *Bp. C. R.* xli. p. 654 (1855).
Phalænopsis leucolaima, *id. t. c.* p. 654.

Adult male. Above dull brown with somewhat of a reddish tint, the back mottled with concealed spots and bars of ochraceous buff; the scapulars shaded with ashy grey, mottled with ochraceous buff spots, the outermost white on the external web, forming a conspicuous longitudinal patch: head rather more rufous than the back and longitudinally streaked with ochraceous buff shaft-stripes; round the hind neck a whitish collar, relieved by black markings, developing into two large patches, one on each side of the hind neck; wing-coverts rather darker than the back, and very sparingly spotted with ochraceous buff, the greater series with large white oval spots bordered with blackish near the tip; quills dark brown banded with rufous, the primaries and inner secondaries chequered with spots of white; upper tail-coverts rufous, crossed with a few obsolete bands of brown; tail dark brown, tipped with fulvous white and crossed with ten bars of rufous; lores and feathers round the eye whitish, the former obscured by blackish bristles; sides of face and ear-coverts dull brown, streaked with buffy white; sides of neck rufous like the head, but mottled with white bases to the feathers; cheeks and chin pure white, the latter succeeded by a band of dark brown across the throat and followed by a large white fore neck; rest of under surface white, streaked with dark brown, the sides of the breast rufous brown, mottled with whitish spots; sides of body ashy brown; thighs ochraceous, varied with small rufous brown bars; tarsal plumes white, spotted with brown; under wing-coverts yellowish white slightly streaked with brown, the greater series yellowish white at base, dark brown at tips, thus resembling the inner lining of the wing, which is ashy brown, barred with white and inclining to yellowish white near the base of the inner web; iris orange. Total length 8 inches, wing 3·85, tail 2·9, tarsus 0·9.

Adult female. Rather larger than the male, shaded with ashy grey on the back and scapulars, this being probably only the sign of the adult bird and existing in both sexes; bars on tail eleven. Total length 8·6 inches, wing 4·4, tail 3·2, tarsus 0·95.

Obs. The full-plumaged birds are much greyer on the back and have the head minutely *spotted* with ochraceous instead of being streaked.

Young. Differs from the adult in being more uniform above and exhibiting less of the streaks on the head; the upper surface rather more grey, and the spots and bars on the hind neck ochre; tail deep rufous, with nine dark brown cross bands, the subterminal one broader. Below, the breast is browner with very few spots, and the streaks on the flanks very broad and merging together.

Hab. Chili and Patagonia.

a. Juv. st.	Port Famine.	Capt. King [C.]. Type of species.
b. Ad. st.	Straits of Magellan.	The Admiralty [P.].
c. ♀ ad. sk.	Province of Colchagua, July 1870.	E. C. Reed, Esq. [C.].
d. ♂ ad. sk.	Rio Negro, Patagonia, May 1871.	W. H. Hudson, Esq. [C.].
e. Ad. sk.	South America.	
f. Juv. st.	South America.	Specimen figured in Gray and Mitchell's 'Genera' (*l. c.*).

2. Glaucidium passerinum.

Strix passerina, *Linn. S. N.* i. p. 133; *Nils. Orn. Suec.* i. p. 69, pl. 3; *Schl. u. Susem. Vög. Eur.* Taf. 43. fig. 2; *Sundev. Sv. Fogl.* pl. xxv. fig. 6.
La Chevêche, *Buff. Pl. Enl.* 439.
Strix tengmalmi, var., *Lath. Ind. Orn. Suppl.* p. 16.
Strix pusilla, *Daud. Traité*, ii. p. 205; *Bree, B. Eur.* i. p. 124, iv. p. 201.

Strix pygmæa, *Bechst. Naturg. Deutschl.* iv. p. 978, t. xxiv.; *Wolf & Meyer, V. D.* ii. p. 54, pl.
Strix acadica (*nec Gm.*), *Temm. Man. d'Orn.* i. p. 96; *Naum. Vog. Deutschl.* i. p. 434, pl. 43. figs. 1, 2; *Werner, Atlas,* Rapaces, pl. 41.
Glaucidium passerinum, *Boie, Isis,* 1826, p. 976; *Bp. Consp.* i. p. 36; *Strickl. Orn. Syn.* p. 158; *Fritsch, Vög. Eur.* tab. 13. fig. 6; *Salvad. Faun. Ital. Ucc.* p. 31; *Sharpe & Dresser, B. Eur.* pt. ii.; *Sharpe, Ibis,* 1875, pp. 38, 259.
Noctua pygmæa, *Steph. Gen. Zool.* xiii. part 2, p. 63.
Noctua acadica, *Cuvier, Règne Anim.* 1829, p. 345.
Surnia passerina, *Bp. Oss. R. A. Cuv.* p. 51; *Keys. & Blas. Wirb. Eur.* p. xxxii; *Degl. & Gerbe, Orn. Eur.* i. p. 120; *Borggr. Vogelf. Norddeutschl.* p. 64.
Noctua passerina, *Gould, B. Eur.* i. pl. 50; *Schl. Mus. P.-B.* Striges, p. 41.
Athene passerina, *Gray, Gen. B.* i. p. 35; *id. Hand-l. B.* i. p. 42.
Micropteryx passerinum, *Kaup, Contr. Orn.* 1852, p. 107; *id. Tr. Z. S.* iv. p. 213.
Glaucidium microrhynchum, *Brehm, Naum.* 1855, p. 273.
Glaucidium pygmæum, *id. l. c.* p. 273.

Adult female. Above umber-brown, with a distinct ashy grey shade on the head, back, and scapulars; the whole of the upper parts varied with spots of yellowish white, inclining to small bars on the back, all the spots somewhat relieved by a continuous dark brown margin, but many of them concealed; occiput more plainly varied with fulvous white spots; the hind neck and wig strongly mottled with dull ochraceous spots, forming an indistinct collar; wing-coverts coloured like the back, washed like the scapulars with rufous brown, the fulvous white spots rather larger and the median coverts pure white, forming a large longitudinal patch of white on the upper part of the wing; quills dark brown, strongly washed with rufous brown from the base upwards, as also are the concealed greater coverts, all barred with lighter brown, inclining to white on outer web and producing a spotted appearance; the secondaries more ashy grey like the back, barred with lighter brown, the white spots larger and forming small bars on the outer web, thus more resembling the back; rump uniform with back, the upper tail-coverts washed with rufous brown and varied with concealed bars of white; tail rufous brown, tipped and crossed with six narrow bars of white, on each margin of which the colour of the tail is a deeper brown; lores and a distinct eyebrow whitish, the former varied with black hair-like bristles; ear-coverts white, crossed with dark brown bars, rather minute in character; cheeks and chin white; throat mottled with brown, which extends on to the sides of the neck, where it is somewhat washed with rufous; below the brown throat-patch the fore neck is conspicuously white, forming a sort of band; rest of under surface of body white, tinged with yellowish on abdomen and vent, the sides of the breast and flanks distinctly streaked with sepia-brown, the breast clouded and mottled with rufous brown, the centre only streaked and the sides of the upper breast entirely rufous brown, mottled with fulvous white spots; legs whitish, the upper parts washed with yellowish brown, the tarsal plumes slightly marked with brown; under tail-coverts white, washed with brown at their tips; under wing-coverts yellowish white, those on the edge of the wing pure white, the centre ones longitudinally streaked with dark brown, the greater series whitish at base, ashy brown at tip, thus resembling the inner lining of the wing, which is ashy brown barred with white; bill yellow; feet brownish yellow, talons blackish brown; iris bright yellow. Total length 6·8 inches, wing 4·15, tail 2·8, tarsus 0·65.

Adult male. Differs from the female only in being smaller. The statement made by some writers that the male has more of an ashy grey shade above than the female does not seem to me to hold good; for the bird before me is quite as grey as any old male. The bars on the tail are seven in number. Total length 6·2 inches, wing 3·85, tail 2·6, tarsus 0·6.

Young male. Much browner than the adults and characterized by an absence of spots, the latter being apparently gradually assumed; the white wing-patch represented only by a few oval spots, and the bands on the quills and tail very narrow and dusky white; underneath the stripes on the flanks are very much broader.

Hab. The whole of Northern and Central Europe.

a. ♂ juv. st.	Germany.	
b. ♂ ad. st.	Carniola.	Purchased.
c. ♀ juv. st.	Carniola.	Purchased.
d. ♂ ad. sk.	Gardsjo, Sweden, March 15, 1863 (*Wheelwright*).	J. Gould, Esq. [P.].
e. ♀ ad. sk.	Stockholm, Sweden, Dec. 10, 1873.	Purchased.
f. ♂ ad. sk.	Falun, Sweden, Jan. 1, 1873.	Purchased.
g. Ad. sk.	Southern Ural (*Dr. Strader*).	R. B. Sharpe, Esq. [P.].

3. Glaucidium siju.

Noctua siju, *D'Orb. in Ramon de la Sagra H. N. Cuba*, Ois. p. 33, pl. iii.
Athene siju, *Gray, Gen. B.* i. p. 35; *Cass. Cat. Strigidæ Philad. Mus.* p. 13; *Gray, Hand-l. B.* i. p. 42.
Nyctale siju, *Bp. Consp.* i. 54; *Strickl. Orn. Syn.* p. 177.
Strix havanensis, *Licht. Mus. Berol., undè*
Glaucidium havanense, *Kaup, Contr. Orn.* 1852, p. 103.
Glaucidium siju, *Cab. J. f. O.* 1855, p. 59; *Ridgw. Pr. Bost. Soc. N. H.* 1873, p. 65; *Scl. & Salv. Nomencl. Av.* p. 117; *Sharpe, Ibis*, 1875, p. 59.

Adult. Above dark brown, varied with small streaks and broader bars of deep ochre, especially distinct on the scapulars, where the bars incline to whitish on their outer margin; wing-coverts dark brown, rather deeper than the back, the least ones with small ochraceous spots, the median and greater series with large ovate spots of white; quills dark brown, barred with lighter brown, on the outer part of which bars are several spots of buffy white, the quills for the most part notched with white on the inner web, more distinctly on the secondaries; upper tail-coverts rather more ashy than the back, streaked with white longitudinally; tail brown, crossed with seven (including the one at the tip) bands of whitish more or less tinged with rufous; head lighter and more earthy brown than the back, with longitudinal fulvous stripes, somewhat dilated towards the tips of the feathers; round the hind neck a collar of orange-rufous; lores and feathers above the eye white, the former obscured by blackish bristles; ear-coverts dull white, with brown cross bars; cheeks, chin, and fore neck whitish, the latter separated by a band of rufous feathers across the lower throat; rest of under surface white, varied with several longitudinal rows of dull rufous brown spots, either oval or heart-shaped, developing into streaks on the abdomen; the sides of the breast rufescent, varied with dark brown bars; under wing-coverts whitish, spotted with brown, the greater series yellowish white at base, brown at tips, thus resembling the inner lining of the wings, which are ashy brown, notched with yellowish white on the inner web. Total length 5·8 inches, wing 3·55, tail 2·4, tarsus 0·8.

Hab. Cuba.

a. Ad. st. Cuba.

4. Glaucidium gnoma.

Glaucidium gnoma, *Wagler, Isis,* 1832, p. 275; *Cass. B. N. Am.* p. 62; *Cab. J. f. O.* 1869, p. 207; *Sharpe, Ibis,* 1875, pp. 38, 259, pl. i. figs. 1, 2.
Strix passerinoides, *Audubon, Orn. Biogr.* v. p. 271.
Athene gnoma, *Gray, Gen. B.* p. 35; *id. Hand-l.* i. p. 42.
Glaucidium infuscatum, *Cass. B. Calif. & Texas,* p. 189.
Glaucidium californicum, *Sclater, P. Z. S.* 1857, pp. 4, 126.
Glaucidium passerinum, var. californicum, *Ridgw. Pr. Bost. Soc. N. H.* xvi. p. 95; *id. N.-Am. B.* iii. p. 81.

Young. Above clear chocolate-brown, perfectly uniform on the interscapulary region, the scapulars distinctly marked with large rounded or oval spots of rufous ochre on their outer web; head grey, slightly tinged with brown and marked on the fore part of the crown with a few narrow ochraceous streaks; the hind neck mottled with white and forming a sort of collar, relieved by two blackish patches on each side of the neck and followed by a second indistinct rufous collar below the white one; wing-coverts rufous chocolate, varied with half-concealed small bars of rufous, the greater coverts deep brown with oval spots of rufous white; primary-coverts uniform deep brown, quills deep brown, sparsely spotted with rufous on the outer web, the secondaries browner like the back and externally margined and spotted with light rufous; upper tail-coverts rufous chocolate, indistinctly barred with light rufous; tail blackish brown, crossed with eight rows of white spots (counting the terminal one), these spots not continuous, but obscured by ochraceous and not meeting in the centre of the feather; lores and feathers round the eye white; ear-coverts greyish, washed with rufous and streaked with white; chin and cheeks white, extending in a distinct patch under the ear-coverts; throat pure white, separated from the chin by a band of brown feathers mottled with rufous; breast rufous, almost uniform, excepting a division of white feathers down the centre which are streaked with black; rest of under surface white, thickly streaked with black, the sides of the body rufous; leg-feathers ochraceous brown, inclining to fulvous on the inner face; under wing-coverts white, the innermost yellowish, with a patch of black near the outer margin, the greater series yellowish at base, blackish at tips, thus resembling the inner lining of the quills, which are greyish brown, plainly spotted with yellowish on the inner web, more largely near the base.

Obs. From the uniform grey-headed plumage above described the present species moults into its second stage, as is evidenced by a specimen in Mr. H. S. Le Strange's collection. The crown in this bird is becoming dark brown like the back, the shafts indicated by rufous lines widening out in an oval form and crossed with other bars of rufous. This specimen also shows some mottlings on the back, thus approaching the Vancouver-Island specimen next to be described, and which I take to be in the second plumage.

Adult male (in second plumage). Above brown with somewhat of a rufous olive tinge, the back varied with half-concealed arrow-shaped spots of rufous ochre, larger on the scapulars and more oval in shape, inclining to white on the outer margin; the hind neck also strongly mottled with rufous ochre, forming a kind of irregular collar, above which is a more clearly defined one of white, slightly varied with blackish and relieved by two patches of black on either side of the hind neck; head rather darker brown than the back, covered with minute spots of rufous ochre, larger and rather more elongated on the forehead; wing-coverts darker brown than the back and spotted in the same way with rufous ochre, taking the form of notches or bands on the greater series, some of the outermost of which are tipped with white, which forms a kind of longitudinal band down the wing; quills deep sepia-brown, indistinctly banded with blackish, externally spotted with rufous ochre, the secondaries browner, spotted and indistinctly barred across with ochraceous buff; upper tail-coverts brown, longitudinally streaked near the base with white and spotted or barred near the tip with ochraceous white; tail blackish with lighter brown bars inclining to white spots on either extremity, but larger always on the inner web, the tail thus appearing to have five bands without counting the terminal white spot on the inner web of each feather; lores and an indistinct eye-brow dusky white, the former obscured by blackish bristles; sides of face dark brown, minutely spotted with ochraceous; chin and cheeks white, extending in a broad patch below the ear-coverts, crossed by a distinct band of brown, mottled with ochre and rufous and extending on to the sides of the neck; fore neck conspicuously white; rest of under surface pure white, narrowly but very distinctly streaked with black, disappearing on the under tail-coverts; the sides of the upper breast mottled with ochraceous spots, the flanks also varied with dark brown; leg-feathers white, slightly streaked with dark brown; under wing-coverts white, streaked near outer margin with blackish, the greater series whitish at base, ashy brown at tip, thus resembling the inner lining of the wing, which is ashy brown barred with white on the inner web. Total length 6·5 inches, wing 3·5, tail 2·75, tarsus 0·7.

Adult female. Above ashy brown, the whole back varied with more or less concealed whitish specks or bars, often tinged with rufous on the back and scapulars, the outermost of the latter having larger spots of dull white on the outer web; head and nape ashy brown, dotted with dull white, some of the spots washed with rufous and inclining to small bars in some instances; round the hind neck the feathers marked with white, forming a kind of collar with the usual black patches on each side, succeeded by a second collar of rufous ochre, with which colour the feathers are marked in the form of bands; wing-coverts a little browner than the back, with small transverse bars of pale rufous, the tips of the outermost median and greater wing-coverts relieved by an ovate spot of white; quills dark brown, with narrow whitish tips to the inner primaries and the secondaries, the quills indistinctly barred with greyish brown, changing to a white spot on the outer margin of the primaries and producing a chequered appearance, not so distinct on the secondaries, which are browner and have irregular transverse bars of white or ochraceous white; tail deep brown, crossed with seven rows of white spots, larger and more like notches on the concealed inner web; lores and feathers above the eye pure white, meeting over the base of the forehead, the former ending in black bristles; ear-coverts ashy brown, varied with small bars of rufous and linear streaks of white; chin and cheeks white, extending in a conspicuous patch under the ear-coverts; lower throat pure white, separated from the chin by a band of dark brown feathers somewhat varied with rufous; rest of under surface white, distinctly streaked with black; the sides of the body ashy brown, spotted with white, these spots more distinct on the sides of the breast, which are washed with rufous; thighs white, mottled with brown on the outer webs; under wing-coverts white, especially the bend of the wing, the inner ones and axillaries yellowish, with a few blackish marks a little way from the edge of the wing, the greater series yellowish at base, ashy brown at tip, thus resembling the inner lining of the quills, which are plainly barred with white on the inner web, less plainly and more spot-like on the outer, the secondaries barred on both webs with pale rufous. Total length 6 inches, wing 3·5, tail 2·7, tarsus 0·65. (*Mus.* H. S. Le Strange.)

Hab. From Vancouver's Island and Oregon along the western coast of America; Mexico southwards to Guatemala.

a. ♂ ad. st.	Vancouver's Island.	J. K. Lord, Esq. [C.].
b. Juv. sk.	W. Mexico.	Purchased.
c. Ad. sk.	Mexico.	M. A. Bouvier [C.].
d. Ad. st.	[Central America.]	J. Gould, Esq.

5. Glaucidium griseiceps.

Glaucidium griseiceps, *Sharpe, Ibis*, 1875, pp. 41, 259, pl. ii. fig. 2.
Glaucidium pumilum *Auct. ex Amer. centr.*

Adult. Head greyish, with minute white dots, each dot relieved by a black terminal line; round the hind neck an ill-defined collar formed of white spots mottled with black markings, the usual black patches on each side of the hind neck well developed; back clear umber-brown, uniform with the exception of a few whitish spots on the outer web of the scapulars; the upper tail-coverts with a few concealed ochraceous spots, often absent; wing-coverts like the back, except the outermost which are blackish, on the least ones a few spots of rufous ochre, larger and whiter on the greater series; primary-coverts blackish, quills blackish, with indications of obsolete brown bars, which show a small ochraceous spot on the external web of the primaries, larger and more distinct on the secondaries, which are browner and resemble the back; tail blackish, crossed with five rows of white spots, inclining to bars or notches on the inner web, and nearly obsolete on the tips of the feathers; frontal plumes, lores, and a narrow eyebrow white, the lores obscured by black bristles; ear-coverts obscurely barred with white and dusky black; sides of crown and sides of neck very thickly dotted with white; cheeks white, extending backwards in a patch below the ear-coverts; chin and fore neck white, separated by a faintly indicated band of dull brown, washed with rufous; rest of under surface white, with broad longitudinal streaks of rufous brown, inclining to dark brown on the abdomen; the sides of the breast entirely rufous brown, mottled with large spots of white; leg-feathers dusky, slightly mottled with white, especially on the inner face of the thighs; under wing-coverts yellowish, with a few black markings near the outer edge of the wing, the greater series yellowish at base, blackish at tip, thus resembling the inner lining of the wing, all the inner webs being notched with yellowish and being entirely yellow near the base of the quills. Total length 5·6 inches, wing 3·55, tail 2·4, tarsus 6·7.

Young. Head deep grey, with a few whitish streaks on the forehead; hind neck mottled with white, forming an indistinct collar with the usual black side-patches, the feathers below this collar slightly inclining to rufous, but not enough to form a second collar; entire back uniform chocolate-brown, the scapulars with a few scattered spots of rufous; wing-coverts like the back, the greater series with a few rufous spots on the outer web; primary-coverts uniformly blackish, quills blackish, externally spotted with rufous, the secondaries browner like the back, but also spotted with rufous on the outer web; tail blackish, with yellowish white tips to the inner webs of the feathers, which have here and there a few yellowish spots, excepting the two centre ones, which have six rows of small ochraceous spots, not counting the one at the tip; lores and feathers round the eye whitish, the former elongated into black bristles; fore part of cheeks also conspicuously white; hinder part of ear-coverts brown; chin and a patch extending below the ear-coverts pure white, separated from the lower throat, which is also white, by a distinct band of brown; rest of under surface of body white, broadly streaked with blackish; the chest and sides of the body rufous chocolate, with no spots or lighter mottlings; leg-feathers yellowish, mottled with brown on their outer face; under wing-coverts yellowish white, the innermost especially yellow, those near the carpal bend tinged with rufous and distinctly marked with brown, the greater series yellowish at base, blackish at tip, thus resembling the inner lining of the wing, which is strongly barred with yellowish, the base of the feathers being entirely yellow, the bars on the inner web of the secondaries washed with rufous.

Hab. Central America, from Guatemala to Panama.

a. Ad. sk.	Guatemala.	Purchased. (Type of species.)
b. Ad. sk.	Guatemala.	M. A. Bouvier.
c Juv. sk.	Guametala.	M. A. Bouvier.
d, Juv. sk.	Veragua.	M. E. Arcé [C.].

6. Glaucidium pumilum.

Strix pumila, *Temm. Pl. Col.* i. pl. 39.
Noctua pumila, *Steph. Gen. Zool.* xiii. pt. 2, p. 68.
Strix minutissima, *Max. Beitr.* iii. p. 242.
Surnia? minutissima, *Bp. Oss. Regn. Anim. Cuv.* p. 57.
Athene pumila, *Gray. Gen. B.* p. 35; *id. Hand-l. B.* i. p. 42.
Athene minutissima, *Bp. Consp.* i. p. 38.
Glaucidium pumilum, *Kaup, Contr. Orn.* 1852, p. 103; *Burm. Th. Bras.* ii. p. 144; *Scl. & Salv. Nomencl.* p. 117; *Ridgw. Pr. Bost. Soc. N. H.* xvi. p. 97; *Sharpe, Ibis,* 1875, pp. 40, 259, pl. ii. fig. 1.
Athene ferox, *Strickl. Orn. Syn.* p. 162 (*nec Vieill.*).
Phalænopsis minutissima, *Bp. Rev. et Mag. de Zool.* 1854, p. 544.
Noctua minutissima, *Schl. Mus. P.-B.* Striges, p. 32.
Glaucidium passerinum, *Saunders, Ibis,* 1869, p. 397 (*lapsu*).

Adult. Above pale rufous brown, the scapulars with concealed oval spots of ochraceous white; the head rather more ashy or olive-brown than the back and covered with minute spots of dull whitish; hind neck mottled with white, several of the feathers being white on the outer web and forming an indistinct collar, relieved by a black patch on each side of the hind neck, this whitish collar succeeded by a very distinct and characteristic band of orange-rufous, forming a conspicuous reddish collar across the upper back; wing-coverts rufous brown like the back, the greater and median series darker brown towards their tips and distinctly spotted with white on the outer web; quills dark brown, faintly crossed with rufous brown bars, shading off into fulvous spots on the outer webs, the secondaries more or less washed with rufous on the outer web; tail-feathers dark brown, washed with rufous near the base, especially of the outer feather; the tail crossed with four rows of white spots, larger on the inner web and partaking here of the character of notches rather than spots, the fourth spot confined tothe inner web, which it tips; lores and feathers in front of and over the eye white, the former entirely obscured by blackish bristles; sides of face dusky brown, mottled with whitish cross bars or rather spots; cheeks white, extending on to the sides of the neck below the ear-coverts; chin white, followed by a line of orange-rufous across the lower throat;

fore neck pure white; rest of under surface white, distinctly streaked with orange-rufous, the upper breast and sides of the body entirely of the latter colour; under tail-coverts white, with a few brown streaks towards their tips; under wing-coverts whitish, marked with brown, thus forming a longitudinal streak down the wing-coverts and inclining to yellowish white on the inner coverts, the greater series yellowish white at base, dark brown at tip, thus resembling the inner lining of the wing, which is dark brown barred with lighter brown and with yellowish white near the base; leg-feathers orange-buff. Total length 6 inches, wing 3·5, tail 2·2, tarsus 0·7.

Young. Differs from the adult in being much more uniform on the head, where there are few or no spots. The back is a shade more chocolate, so that the rufous collar does not stand out so conspicuously; quills crossed with eight rufous bars, paler on the outer web, but not showing such distinct white spots as in the adult.

Hab. Brazil.

a. Juv. st.	South America.	
b. Ad. st.	South America.	
c, *d*. Ad. sk.	Brazil.	Purchased.
e. Ad. sk.	Brazil.	M. Claussen [P.].
f. Ad. sk.	Bahia.	Dr. Wucherer [C.].
g. Juv. sk.	Bahia.	Dr. Wucherer [C.].

7. Glaucidium cobanense. (Plate XIII. fig. 1.)

Glaucidium ridgwayi, jun., *Sharpe, Ibis*, 1875, p. 47.
Glaucidium cobanense, *Sharpe, Ibis*, 1875, pp. 259, 260.

Adult. Above dull rufous brown, the head with obscure bars of rufous; on each side of the hind neck a concealed patch of white feathers, tipped with black; scapulars rufous brown, with concealed bars of rufous; primary-coverts blackish, with narrow rufous margins; quills blackish, with lighter brown bars, terminating in rounded white spots on the inner web, the outer one minutely spotted with rufous, inclining to buff on the primaries; tail blackish, with seven spots of white on the inner web, those on the middle feathers washed with rusty; lores white, washed with rusty, the shafts developed into long hair-like bristles; feathers round the eye buffy white; ear-coverts and sides of face brighter rufous, mottled with dark brown; chin and cheeks buffy white, the fore neck and chest pure white, separated from the chin by a band of bright rufous; the rest of the under surface of the body white, broadly streaked with rufous brown, the flanks entirely rufous; leg-feathers white, streaked with dark brown and washed with rufous; under wing-coverts white, as also the edge of the wing, the inner ones washed with yellow, those near the edge of the wing dark brown, the lower series white, tipped with brown, resembling the inner lining of the wings, which are ashy brown, barred with white on the inner web; bill yellow, dusky at base of both mandibles; legs yellow, claws black. Total length 5 inches, wing 3·45, tail 2·55, tarsus 0·75.

Hab. Coban, Vera Paz, Guatemala.

a. Ad. sk.	Coban, Guatemala.	M. A. Bouvier.

8. Glaucidium ferox.

Under the heading of this species three subspecies or races are included, as follows:—

Subsp. *a.* Glaucidium ferox.

a. Brown phase.

Caburé, *Azara, Apunt.* i. p. 225.
Strix ferox, *Vieill. N. Dict. d'Hist. N.* vii. p. 22.
Strix passerinoides, *Temm. Pl. Col.* 344; *Max. Beitr.* iii. p. 239.
Noctua passerinoides, *Less. Man. d'Orn.* i. p. 112.
Surnia passerinoides, *Bp. Oss. Regn. Anim. Cuv.* p. 57.
Strix infuscata, *Temm. Ind. Gen. Pl. Col.* p. 7.
Noctua ferox, *D'Orb. Voy. Am. Mér.* Ois. p. 127.
Athene passerinoides, *Gray, Gen. B.* i. p. 35.
Athene infuscata, *Bp. Consp.* i. p. 37; *Strickl. Orn. Syn.* p. 163; *Gray, Hand-l. B.* i. p. 42.
Glaucidium infuscatum, *Kaup, Contr. Orn.* 1852, p. 103; *Cab. J. f. O.* 1869, p. 207; *Scl. P. Z. S.* 1860, p. 289; *Ridgw. Pr. Bost. Soc. N. H.* xvi. p. 102.
Phalænopsis infuscata, *Bp. Rev. et Mag. de Zool.* 1854, p. 544.
Phalænopsis ferruginea, *Bp. l. c.* p. 544.
Glaucidium passerinoides, *Burm. Th. Bras.* ii. p. 143; *id. Reise La Plata St.* ii. p. 440.
Glaucidium ferrugineum, *Scl. P. Z. S.* 1860, p. 289.
Noctua infuscata, *Schl. Mus. P.-B.* Striges, p. 31.
Athene ferruginea, *Pelz. Orn. Bras.* p. 399.
Athene minutissima, *Pelz. Orn. Bras.* pp. 9, 399.
Glaucidium phalænoides (pt.), *Scl. & Salv. Nomencl.* p. 117.

b. Rufous phase.

Strix ferruginea, *Max. Reis. Bras.* i. p. 105; *Temm. Pl. Col.* 199.
Noctua ferruginea, *Steph. Gen. Zool.* xiii. pt. 2, p. 69; *Tsch. Faun. Per.* pp. 19, 117; *Schl. Mus. P.-B.* Striges, p. 30.
Surnia ferruginea, *Bp. Oss. Regn. An. Cuv.* p. 56.
Athene ferruginea, *Gray, Gen. B.* i. p. 35; *Bp. Consp.* i. p. 38; *Strickl. Orn. Syn.* p. 162; *Gray, Hand-l. B.* i. p. 42.
Glaucidium ferrugineum, *Kaup, Contr. Orn.* 1852, p. 104; *Burm. Th. Bras.* ii. p. 141; *Cab. J. f. O.* 1869, p. 207.

Brown phase. *Adult.* Above ashy brown, with concealed diamond-shaped spots of white on the back, extending to the upper tail-coverts, and changing to bars on the scapulars, which have also large ovate spots of white on the outermost feathers; head rather more earthy brown than the back, thickly covered with tiny ovate dots of pale ochraceous, sometimes joined together by a central shaft-line, and thereby producing a streak; round the hind neck a sort of collar, produced by mixed black and white feathers, some of them washed with ochre, but not producing a second collar, the usual black neck-patches tolerably well pronounced; wing-coverts coloured like the back, and spotted in the same manner, the outer median and the greater coverts with well-pronounced ovate spots of white; primary-coverts uniform ashy brown; quills ashy brown, rather darker than the back, excepting near the base of outer web of primaries, which is clear ashy, all of them externally chequered with white spots or notches, the latter much larger on the inner web, and especially on the secondaries, where the notches sometimes nearly reach across the feather and form a bar; tail blackish, darker even than the wings, and crossed with seven rows of white spots, larger and taking the form of bars on the two centre ones and on the inner webs of the others, but never reaching to the shaft of the feather or forming a complete band; feathers in front of and over the eye white, forming a tolerably distinct eyebrow, the lores terminating in black bristles; ear-coverts ashy grey, with broad streaks of ochraceous; chin and lower throat pure white, extending backwards under the ear-coverts, and separated from each other by a blackish grey band; rest of under surface white, streaked with ashy grey, some of the stripes on the abdomen blackish; the sides of the body and the upper breast greyish brown, the latter spotted irregularly with white; thighs and leg-feathers white on the inner surface, and strongly mottled with ashy brown; under tail-coverts pure white, with distinct central streaks of black; under wing-coverts yellowish, the outermost pure white, especially round the bend of the wing, these last strongly marked with dark brown; the greater series yellowish at base, ashy brown at tip, thus resembling the inner lining of the quills, which have the inner webs notched with white, inclining to yellow near the base of the feathers; iris chrome-yellow. Total length 7·5 inches, wing 4, tail 3·2, tarsus 0·85. (*Peru*: *Mus.* Salvin & Godman.)

Female (?). Above ochraceous brown, the back varied with more or less concealed star-like spots of white, rather larger on the scapulars, which have bars of white, the outermost with large ovate spots of the same; wing-coverts coloured like the back, with the same tiny spots of white, the median and some of the greater coverts with large oval white spots, smaller on the greater coverts, which are externally washed with rufous, the outer ones blackish; primary-coverts brown, washed with rufous and obscurely barred with dull ochraceous; quills dark brown, banded with pale rufous, the bars inclining to white on their outer webs, and producing a chequered appearance, especially on the secondaries, where the white spots are more distinct, the innermost barred with white nearly across the feather, rufous only in the middle of the band; upper tail-coverts rather more sandy rufous than the back, distinctly spotted with white; tail dark brown, crossed with eight bars of white, including the terminal one, which is very narrow, the bars as a rule not meeting across the shaft, but doing this in many cases, the interspaces washed with rufous; head ochraceous brown, a little duller than the back, streaked with white, with a few spots of the same, some of the streaks dilated into spots, sometimes rufous, sometimes white; round the neck an indistinct collar of white, here and there inclining to ochre, the usual black patches on the hind neck very feebly developed; below this white collar a few feathers of rufous ochre are present, but not strongly enough developed to form a second collar; lores as well as feathers over and round the eye white, the former terminating in blackish bristles; sides of face dark brown, the ear-coverts very distinctly streaked with ochraceous white; chin and cheeks pure white, extending backwards under the ear-coverts; throat also pure white, separated from the chin by a band of ochraceous brown feathers, joining the sides of the neck, which are minutely spotted with white; rest of under surface of body pure white, sparsely streaked with dull brown, the flanks also white, with the same brown streaks; the sides of the breast ochraceous brown, with a few white spots; under wing-coverts yellowish, the outer ones white as well as the bend of the wing, with a few dark brown spots a little way in; the greater coverts yellowish at base, dark brown at tip, thus resembling the inner lining of the wing, which is dark brown, barred with white on the inner web, inclining to yellow near the base. Total length 7·5 inches, wing 4, tail 2·85, tarsus 0·8. (*Santa Martha*: *Mus. Brit.*)

Rufous phase. *Adult female.* Above uniform cinnamon-rufous, the wing-coverts a little darker and browner than the back, both these and the scapulars varied with a few half-concealed oval spots of white; quills of the same colour as the back, indistinctly banded across with dark brown, more plainly on the primaries, which are dusky brown towards their tips; the two outer ones externally dusky brown; tail entirely cinnamon-rufous; lores and a distinct eyebrow white, the former varied with black bristles; ear-coverts cinnamon, with white shaft-streaks; cheeks white, extending backwards on to the sides of neck below the ear-coverts; chin white; lower throat cinnamon, followed by a large patch of white on the fore neck; rest of under surface of body fulvous white, streaked with cinnamon, the stripes having a shade of brownish on the sides of the body; sides of the body entirely cinnamon-rufous, as also the under wing-coverts, which are rather paler, the greater series yellowish buff, tipped with dark brown, the inner lining of the quills rufous, barred with brown, and inclining to yellowish buff near the base of the wing; edge of the wing whitish; leg-feathers pale cinnamon; under tail-coverts yellowish buff, streaked with cinnamon. Total length 9·7 inches, wing 4·6, tail 3·5, tarsus 0·9.

Adult male. Similar to the female in colour, but smaller. Total length 8 inches, wing 4 2, tail 2·9, tarsus 0·85.

Hab. South America, Brazil, Peru, and Bolivia; northwards through Ecuador to New Granada.

a. ♀ ad. st.	Brazil.	Dr. Such [C.].
b. ♂ ad. st.	Brazil.	M. Claussen [P.].
c. Juv. st.	Brazil.	M. Claussen [P.].
d. Ad. sk.	Brazil.	Purchased.
e, f. Ad. sk.	Bahia.	Dr. Wucherer [C.].
g, h. Ad. st.	Bolivia.	Mr. Brydges [C.].
i, k. Ad. sk.	Santa Martha.	Purchased.
l. Ad. sk.	Santa Martha.	Purchased.

Subsp. β. Glaucidium phalænoides.

Strix phalænoides, *Daud. Traité*, ii. p. 206.
Phalænopsis phalænoides, *Bp. Rev. et Mag. de Zool.* 1854, p. 544.
Glaucidium ferrugineum, *Taylor, Ibis*, 1864, p. 80.
Athene infuscata, *Finsch, P. Z. S.* 1870, p. 557.
Athene phalænoides, *Gray, Hand-l.* i. p. 42.

Brown phase. *Adult.* Above deep sepia- (almost blackish) brown, the head profusely marked with small rounded or minute triangular dots of fulvous white; the back uniform brown, with scarcely any concealed white dots; round the hind neck an indistinct collar of ochraceous buff, plentifully intermingled with black, with the usual two nape-patches of the latter colour; scapulars scarcely spotted at all, but exhibiting a few large white spots near their tips; wing-coverts a little browner than back and nearly uniform, only the greater series with a few distinct white spots towards the tip of the feather; the primary-coverts uniform brown; quills dark brown, indistinctly barred with ashy brown, the bars being indicated on the outer web by a small white spot, inclining to bars on the secondaries; upper tail-coverts uniform with back, and indistinctly spotted with white; tail blackish, crossed with six rows of yellowish spots; lores whitish, obscured somewhat by blackish shaft-stripes; white spots on side of crown rather larger, and forming a narrow eyebrow; sides of face brownish, with a few streaks and bars of dull fulvous; chin and cheeks white, separated from the white fore neck by a band of dark brown feathers, plentifully mottled with fulvous spots; rest of under surface white, broadly streaked with chocolate-brown, the white markings on the flanks rather larger and more oval; the sides of the body, and especially of the upper breast, more uniform chocolate-brown, with very distinct white spots, rather ovate in form; under tail-coverts white, with narrow streaks of dark brown; under wing-coverts yellowish, strongly mottled with dark brown round the bend and near the edge of the wing, the greater series yellowish at base, dark brown at tip, resembling more the inner lining of the wing, which is dark brown, with yellowish white bars on the inner webs of the quills; bill yellowish white; feet yellowish white; iris yellow. Total length 6·5 inches, wing 3·9, tail 2·65, tarsus 0·75.

Rufous phase. *Adult.* Above dusky ferruginous chestnut, the back and scapulars margined with more or less concealed spots of light ochraceous, the head with numerous narrow stripes of rufous ochre; nuchal collar whitish, mottled more or less with black and relieved by the usual patches on each side of the hind neck, these being more indistinct than in some of the allied species, the upper back mottled with large ovate spots of yellowish ochre below the nuchal collars; scapulars coloured like the back and spotted in the same manner and exhibiting very distinct oval spots of white on their outer margin; upper wing-coverts also uniform with the back and minutely spotted with lighter rufous, the greater series having large terminal spots of white like the scapulars; primary-coverts brown, with a few rufous spots; quills dark brown, barred across with light rufous, these lighter bars terminating externally in a fulvescent spot; tail rufous brown, crossed with eight distinct bars of light rufous, sometimes but not always completely traversing the feather and here and there inclining to whitish; lores whitish at base, terminating in yellowish bristly hairs; sides of face rufous brown, narrowly streaked with buffy white; chin and fore part of cheeks whitish, separated from the fore neck, which is also white, by a broad band of rufous on which slight ochraceous mottlings appear; rest of under surface white, blotched with broad stripes of bright rufous, the sides of the body almost uniformly of this colour, the sides of the breast distinctly marked with oval spots of ochraceous white; under wing-coverts whitish, marked with brown near the bend and edge of the wing, the greater series yellowish at base, brown at tip, resembling the inner lining of the wing, which is brown barred with yellowish white on the inner webs of the quills, these bars inclining to pale rufous near the ends of the quills. Total length 6·5 inches, wings 4·1, tail 2·55, tarsus 0·75.

Another rufous specimen has a more uniform upper surface with less distinct stripes on the head, which is obscurely spotted with paler rufous; the tail rufous, barred with dusky brown, eight bars being perceptible on the centre feathers, but these irregular in shape and more or less connected together by a longitudinal mark along the shaft; these bars more or less obliterated on the other feathers, which exhibit the centre part for the most part dusky brown, from which branch on either side irregular and ill-defined bars of the same colour.

Another specimen in the Museum represents what I believe to be the young bird of the rufous phase. It is of a deeper and more maroon-chestnut colour, and is nearly uniform everywhere on the upper surface; the crown is dusky rufous brown, with only a few narrow fulvous stripes on the forehead; the tail uniform deep rufous, with the faintest indications of obsolete bars near the tips of some of the feathers.

Obs. The Trinidad form of *G. ferox*, which I keep separate from the Central-American and Brazilian forms, holds an intermediate position between the two latter, not being exactly similar to either and partaking partially of the characters of both. The brown phase is considerably darker than the corresponding plumage of either *G. ferox* or *G. ridgwayi*, and is more of a chocolate-brown; the spots on the tail are six in number and smaller than in Brazilian birds. The rufous plumage is also much darker and more chestnut than in either of the two allied birds, and is separated by its distinctly spotted back from the corresponding stage of *G. ridgwayi*, though resembling the latter in the characters of its tail when adult. Until, therefore, we know more of the species of *Glaucidium* from Venezuela and Trinidad, it seems better to keep this as an intermediate and at present geographically isolated form.

Hab. Trinidad [? Venezuela and Guiana].

a, *b*, *c*. Ad. st. Trinidad.

d. Juv. st.	Trinidad.	
e. Ad. sk.	Trinidad.	M. A. Bouvier.
f, *g*. Ad. sk.	Trinidad.	M. A. Bouvier.
h. Juv. sk.	Trinidad.	M. A. Bouvier.

Subsp. γ. **Glaucidium ridgwayi.**

Athene infuscata, *Sclater, P. Z. S.* 1856, p. 285.
Glaucidium infuscatum, *Scl. & Salv. Ibis,* 1859, p. 220; *G. C. Taylor, op. cit.* 1860, p. 226; *Lawr. Ann. Lyc. N. Y.* ix. p. 207.
Glaucidium jardinii, *Scl. & Salv. Ibis,* 1859, p. 220.
Glaucidium gnoma, *Lawr. Ann. Lyc. N. Y.* ix. p. 132.
Glaucidium phalænoides, *Salvin, P. Z. S.* 1870, p. 217.
Glaucidium infuscatum, var. gnoma, *Ridgw. Pr. Bost. Soc. N. H.* xvi. p. 103.
Glaucidium ridgwayi, *Sharpe, Ibis,* 1875, pp. 55, 259.

BROWN PHASE. *Adult.* Above greyish umber, the back apparently uniform, but exhibiting a few concealed spots of whitish on the feathers being disarranged; the head striped with whitish, rather more indistinctly towards the nape; round the hind neck a whitish collar, the markings being somewhat oval in shape and mottled with blackish, forming two large black patches on either side of the hind neck; scapulars coloured like the back, but having large oval spots of white near their outer margin; wing-coverts slightly darker brown than the back, the least ones unspotted, the median series with rather larger spots and bars of white or yellowish white, these markings much larger and more oval on the greater series; quills dark brown, narrowly barred across with greyish brown, becoming gradually lighter on the outer web and terminating externally in a white spot; upper tail-coverts rather more rufescent than the back, with a few concealed whitish spots; tail black, crossed with six rows of white spots, inclining to bars which are not continued across the feather; lores whitish, slightly varied with blackish bristles; ear-coverts greyish umber like the head, slightly streaked or mottled with a few yellowish white cross bars; fore part of cheeks whitish, as also the chin, which is separated from the white fore neck by a band of dark brown; rest of under surface white, streaked with greyish umber, a very little, if any, darker than the back, the sides of the upper breast more uniform, varied with a few whitish spots; leg-feathers dark brown on the outer aspect and mottled with white, the inner aspect much whiter; under tail-coverts white, with very broad streaks of dark brown; under wing-coverts yellowish, with small dark brown markings near the edge of the wing, the greater series yellowish at base, dark brown at tip, resembling the inner lining of the quills, which are dark brown with yellowish bars on the inner web. Total length 6 inches, wing 3·6, tail 2·7, tarsus 0·8.

Less adult (? female). Browner in all respects than the foregoing description, the head being partly streaked and partly spotted with dull white or pale rufous; tail rufous, crossed with nine bands of brown of about equal width with the rufous interspaces. These tail-bands are sometimes eight and sometimes nine in number, and vary much in breadth. On the subject of this variation see my paper on *Glaucidium* (Ibis, 1875, p. 55).

RUFOUS PHASE. *Immature female.* Above bright cinnamon, quite uniform on the head and back, the scapulars with no light-coloured oval spots, but have a few dusky bars on the outer web; the upper tail-coverts perfectly uniform and bright cinnamon like the back; round the hind neck a faint indication of an ochraceous buff collar, relieved by two tolerably distinct neck-spots; wing-coverts darker and more bay than the back, the greater series mottled with lighter rufous on their outer webs; quills blackish, notched with cinnamon on the outer web, terminating a faintly indicated dusky band traversing the quills; the inner secondaries cinnamon like the back, spotted externally with brighter rufous, but very distinctly notched on the inner web with ochraceous buff; tail dull brown, washed more or less with rufous, especially on the centre feathers, and having eight fulvous bars not meeting across the feathers but inclining to buffy white on the exterior margin of some of them, the outer feathers internally barred with whitish, but not showing more than a tiny fulvous spot externally; over the eye an eyebrow of buffy white feathers; lores and sides of the face bright cinnamon, the lores thickly obscured with black bristles, the cheeks slightly streaked with fulvous; a narrow chin and cheeks which are drawn backwards below the ear-coverts pure white, separated from the white fore neck by a band of cinnamon; rest of under surface white, streaked with cinnamon, deepening into dark brown on a few of the abdominal plumes; the sides of the body bright cinnamon, especially the sides of the upper breast which converge upon the chest; thighs cinnamon, lighter on the inner surface; under wing-coverts yellowish, with a few cinnamon feathers near the edge of the wing, the outer median series marked with dark brown, the greater series yellowish at base, dark brown at tip, resembling the inner lining of the wing-feathers, which are broadly barred with yellowish white on the inner web. Total length 6 inches, wing 3·7, tail 2·5, tarsus 0·8. (Guatemala: *Mus.* Salvin and Godman.)

More adult bird. Differs very little from the foregoing bird, excepting that the upper surface is slightly browner and the head is mottled with rufous spots and the upper tail-coverts with ochraceous buff; the tail is blackish, the spots more distinct and pronounced in shape.

Adult. Above dull ferruginous, with concealed spots of lighter rufous on the back, a little plainer on the scapulars, the outermost of which have large ovate spots of white; head narrowly streaked with rufous ochre, some of the feathers showing slight spots; round the hind neck a whitish collar slightly mottled with rufous, the usual black patches on each side of the hind neck feebly developed; wing-coverts coloured like the back, with minute spots of pale ochraceous buff, some large oval spots of white appearing both on the outer coverts of the median and greater series; quills dark brown, transversely banded with rufous, somewhat paler on the secondaries and towards the outer web of the primaries; upper tail-coverts like the back, but mottled with distinct streaks and bars of pale rufous; tail rufous, crossed with eight bars of clear brown; lores brown, the shafts terminating in blackish bristles; sides of face rufous, narrowly streaked with buffy white; fore part of cheeks and chin pure white, separated from the white fore neck by a rufous band not very distinct but slightly mottled with pale ochraceous spots; rest of under surface white, broadly streaked with rufous, the streaks on the sides of the body broader and more obscure; the sides of the breast nearly uniform, with a few ochraceous spots; leg-feathers clear rufous externally, white internally; under tail-coverts white, streaked with rufous; under wing-coverts yellowish, those along the outer edge of the wing white, with a few brown markings, the greater series yellowish at base, dull brown at tip, not distinguishable from the inner lining of the wing, the inner webs of the feathers being barred with yellowish inclining to rufous near the tips of the feathers.

Hab. From the Rio Grande southwards through Central America to Veragua.

a. Ad. sk.	Mexico.	Purchased.
b. Ad. sk.	Mexico.	Purchased.
c. Ad. st.	Guatemala.	John Gould, Esq.
d, e. Ad. sk.	Costa Rica.	Dr. van Patten [C.].
f, g, h, i. Ad. sk.	Costa Rica.	Purchased.
k. Juv. sk.	Honduras.	Purchased.
l. Ad. sk.	Veragua.	M. E. Arcé [C.].

9. Glaucidium jardinii.

Phalænopsis, sp., *Sclater, P. Z. S.* 1855, p. 134.
Phalænopsis jardinii, *Bp. C. R.* xli. p. 654.
Glaucidium jardinii, *Cab. J. f. O.* 1869, p. 208; *Ridgw. Pr. Bost. Soc. N. H.* xvi. p. 99; *Sharpe, Ibis,* 1875, pp. 43, 259.
Noctua jardinii, *Schl. Mus P.-B. Revue Accipitr.* p. 29.
Glaucidium langsbergii, *Ridgw. Pr. Bost. Soc.* xvi. p. 98 (*nec Bp.*).

Brown phase. *Young.* Above reddish brown, the head more earthy brown, but the whole of the upper surface almost entirely uniform, excepting a few concealed rufous ochre spots on the interscapulary region, changing to fulvous bars on the outer scapulars; round the hind neck an indistinct collar of ochraceous buff, with some of the feathers whitish, nearly all of them tipped with black, which is more distinct on the sides of the hind neck; wing-coverts rather darker than the back, minutely spotted with rufous ochre, the greater series plainly spotted and barred with ochraceous buff; quills blackish, barred across with the same reddish brown as the back, paler and more ochraceous on the inner web and exhibiting paler spots of the same on the outer secondaries, whitish on the primaries; tail black, crossed with five (one concealed) rows of rounded white spots, without including the one at the tip of the inner web; lores and a distinct eyebrow whitish, the latter washed with ochre, the former obscured by blackish bristles in front of the eye; ear-coverts brown, with narrow whitish shaft-lines and barred with pale ochre; chin white, extending in a large white patch below the ear-coverts on to the sides of the neck; across the lower throat a bar of plain dark brown extending on to the sides of the neck and followed by a large white patch on the fore neck; rest of under surface of body ochraceous, the centre of the abdomen white, the under tail-coverts brownish at tip; the lower surface of the body broadly but irregularly streaked with dull reddish brown, the sides of the upper breast and the flanks uniform reddish brown; under wing-coverts deep ochre, with a longitudinal patch of dark brown near the outer edge of the wing, which is whitish round the bend, the greater series yellowish white at base, blackish at tip, thus resembling the inner lining of the wing, which is blackish, barred with lighter brown towards the tip, inclining to yellowish white near the base.

Adult. Above dark brown, with a slight tinge of reddish on the back, which is varied with more or less concealed bars of deep ochre, each bar having a conterminous line of black above and below, the outer scapulars and greater wing-coverts relieved with large spots of ochre or white; head more greyish brown, contrasting with the back and covered with numerous rounded or elongated spots of ochraceous white, extending on to the sides of the neck, which are coloured like the head; round the hind neck a whitish collar, here and there varied with deep ochre and black, a large patch of the latter being visible on each side of the hind neck; wings coloured as in foregoing young specimen, excepting that the rufous bands are more distinct and the external white spots are larger; tail black, with five distinct rows of white spots (not including the terminal spot on the inner web), the two centre feathers with a subterminal spot of white on either web; facial features and under surface of body much as in foregoing description, excepting that the stripes on the lower parts are more distinct and the sides of the breast are minutely spotted with ochraceous white like the head; under wing-coverts and bars on inner lining of wing whiter and not so rich ochre as in the young bird. Total length 6 inches, wing 4·15, tail 2·75, tarsus 0·7.

Rufous phase. *Adult.* Above rufous brown, the head mottled with rufous chestnut, which takes the form of broad shaft-streaks, widening out here and there into large spots or lateral bars; round the hind neck encircled with an indistinct collar of rufous ochre, somewhat varied with white spots and having the usual black neck-patches well defined; back rufous brown, thickly mottled and barred across with rufous, most of the feathers with distinct transverse bars; wing-coverts dark brown, barred with rufous; quills blackish brown, barred with lighter and more rufous brown, changing to clear rufous on the outer web of the primaries and thus producing a chequered appearance, the inner secondaries regularly barred with rufous and black; tail bright rufous, with indications of seven black cross bands, irregular in shape and sometimes connected by a mesial streak along the shaft; feathers round the eye, forming an indistinct eyebrow, rich ochre; the lores whiter, and terminating in black bristles; ear-coverts bright rufous, streaked with narrow white shaft-lines; chin and cheeks white, extending in a patch backwards under the ear-coverts; lower throat pure white, separated from the chin by a band of rufous; remainder of under surface rich ochre, with the exception of a few white feathers near the centre of the body; the sides of the upper breast rufous brown, mottled and barred with rufous, like the back; the sides of the body also bright rufous, the flanks mottled with patches and bars of dark brown and slightly streaked here and there with the same colour; leg-feathers bright ochre, minutely barred with brown; under wing-coverts very deep ochre, as also the edge of the wing, a short distance from which are several blackish spots, the greater series ochraceous at base, ashy black at tip, thus resembling the inner lining of the wing, which is blackish, broadly barred with rufous on the inner web, inclining to deep ochre near the base of the inner web. Total length 6 inches, wing 3·9, tail 2·55.

Young. Differs from the adult in being more uniform on the back and in not showing the bars which produce the mottled back in the latter; the bands on the quills and tail are not perfect, but the dark brown colour occupies the greater part of the inner webs of the secondaries, extending on to the outer one in a few dusky blackish notches; the tail uniform rufous, with a few indistinct blackish markings, principally longitudinal and close to the shaft of the feather. (New Granada: *Mus.* Salvin and Godman.)

Obs. Another young bird in the Museum, obtained by Goering in the Sierra Nevada, Venezuela, is, in some respects, different from the one above noticed; for it has the transverse bars on the secondaries complete, and the tail as strongly barred with black, though as irregularly, as in the adult; but the rest of the upper surface is dull rufous chocolate, almost uniform, with the exception of a few narrow and indistinct rufous shaft-lines on the crown; underneath the bird is ochraceous, with no white, but streaked with rufous on the lower parts, the sides of the upper breast and flanks being rufous mottled with brown.

Hab. New Granada and Venezuela.

a. Ad. st.	Columbia.	J. Gould, Esq.
b. Ad. st.	Columbia.	M. Verreaux.
c. Juv. st.	Columbia.	M. Verreaux.
d, e, f, g. Ad. sk.	Bogotá.	Purchased.
h. Juv. sk.	Bogotá	Purchased.
i. Juv. sk.	Sierra Nevada, Venezuela.	M. A. Goering [C.].

10. Glaucidium perlatum.

La Chevêchette perlée, *Levaill. Ois. d'Afr.* vi. pl. 284; *Sundev. Crit. om Levaill.* p. 55.
Pearl Owl, *Lath. Gen. Hist.* i. p. 366.
Strix perlata, *Vieill. N. Dict. d'Hist. Nat.* vii. p. 26.
Strix occipitalis, *Temm Pl. Col.* i. pl. 34.
Noctua occipitalis, *Steph. Gen. Zool.* xiii. pt. 2, p. 632.
Surnia perlata, *Bp. Oss. R. A. Cuv* p. 58.

Noctua perlata, *Smith, S.-Afr. Q. Journ.* 2nd ser. no. 4, pt. 1, p. 313; *Schl. Mus. P.-B.* Striges, p. 36; *id. Revue Accipitr.* p. 30: *Heugl. Orn. N.O.-Afr.* i. p. 120.
Nyctipetes perlatus, *Swains. Classif. B.* ii. p. 218.
Scotophilus perlatus, *Swains. B. W. Afr.* i. p. 130.
Strix licua, *Licht. Verz. Saug. u. Vog. Kaffernl.* p. 12.
Athene perlata, *Gray, Gen. B.* i. p. 35; *Cass. Cat. Strigidæ Philad. Mus.* p. 15; *Bp. Consp.* i. p. 37; *Strickl. Orn. Syn.* p. 163; *Hartl. Orn. W.-Afr.* p. 17; *Gray, Hand-l. B.* i. p. 40; *Blanf. Geol. & Zool. Abyss.* p. 303; *Finsch, Tr. Z. S.* vii. p. 209; *Gurney in Anderss. B. Dam. Ld.* p. 37; *Antin. & Salvad. Viagg. Bogos.* p. 37.
Microglaux licua, *Kaup, Isis,* 1848, p. 769; *id. Contr. Orn.* 1852, p. 104.
Microglaux perlata, *Kaup, ut suprà.*
Athene licua, *Strickl. & Sclater, Contr. Orn.* 1852, p. 142; *Strickl. Orn. Syn.* p. 164; *Bp Consp.* i. p. 37; *Layard, B. S. Afr.* p. 38.
Athene pusilla, *Heugl. Syst. Uebers.* p. 14.
Athene occipitalis, *Heugl. Syst. Uebers.* p. 14; *Gray, Hand-l. B.* i. p. 40.
Noctua perlata capensis, *Schl. Mus. P.-B.* Striges, p. 37.
Strix senegalensis, *Chapm. Trav. S. Afr.* ii. App. p. 393.
Carine perlata, *Sharpe, ed. Layard, B. S. Afr.* p. 77.
Glaucidium occipitale, *Sharpe, Ibis,* 1875, p. 259.

Adult male. Above ashy brown, on the back spotted with white, more or less concealed, the more conspicuous spots relieved by a small conterminous black mark; scapulars with a large white patch on their outer webs, forming a conspicuous longitudinal mark on each side of the back; head and nape rather more ochraceous brown than the back, the forehead spotted, but the crown and wig barred with white, each bar having a conterminous one of black; hind neck silvery white, mottled with black near the bases of the feathers, forming a kind of irregular collar, which is succeeded by another one of orange-buff, which obscures the white spotting on the feathers; wing-coverts coloured like the back, but slightly washed with ochraceous, largely spotted with white on the median and greater series; quills dark brown, the primaries inclining to ashy at their tips, the secondaries narrowly tipped with white, all the quills barred across with light rufous brown, changing to white on the outer web, and thus producing a spotted appearance, the bars becoming fulvous white on the inner web and banding the innermost secondaries with white; tail-feathers dark brown, washed with rufous, tipped on the inner web with white, and conspicuously spotted with white on both webs, the spots on the inner one larger and more oval and six in number; loral plumes, feathers above and below the eye, as well as the sides of the face white, the ear-coverts obsoletely marked with blackish; chin white; rest of under surface white, the throat marked with ochraceous brown, the breast almost entirely of the latter colour on the sides, but mottled with more or less concealed spots and bars of white; the belly, flanks, and sides of body thickly streaked with dark brown, the centre of the body white; leg-feathers white, slightly mottled with dark brown; under tail-coverts white, excepting a slight mark of dark brown near the tips of the middle ones; under wing-coverts ochraceous buff, here and there marked with dark brown, the greater series dark brown at tip, thus resembling the lining of the quills, which have the inner web ochraceous near the base; small coverts round the bend of the wing white; bill yellow; iris yellow. Total length 7 inches, wing 4·15, tail 3·2, tarsus 0·95.

Adult female. Similar to the male. Total length 8 inches, wing 4·4, tail 3·2, tarsus 1.

Young. Nothing is as yet precisely known regarding the young plumage of this Owl; but I fancy that the more spotted birds are the oldest, although it is almost impossible to say, as in some specimens the spots and bars are about equally distributed.

Two Abyssinian specimens, collected by Captain Sturt during the British expedition, have more of an orange-buff shade on the head than those from Southern Africa, and they appear to have about one row of spots more on the tail. At present, however, my material is not extensive enough to separate them specifically. The Gambian specimen agrees best with the Abyssinian.

Hab. The whole of Africa south of the Sahara, excepting the forest regions of the west and east coasts.

a. Ad. st.	South Africa.	Earl of Derby [P.].
b. ♀ ad. sk.	Griqua Land, May 1871 (*T. C. Atmore*).	R. B. Sharpe, Esq. [P.].
c. ♂ ad sk.	Griqua Land, May 1871 (*T. C. Atmore*).	T. C. Atmore, Esq. [C.].
d. Ad. sk.	Tette, Zambesi.	Dr. Kirk [C.].
e. Ad. sk.	Tette, Zambesi.	Dr. Kirk [C.].
f. ♂ ad. sk.	Tatti, Matabili.	T. E. Buckley, Esq. [P.].
g, h, i, k. Ad. sk.	Damara Land.	C. J. Andersson, Esq. [C.].
l. Ad. st.	River Gambia.	Purchased.

11. **Glaucidium tephronotum.** (Plate XIII. fig. 2.)

Glaucidium tephronotum, *Sharpe, Ibis*, 1875, p. 260.

Adult. Above dull bluish grey, the upper surface almost entirely uniform, with the exception of several white spots at the base of the feathers on the hind neck; wing-coverts rather blacker than the back; quills brown, the secondaries greyish brown, much duller than the back; tail black, with three large notches of white on the inner webs of all the feathers, the centre ones with four ovate white spots, the subterminal one very small, also confined to the inner web; lores and an ill-defined eyebrow whitish, the bristles overhanging the former black; ear-coverts dull greyish brown; cheeks whitish, inclining to brown on the hinder parts; chin white; under surface of body white, the breast and sides vinous chestnut, all the plumes in the centre of the lower breast and abdomen dark brown, causing a distinctly streaked appearance; under tail-coverts white; leg-feathers rich ochre; under wing-coverts vinous chestnut, streaked with brown, the innermost ones and the axillaries yellowish, the edge of the wing white, the lower coverts ashy brown, yellowish white at base, resembling the lining of the quills, which are ashy brown, yellowish at base, the inner webs also barred with yellowish; bill yellowish, greenish towards the base; feet yellow. Total length 7·5 inches, wing 4·05, tail 3, tarsus 0·75.

Hab. South America [?].

a. Ad. sk. "South America." W. Wilson Saunders, Esq. [P.]. Type of species.

12. **Glaucidium brodiei.**

Noctua brodiei, *Burton, P. Z. S.* 1835, p. 152; *Schl. Mus. P.-B.* Striges, p. 35.

Noctua tubiger, *Hodgs. As. Res.* xix. p. 175; *id. J. A. S. B.* vi. p. 369.

Athene brodiei, *Blyth, J. A. S. B.* xi. p. 163; *Gray, Gen. B.* i. p. 35; *Blyth, Cat. B. Mus. A. S. B.* p. 40; *Bp. Consp.* i. p. 40; *Horsf. & Moore, Cat. B. Mus. E. I. Co.* i. p. 66; *Strickl. Orn. Syn.* p. 169; *Gray, Hand-l. B.* i. p. 40; *Gould, B. Asia,* part xxii.; *Swinh. P. Z. S.* 1871, p. 343.

Athene badia, *Hodgs. in Gray's Zool. Misc.* p. 82.

Tænioptynx brodiei, *Kaup, Isis,* 1848, p. 768; *id. Contr. Orn.* 1852, p. 104; *id. Tr. Z. S.* iv. p. 206; *Hume, Nests and Eggs Ind. B.* i. p. 72.

Glaucidium brodiei, *Jerdon, B. Ind.* i. p. 146; *Hume, Rough Notes,* ii. p. 417; *Sharpe, Ibis,* 1875, p. 259.

Athene minutilla, *Gould, B. Asia,* part xxii.

Adult (grey phase). Above dull greyish brown, everywhere narrowly banded with white; the head greyer than the back, and rather more broadly banded, the forehead and sides of the latter very minutely dotted with white; hind neck surrounded by a kind of fulvous collar, the feathers being edged with fulvous; on each side of the hind neck a black patch; scapulars rather largely marked with white on the outer web, forming a tolerably distinct shoulder-patch; wing-coverts coloured like the back, the white bars rather wider apart; primary-coverts uniform blackish brown; quills blackish brown, the secondaries tipped with whitish, the primaries with tiny notches of white on the outer web, these being larger on the secondaries, and taking the form of bars on the innermost; upper tail-coverts dark brown, barred and spotted with white; tail dark brown, very narrowly tipped with whitish, and crossed with seven other bands of white; lores and a distinct eyebrow white, the former obscured by black hair-like bristles; ear-coverts dark brown, barred and spotted with whitish; cheeks white, extending backwards behind the ear; chin and fore neck white, separated by a band of dark brown, minutely barred with white, drawn across to the sides of the neck, which the band resembles in colour; rest of under surface white, with large terminal spots of brown, the flanks banded with brown; sides of breast brown, banded with white; under tail-coverts spotted with brown at the tip; leg-feathers brown, with a few minute bars of white, the anterior face of the tarsus white; under wing-coverts white, the inner ones yellowish, with a few brown markings a little distance from the edge of the wing; under surface of wing dull brown, the inner web barred with white, the bases almost entirely whitish; bill pale greenish yellow; toes the same, but more yellow; irides bright pale yellow. Total length 7 inches, wing 3·45, tail 2·8, tarsus 0·8.

Rufous phase. Similar to the grey phase in character of markings, but of a dark brown colour, with all the cross markings ochraceous; under surface of body white, with all the terminal spots ochraceous as well as the bars on the sides of the body; bands on tail eight, not counting the one at the tips of the feathers.

Young. Nearly uniform above, the head with a few ochraceous spots and streaks on the crown; collar round the neck and nape-patches of black as in adult; back brown, margined with rufous brown; wing-coverts blackish, with ochraceous margins, the greater series with notches of ochraceous; scapulars with distinct white spots on the outer webs; primary-coverts uniform blackish; quills blackish externally, notched with deep ochre, these bars larger and more like bars on the secondaries; tail black, with seven ochraceous bands; a small eyebrow whitish; ear-coverts dark brown, banded with ochraceous; chin, cheeks, and fore neck white, with a band of rufous brown across the throat; remainder of under surface white, with longitudinal brown markings at the tips of the feathers; the sides of the breast nearly uniform, with a few ochraceous spots.

Obs. The type of Mr. Gould's *Athene minutilla* is now before me, and I am fain to confess that it seems to me to be only the male of the ordinary *G. brodiei*. It is a young bird, with nearly uniform upper surface, and measures as follows—total length 5·8 inches, wing 3·4, tail 2·4, tarsus 0·75. The Museum contains other specimens of this small bird, evidently the same species, and exhibiting both rufous and grey phases. The following are the measurements:—

	Total length in	Wing in.	Tail in.	Tarsus in.
a. Ad. Himalayas	5·5	3·3	2·45	0·75
b. Ad. Himalayas	6	3·4	2·5	0·7
c. Ad. Darjiling	6	3·4	2·45	0·8
d. Ad. Nepal (*Hodgson*)	5·8	3·2	2·4	0·7

To compare with the above I add the dimensions of three Nepalese specimens of the larger bird, which I consider to be the female. It is the true *Glaucidium brodiei* of Gould.

	Total length. in.	Wing. in.	Tail. in.	Tarsus in.
a. Ad. Nepal (*Hodgson*)	7	3·8	2·75	0·75
b. Ad. Nepal (*Hodgson*)	6·5	3·6	2·6	0·7
c. Ad. Nepal (*Hodgson*)	7	3·8	2·7	0·7

After allowance has been made for the difficulty of measuring these small Owls, I cannot see any reason for supposing that those measured above are any thing more than the male and female of one and the same species.

Hab. Himalayas.

a. Ad. st.	Himalaya Mountains.	Zoological Society.
b, c. Ad. sk.	Himalaya Mountains.	Purchased.
d. Juv. sk.	Himalaya Mountains.	Purchased.
e. Ad. sk.	Darjiling.	J. Gould, Esq. [P.].
f. Ad. sk.	Nepal (*Eccles*).	J. Gould, Esq. [P.].
g, h, i. Ad. st.	Nepal.	B. H. Hodgson, Esq. [P.].
k. Ad. st.	Nepal.	B. H. Hodgson, Esq. [P.].
l, m. Ad. sk.	Nepal.	B. H. Hodgson, Esq. [P.].
n. Juv. sk.	Assam.	J. Gould, Esq. [P.]. Type of *Athene minutilla.*

13. Glaucidium pardalotum.

Athene pardalota, *Swinhoe, Ibis,* 1863, p. 216; *Gray, Hand-l. B.* i. p. 40; *Swinh. P. Z. S.* 1871, p. 343.

Glaucidium pardalotum, *Sharpe, Ibis,* 1875, p. 259.

Adult. Head umber-brown, thickly dotted with ochraceous spots on the fore part of the crown, the hind part and the nape with ochraceous bars; round the hind neck a tolerably distinct collar of clear ochraceous feathers barred with black, the latter colour being very distinct on the sides of the hind neck; rest of the upper surface ochraceous brown, everywhere distinctly barred with bright ochraceous, most of these bars with a conterminous black margin; upper tail-coverts spotted with ochraceous; tail blackish brown, tipped with ochraceous and crossed with six other bars of the same colour; scapulars brown like the back, distinctly barred across with bright ochraceous, the outermost white for apical half, with a terminal bar of black; quills dark brown, barred with ochraceous and slightly washed with the same colour, the bars very much broader on the greater series; quills blackish, the primaries with a few tiny spots of ochraceous, the secondaries narrowly tipped with ochraceous, the spots on the outer web larger than on the primaries; lores and a small eyebrow white, the shafts of the former produced into long hair-like bristles; sides of face blackish brown, spotted with ochraceous; chin and cheeks white, the latter forming a large triangular patch on the sides of the neck, and separated from the white fore neck by a band of dark brown feathers spotted with light ochraceous; centre of the body pure white, the belly streaked with light brown, the sides of the body ochraceous brown, the sides of the upper breast barred with light ochraceous; leg-feathers dark brown; under tail-coverts white, with a few narrow brown streaks; under wing-coverts white, the inner ones washed with yellow; a few brown spots near the outer edge of the wing, the greater series yellowish at base, ashy brown at tip, like lining of the quills, which are ashy brown, barred on the inner web with yellowish white. Total length 6 inches, wing 3·45, tail 2·5, tarsus 0·75. (*Mus.* R. Swinhoe.)

Obs. This species is extremely close to *G. brodiei*, and is said by Mr. Swinhoe to differ from that bird by having the belly spotted with black. It certainly does not show so much white below and is more spotted; but the characters are very slight; the bars on the scapulars are rather more sagittate.

Hab. Formosa.

14. Glaucidium sylvaticum.

Athene sylvatica, *Bp. Consp. Av.* i. p 40 (*ex Mull. MS. in Mus. Lugd.*); *Wall. Ibis*, 1868, p. 23; *Gray, Hand-l. B.* i. p. 40.
Tænioptynx sylvatica, *Bp. Rev. et Mag. de Zool.* 1854, p. 544.
Noctua sylvatica, *Schl. Mus. P.-B.* Striges, p. 36.
Glaucidium sylvaticum, *Sharpe, Ibis*, 1875, p. 259.

Adult (type of species). Above deep rufous brown, crossed with transverse bars of ochraceous buff, the scapulars with large white spots on the outer web, forming a more or less distinct wing-bar; the wing-coverts coloured like the back, the fulvous cross bars a little broader and more distinct; quills and tail blackish brown, with distinct bars of ochraceous buff or whitish, not strictly continuous across the wing, the inner primaries and all the secondaries with large white bars on the inner webs, the tail-feathers with whitish bars on the inner web, more distinct below; head rather more ashy brown than the back, spotted and barred with dull white; the sides of the face like the head, but the lores and feathers behind the eye white, the hinder neck encircled with a not very distinct collar of ochraceous buff; throat and sides of neck pure white, as also the chest and the entire centre of the body, the lower part of the throat and sides of the breast rufous brown, the flank-feathers somewhat margined with white, producing a striped appearance; under wing-coverts white, with a bar of brown feathers across the middle. Total length 6·5 inches, culmen about 0·65, wing 3·8, tail 2·1, tarsus 0·8, middle toe 0·7. (*Mus. Lugd.*)

Hab. Sumatra.

Obs. The note that I made at the time of examining the type was that it was "closely allied to *G. brodiei*, but differing in its rufous brown back, dusky sides of face, and especially in the uniform chestnut sides of the body."

15. Glaucidium castanonotum.

Athene castanonota, *Blyth, Cat. B. Mus. A. S. B.* p. 39; *Kelaart, Prodr. Faun. Zeyl.* p. 27.
Tænioglaux castanonotus, *Bp. Rev. et Mag. de Zool.* 1854, p. 544.
Noctua castanonota, *Schl. Mus. P.-B.* Striges, p. 34.
Athene castaneonota, *Gray, Hand-l. B.* i. p. 39 (1869); *Hume, Rough Notes*, ii. p. 412; *Holdsw. P. Z. S.* 1872, p. 418.
Glaucidium castanonotum, *Sharpe, Ibis*, 1875, p. 259.

Adult. Upper surface deep rufous chestnut, with remains of blackish cross bars more or less visible on the back and greater wing-coverts, some of the outer median coverts slightly notched with fulvous, all the rest uniform chestnut like the back; primary-coverts dull brown, with a few rufous bands externally; primaries dull brown, barred with rufous brown on the inner web, the outer web having a corresponding spot of rufous, inclining to fulvous on the outermost primaries; secondaries also dark brown, banded with rufous on the outer and with fulvous on the inner web, these bands much less distinct on the inner secondaries; upper tail-coverts blackish brown, narrowly banded with fulvous, exactly resembling the tail, which has eight narrow fulvous bands including the terminal one; entire head and neck dark brown, everywhere narrowly banded with ochraceous, the hind neck banded and spotted with white; ear-coverts rather darker than the head, but barred in exactly the same manner; lores brown, barred with ochraceous, the bases being white; cheeks pure white, forming a triangular patch; chin whitish; throat crossed with a band of dark brown feathers, barred with ochraceous like the sides of the neck, with which it is continuous; sides of the breast blackish brown, narrowly barred with ochraceous; the rest of the under surface of the body white, streaked down the centre with brown, with which colour the flanks are also barred; under tail-coverts whitish; leg-feathers fulvous, barred with brown; under wing-coverts whitish, with a few brown spots near the outer edge of the wing, the greater series whitish at base, brown at tip; inner lining of the wing dull and slightly rufous brown, spotted on the outer and barred on the inner web with pale rufous, inclining to fulvous on the latter; bill yellow; feet greenish brown. Total length 8·7 inches, wing 5·5, tail 2·7, tarsus 0·9.

Obs. The specimen presented to the Museum by Mr. Gould is not in very good order, but shows that the amount of cross barring on the back varies; for whereas in the subject of our description the back is nearly uniform, in the other bird the blackish cross bars can be traced very plainly all over the upper surface.

Hab. Ceylon.

a. Ad. st.	Ceylon.	Hugh Cuming, Esq. [C.].
b. Ad. sk.	Ceylon.	J. Gould, Esq. [P.].

16. Glaucidium castanopterum.

Strix spadicea, *Reinw. MS.*
Strix castanoptera, *Horsf. Tr. Linn. Soc.* xiii. p. 140; *Temm. Pl. Col.* i. pl. 98.
Noctua castanoptera, *Steph. Gen. Zool.* xiii. pt. 2, p. 60; *Less. Traité*, p. 105; *Schl. Mus. P.-B.* Striges, p. 34.
Athene castanoptera, *Gray, Gen. B.* i. p. 35; *Blyth, J. A. S. B.* xv. p. 280; *Cass. Cat. Strigidæ Philad. Mus.* p. 14; *Bp. Consp.* i. p. 44; *Horsf. & Moore, Cat. B. Mus. E. I. Co.* i. p. 67; *Strickl. Orn. Syn.* p. 168; *Gray, Hand-l. B.* i. p. 39; *Hume, Rough Notes*, ii. p. 413.
Tænioglaux castanoptera, *Kaup, Contr. Orn.* 1852, p. 106; *id. Tr. Z. S.* iv. p. 211; *Bp. Rev. et Mag. de Zool.* 1854, p. 544.
Glaucidium castanopterum, *Sharpe, Ibis*, 1875, p. 259.

Adult. Above uniform rufous chestnut, the outer web of the scapulars white, forming a broad longitudinal band; wing-coverts chestnut like the back, only the outer greater wing-coverts externally white; primary-coverts brown, externally washed with rufous; quills brown, barred on the inner web with ochraceous, the outer webs notched with dull ochraceous and terminally margined with the same colour; upper tail-coverts brown, narrowly barred with ochraceous; tail dark brown, crossed with seven rather narrow ochraceous bands including the terminal one; entire head and neck barred with dark brown and ochraceous, the sides of the neck somewhat spotted with white; the ear-coverts and sides of the neck barred exactly like head, the loral plumes and ear-coverts with white shaft-streaks, the cheeks white, all the feathers narrowly tipped with brown; chin whitish, the throat-feathers banded with dark brown and ochraceous like the head; fore neck white; sides of upper breast brown, barred with ochraceous like the sides of the neck and head; the rest of the under surface white, streaked with bright rufous chestnut in about equal proportions; the feathers of the lower breast rufous, broadly margined with white on both webs; the flank-feathers chestnut on the outer web, white on the inner one; leg-feathers ochraceous, minutely marked with brown; under tail-coverts dull whitish; under wing-coverts also white as well as the edge of the wing, the feathers adjoining the latter spotted with brown, the greater series brown at tip, yellowish at base; the inner lining of the quills brown, washed with rufous on the outer web and barred with ochraceous on the inner; the secondaries yellowish at base of inner web, shading into ochraceous, with which they are deeply notched on the inner web. Total length 10·5 inches, wing 5·9, tail 3·8, tarsus 0·9.

Another specimen in the Museum is rather smaller than the one described; but the shoulder-patch is not perfectly white, many of the feathers being chestnut like the back, barred externally with dark brown and fulvous. Total length 9 inches, wing 5·7, tail 3, tarsus 0·85.

Hab. Java.

a. Ad. st.	Java.	Hon. East-India Co. [P.].
b. Ad. st.	Java.	J. Baird, Esq. [P.].

17. Glaucidium radiatum.

Strix radiata, *Tickell, J. A. S. B.* ii. p. 572.
Noctua perlineata, *Hodgs. J. A. S. B.* xi. p. 269.
Athene erythroptera, *Gould, P. Z. S.* 1837, p. 136; *Gray, Gen. B.* i. p. 34.
Athene undulata, *Blyth, J. A. S. B.* xi. p. 457.
Athene perlineata, *Hodgs. in Gray's Zool. Misc.* p. 82.
Athene radiata, *Blyth, J. A. S. B.* xv. p. 281; *id. Cat. B. Mus. A. S. B.* p. 39; *Cass. Cat. Strigidæ Philad. Mus.* p. 14; *Bp. Consp.* i. p. 40; *Horsf. & Moore, Cat. B. Mus. E. I. Co.* i. p. 67; *Strickl. Orn. Syn.* p. 167; *Jerd. B. Ind.* i. p. 143; *Gray, Hand-l. B.* i. p. 39; *Hume, Rough Notes,* ii. p. 409; *id. Nests & Eggs Ind. B.* i. p. 70; *Ball, Str. F.* ii. p. 383.
Tænioglaux erythroptera, *Kaup, Isis,* 1848, p. 769; *id. Contr. Orn.* 1852, p. 106; *id. Tr. Z. S.* iv. p. 212.
Tænioglaux radiata, *Bp. Rev. et Mag. de Zool.* 1854, p. 544.
Noctua radiata, *Schl. Mus. P.-B.* Striges, p. 34.
Glaucidium radiatum, *Sharpe, Ibis,* 1875, p. 259.

Adult. Above dusky brown, everywhere narrowly banded with

rufous buff, those on the hind neck broader and lighter-coloured, and inclining to white on the scapulars, the outer web of which is almost entirely white with a few dusky brown spots or bars; wing-coverts dark brown, the cross bars more rufous and broader than on the back, those on the greater series more broadly barred with white, the greater series externally marked with large white spots, the edge of the wing also white; primary-coverts dark brown, barred with rufous; quills dark brown, broadly banded with rufous, inclining to fulvous on the outer web and causing a slightly spotted appearance; the bands on the secondaries much narrower, especially on the innermost, where they resemble those of the scapulars; tail-feathers blackish brown, narrowly tipped with whitish and crossed with eight other narrow bands of white, these bars not being continuous; ear-coverts dusky brown, minutely spotted with fulvous; chin and cheeks white, as also the fore neck, separated from the two latter by a band of dusky blackish feathers, narrowly banded with rufous; rest of under surface white in the centre of the body, the flanks broadly barred with white and brown; sides of upper breast thickly barred with dusky brown and fulvous; under tail-coverts white, spotted with brown; leg-feathers fulvous, with a few small spots of brown; under wing-coverts white, washed with rufous, with a few small spots of brown, the lower series rufous white, tipped with brown; inner lining of quills deep brown, barred with dull rufous on the inner web, much brighter near the base of the latter, the bars gradually broken up; bill greenish horn-colour; feet greenish horn-colour; claws dusky; iris golden yellow. Total length 8 inches, wing 4·95, tail 2·85, tarsus 3·9.

Hab. India, from the Himalayas southwards through the peninsula, but not extending far below the Neilgherries.

a. Ad. sk.	India.	J. Gould, Esq. [P.].
b. Ad. st.	Himalaya Mts.	Major-Gen. Hardwicke [P.].
c. Ad. st.	Himalaya Mts.	Hon. East-India Co. [P.].
d. Ad. st.	Nepal.	B. H. Hodgson, Esq. [P.].
e, f. Ad. sk.	Nepal.	B. H. Hodgson, Esq. [P.].
g. Ad. sk.	Behar.	B. H. Hodgson, Esq. [P.].
h, i. Ad. sk.	Darjiling.	Purchased.

Subsp. *α*. **Glaucidium malabaricum.**

Noctua cuculoides, *Jerd. Madr. Journ.* x. p. 91.
Noctua spadicea, *Jerd. Madr. Journ.* xiv. p. 120.
Athene castanoptera, *Blyth, J. A. S. B.* xiv. p. 184.
Athene malabarica, *Blyth, J. A. S. B.* xv. p. 280, et xix. p. 500; *id. Cat. B. Mus. A. S. B.* p. 39; *Strickl. Orn. Syn.* p. 168; *Jerd. B. Ind.* i. p. 144; *Gray, Hand-l. B.* i. p. 40; *Hume, Rough Notes,* ii. p. 413.
Glaucidium malabaricum, *Sharpe, Ibis,* 1875, p. 259.

Adult. Head, neck, and interscapulary region tawny rufous, narrowly barred across with black, the scapulars and hinder portion of the back and rump fulvous or greyish buff, with much broader black bars, the outer scapulars more or less uniformly white externally, exhibiting a tolerably distinct shoulder-patch; wing-coverts tawny rufous, barred with black, the greater series white on the outer web, with a few irregular bars of black, forming a sort of alar bar; quills deep tawny or chestnut, broadly barred across with black, the bars replaced by spots, and gradually disappearing towards the base of the feathers; the innermost secondaries fulvous, barred with black, and resembling the scapulars; upper tail-coverts whitish, barred with black, and resembling the tail; the latter black, tipped with white, and crossed with nine other bars of the same colour; lores whitish, slightly washed with brown, the shafts developing into long hair-like bristles; round the eye a fringe of buffy white feathers; sides of face rufous, barred with brown; chin and cheeks white, the latter forming a triangular patch on the sides of the face, and separated from the fore neck, which is also white, by a broad band of tawny feathers, narrowly barred with black; rest of under surface of body white, the sides of the body thickly barred with black, these becoming more distinct on the flanks; the upper breast rufous or fulvescent, barred with black, and resembling the sides of the neck; leg-feathers white, tinged with rufous on the thighs, which have a few remains of blackish bars; under tail-coverts buffy white; under wing-coverts white, with a few brown markings near the edge of the wing, where there is also a tinge of rufous, the greater series with dusky brown tips. Total length 7·8 inches, wing 4·9, tail 2·9, tarsus 0·9.

Obs. Whether *G. malabaricum* be specifically distinct from *G. radiatum* will be a matter for Indian ornithologists to decide; but the general appearance of Malabar birds is very different from those of Northern India. The rufous head and wings, and the greater breadth of the white bars on the tail, are very conspicuous in the southern form. Mr. Hume (Rough Notes, ii. p. 411) notices certain differences between these little Owls from Northern and Southern India, but does not consider the difference to be specific; the Museum also contains both grey and rufescent birds from the Himalayas, but none so chestnut in appearance as the ones from Malabar.

Hab. South India.

a, b. Ad. sk.	South India.	J. Gould, Esq. [P.].
c. Ad. st.	Madras.	Dr. Jerdon [P.].
d, e. Ad. st.	Madras.	Rev. H. H. Baber [P.].
f, g. Ad. sk.	Malabar.	Purchased.

18. Glaucidium cuculoides.

Noctua cuculoides, *Gould, Cent. Himal. B.* pl. 4; *Schl. Mus. P.-B.* Striges, p. 33; *id. Revue Accipitr.* p. 28.
Noctua auribarbis, *Hodgs. J. A. S. B.* vi. p. 369.
Athene auribarbis, *Gray, Gen. B.* i. p. 35.
Athene cuculoides, *Gray, Gen. B.* i. p. 35; *Blyth, Cat. B. Mus. A. S. B.* p. 38; *Cass. Cat. Strigidæ Philad. Mus.* p. 14; *Bp. Consp.* i. p. 40; *Horsf. & Moore, Cat. B. Mus. E. I. Co.* i. p. 66; *Strickl. Orn. Syn.* p. 167; *Jerd. B. Ind.* i. p. 145; *Gray, Hand-l. B.* i. p. 39; *Hume, Rough Notes,* ii. p. 414; *id. Nests and Eggs Ind. B.* i. p. 71.

Tænioglaux cuculoides, *Kaup, Isis*, 1848, p. 769; *id. Contr. Orn.* 1852, p. 106; *id. Tr. Z. S.* iv. p. 212.
Glaucidium cuculoides, *Sharpe, Ibis*, 1875, p. 259.

Adult. Above brown, the feathers all slightly tipped and distinctly barred across with ochraceous, the bars on the hind neck wider and here and there inclining to white spots, forming an indistinct kind of collar; scapulars externally white, showing a broad longitudinal shoulder-patch; wing-coverts marked exactly like the back, but the bars rather more rufous ochre, the outer median and greater coverts with large and distinct spots of white; primary-coverts dark brown; quills dark brown, tipped with whitish, the primaries obsoletely barred on the inner web with lighter brown, and notched on the outer one with fulvous, more or less inclining to rufous; the secondaries barred with fulvous on the inner web, the bars nearly traversing the innermost secondaries; upper tail-coverts darker brown than the back, barred with ochraceous, the lower ones with white; tail-feathers blackish brown, narrowly tipped with whitish and crossed with six other bands of white; lores and an indistinct eyebrow dull white, the former obscured with blackish shaft-lines; ear-coverts fulvous, barred with blackish; cheeks white, forming a triangular patch on the sides of the neck; chin and fore neck whitish, separated from each other by a tolerably broad band of dark brown feathers, somewhat mottled with ochraceous bars; the sides of the upper part of the breast brown, distinctly barred across with ochraceous, these bars dissolving into lateral spots on the lower sides of breast; rest of under surface of body white, very broadly streaked with brown, the flanks barred with ochraceous buff, but less distinctly than the breast; under tail-coverts white; leg-feathers fulvous, with dusky brown cross bars; under wing-coverts yellowish, inclining to white near the edge of the wing, the greater series yellowish at base, brown at tip, the wing-lining being dark brown, barred with yellowish near the base of the inner web and inclining more to rufous. Total length 11 inches, wing 6·2, tail 3·8, tarsus 1.

Obs. None of the Museum specimens are sexed, and I therefore cannot define the relative size of the sexes. Specimens vary much in colouring, some being very much darker than others.

Young. In general similar to the adults, but having the upper surface much more spotted, none of the bars being so distinct, and the whole appearance of the upper surface more rufous; on the hind neck a few whitish triangular spots; patch of white on scapulars absent; wing-coverts with distinct whitish external marks, but these somewhat washed with ochraceous; underneath more dingy and obscured than the adult, the bars on the sides of the breast very faintly indicated.

Hab. Himalaya Mountains, extending eastwards to Burmah and Siam.

a. Juv. st.	India.	H. Drummond, Esq. [P.].
b. Juv. sk.	India.	Purchased.
c. Ad. st.	Himalaya Mountains.	Major-Gen. Hardwicke [P.].
d. Juv. sk.	Darjiling.	J. Gould, Esq. [P.].

e, f. Ad. st.	Nepal.	B. H. Hodgson, Esq. [P.].
g. Juv. st.	Nepal.	B. H. Hodgson, Esq. [P.].
h, i, k, l. Ad. sk.	Nepal.	B. H. Hodgson, Esq. [P.].
m. Juv. sk.	Nepal.	B. H. Hodgson, Esq. [P.].
n. Ad. st.	Moalmaza.	
o. Ad. sk.	Tonghoo, Burmah.	Purchased.
p. Ad. sk.	Siam.	M. Mouhot [C.].

Obs. The specimen from "Moalmaza" is thus marked by Mr. G. R. Gray; at present I have not found this name in an atlas. He calls it "*Athene cuculoides*, var.;" but I cannot see that it is different from the ordinary species. It is more rufous underneath, and has the stripes on the abdomen broken up into bars; but I find this appearance on some of the Nepal examples, though scarcely to the same extent. The head is uniformly coloured with the back. I was at one time inclined to separate the Siamese bird as distinct on account of its grey head and orange-striped under surface. The Museum contains a specimen collected by the late M. Mouhot in Siam, of which I append an exact description; but I can scarcely believe it to be a species, as I saw specimens in the Paris Museum, obtained by M. Bocourt in Siam, which seemed to me inseparable from Himalayan examples.

Adult. Upper surface brown, everywhere crossed with narrow ochraceous bars, inclining to whitish on the hind neck, scapulars, and upper tail-coverts, the bars on the latter rather broader; outer web of scapulars pure white, forming a narrow longitudinal shoulder-patch; wing-coverts more rufous brown than the back, barred across with rufous ochre, the median series externally white, forming a broad but imperfect wing-bar; primary-coverts brown, with obsolete rufous cross bars; quills dark brown, barred with rufous on the inner web, these bars being represented on the outer web by a large fulvescent notch; the secondaries completely barred across with rufous, inclining to fulvous on the outer web, the bars on the innermost secondaries rather narrower; tail brown, slightly washed with rufous, and crossed with seven bands of white, nowhere very broad, and tinged with fulvous near the base; head and neck much greyer than the rest of the back, everywhere barred across with whitish, more broadly just behind the ear and on the sides of the neck; lores dusky, the shafts white near the base; ear-coverts dull brown, barred across with buffy white; cheeks white, forming a triangular patch; chin whitish, separated from the fore neck, which is also white, by a band across the throat, consisting of brown feathers barred with fulvous like the adjoining sides of the neck; the sides of the upper breast dark brown, barred with buffy white, the brown bars gradually becoming light rufous towards the lower breast; rest of under surface white, broadly streaked with pale rufous, the flanks barred with dusky brown and pale rufous; under tail-coverts white; leg-feathers fulvous white, slightly spotted with brown. the plumes at the bend of the tarsus light rufous, slightly marked with brown spots; under wing-coverts white, the innermost and the axillaries yellowish, the greater ones washed with rufous and tipped with ashy brown, appearing as if part of the lining of the quills, which are ashy brown below, barred across with yellowish, inclining to pale rufous towards the tips of the feathers. Total length 10·5 inches, wing 5·8, tail 3·6, tarsus 1·1.

19. Glaucidium whitelyi.

Athene cuculoides, *Swinh. Ibis*, 1861, pp. 25, 265; *id. P. Z. S.* 1863, p. 262; *id. Ibis*, 1867, p. 406.
Athene whitelyi, *Blyth, Ibis*, 1867, p. 313; *Gray, Hand-l. B.* i. p. 40; *Hume, Rough Notes*, ii. p. 415; *Jerd. Ibis*, 1871, p. 349; *Swinh. P. Z. S.* 1871, p. 343.
Glaucidium whitelii, *Sharpe, Ibis*, 1875, p. 259.

Adult male. Above brown, everywhere broadly barred across with ochraceous, the bars broader and a little more rufescent on the scapulars, the bars on the hind neck rather whiter, especially towards the base of the feathers; wing-coverts brown, the bars deeper ochraceous and wider apart than on the back; the outer scapulars with large white spots externally, forming a distinct shoulder-patch, the outer and median wing-coverts also with very large white terminal spots, forming a longitudinal band down the wing; primary-coverts dark brown, slightly tipped with whitish, and crossed with a few nearly obsolete bars of lighter brown; quills dark brown, tipped with white and crossed with bars of lighter brown, these bars represented on the outer web by fulvous or fulvous brown notches, these notches widening into bars on the secondaries, extending right across the innermost of them; tail blackish, broadly tipped with white, and crossed, moreover, with five other bands of white; lores and a tolerably distinct eyebrow white, the former obscured by central shaft-streaks of black; ear-coverts barred with ochraceous and black; cheeks white, extending backwards and forming a triangular white patch behind the ear-coverts; throat pure white, as also the fore neck, these being separated from each other by a band of ochraceous feathers barred with brown, extended on both sides and joining the sides of the neck, which are similarly barred; sides of breast broadly barred with ochraceous and dark brown; rest of under surface white, the abdomen broadly streaked with rather pale brown, the flanks thickly mottled with brown bars; thighs for the most part white, washed with fulvous on the upper part, and barred externally with brown; under tail-coverts white, with a few narrow brown shaft-streaks near the tips of the feathers; under wing-coverts and edge of wing white, the inner ones washed with yellowish, the lower series yellowish at base, ashy brown at tip, thus resembling the inner lining of the quills, which are ashy brown below, broadly barred with yellowish on the inner web, much more distinctly on the secondaries. Total length 11·5 inches, wing 6·4, tail 4·3, tarsus 1·25.

Adult female. Rather browner than the male, and not so closely mottled with ochraceous bars; bands on tail six, not including the terminal one. Total length 12 inches, wing 6·3, tail 4·5, tarsus 1·3.

Obs. Both the Fokien birds in the collection have five bars on the tail besides the terminal one; they are, however, much more rufous than the specimens described, all the cross bars of the upper surface being rufous, and in general appearance they look much more like Indian specimens of *G. cuculoides*. *G. whitelii*, however, is always larger than the latter, and never has more than six bars on the tail, instead of seven as in the Himalayan species.

Hab. China.

a, b. Ad. sk.	Fokien, China.	J. Gould, Esq. [P.].
c. ♀ ad. sk.	Ningpo, Dec. 24, 1871.	R. Swinhoe, Esq. [C.].
d. ♂ ad. sk.	Ningpo, Dec. 31, 1871.	R. Swinhoe, Esq. [C.].

20. Glaucidium capense.

Athene capensis, *Smith, S. Afr. Quart. Journ.* ser. 2, no. 4, pt. 1, p. 313; *id. Ill. Zool. S. Afr.* Aves, pl. 33; *Gray, Gen. B.* i. p. 35; *Bp. Consp.* i. p. 37; *Strickl. Orn. Syn.* p. 169; *Layard, B. S. Afr.* p. 37; *Gray, Hand-l. B.* i. p. 40; *Finsch & Hartl. Vög. Ostafr.* p. 98.

Tænioglaux capensis, *Kaup, Isis*, 1848, p. 769; *id. Tr. Z. S.* iv. p. 212; *Gurney, Ibis*, 1868, p. 147; *id. in Anderss. B. Dam. L.* p. 38.

Smithiglaux capensis, *Bp. Rev. et Mag. de Zool.* 1854, p. 544.

Noctua capensis, *Schl. Revue Accipitr. Mus. P.-B.* p. 28.

Carine capensis, *Sharpe, ed. Layard, B. S. Afr.* p. 77, pl. 3.

Glaucidium capense, *Sharpe, Ibis*, 1875, p. 259.

Adult. General colour above deep brown, narrowly barred with rufous ochre; the head lighter brown, minutely dotted with whitish, rather inclining to small bars on the hind neck; scapulars white, tipped with brown, forming a very distinct longitudinal band down the body; wing-coverts brown, narrowly barred with rufous ochre like the back, the outer median and greater coverts with a large spot of pure white on the outer web; primary-coverts uniform sepia-brown; quills dark brown, narrowly banded with rufous ochre, the primaries distinctly spotted with white on the outer web, all the quills more broadly barred with white on the inner web, plainer on the secondaries, especially the innermost; tail brown, crossed with fourteen narrow bars of rufous ochre, including the terminal one; loral plumes whitish, with brown shafts; ear-coverts brown, streaked with buffy white; cheeks white, forming a triangular patch; sides of neck brown, spotted with minute whitish bars like the hind neck; chin dusky white; throat brownish; fore neck white, the feathers tipped with pale brown; chest white in the centre, barred with brown, the sides of the breast barred with fulvous and brown; rest of under surface pure white, with distinct terminal spots and bars of brown; under tail-coverts whitish; leg-feathers yellowish buff, with a few brown bars; under wing-coverts yellowish white, the edge of the wing pure white, the adjoining feathers brown, forming an irregular longitudinal band just inside the wing, the lower series brown, yellowish at base; inner lining of wing brown, the primaries barred on the inner web with ochraceous, the secondaries yellowish at base of inner web, barred with the same colour for the greater part of this web. Total length 11 inches, wing 5·9, tail 4, tarsus 1.

Hab. South Africa.

a. Ad. st.	South Africa.	Sir A. Smith [C.].	Type of species.

14. MICRATHENE.

Type.
Micrathene, *Coues, Pr. Philad. Acad.* 1866, p. 15 .... M. whitneyi.

Range. Colorado region of the United States and Western Mexico.

1. Micrathene whitneyi.

Athene whitneyi, *Cooper, Pr. Calif. Acad. Sci.* 1861, p. 118.
Micrathene whitneyi, *Coues, Pr. Philad. Acad.* 1866, p. 15; *Elliot, B. N. Am.* pl. xxix.; *Coues, Key N. Am. B.* p. 207; *Lawr. Mem. Bost. Soc. N. H.* ii. p. 297; *Ridgw. B. N. Amer.* iii. p. 87; *Coues, B. N. W.* p. 320.

Above umber-brown (less pure and uniform than in *Glaucidium*), each feather with an irregular transversely elliptical spot of pale rufous, these largest on the forehead bordering the white eyebrows; the feathers everywhere minutely mottled transversely with darker, this being most noticeable where bordering the yellowish spots. Scapulars with their outer webs almost wholly white. Wings with the ground-colour a little darker than the back; lesser coverts with numerous spots of light rufous, there being two on each feather, one concealed; middle and secondary coverts with a very large oval spot of pure white terminating the outer webs, the white spot on the latter preceded by a pale rufous one; secondaries with five (exposed) bands of pale ochraceous (the last terminal), these passing into white on the edge; primary-coverts with three large ochraceous spots; primaries with about six (including the terminal) conspicuous spots of the same, those anterior to the emargination on the third, fourth, and fifth quills almost white; tail like the wings, but more uniform, crossed by six irregular narrow bands of pale ochraceous, the last or terminal of which is not well defined; these do not touch the shaft, and on the inner webs they are pure white; lores and eyebrows, cheeks, lining of the wings, and ground-colour of the lower parts white; ear-coverts and suborbital space like the crown, but more rusty; lateral lower parts much washed with plumbeous, this especially prevalent on the flanks; behind the sharply defined white of the cheeks is a black transverse wash; throat, jugulum, breast, and abdomen with each feather having a median longitudinal ragged-edged blotch of pale rufous, these blotches most clearly defined on the abdomen, more confused anteriorly; anal region and tibiæ almost immaculate; tibiæ with numerous transverse narrow blackish bars on a pale ochraceous ground; lining of the wing faintly spotted at the bend and on the primary-coverts, the terminal half of which is plain dusky; under surface of primaries blackish, with obscure transverse paler spots, those anterior to the emargination almost white, those beyond darker, the last being scarcely distinguishable, on the longest quill eight can be detected; bill pale greenish; iris yellow. Length "6·25" inches, extent "15·25," wing 4·40, tail 2·30, culmen 0·35, tarsus 0·80, middle toe 0·60. (*Ridgway.*)

A male from Socorro Island (Colonel A. J. Grayson) is less adult than the preceding. The upper plumage is more brownish and more mottled; the rufous spots, though deeper and larger, are less sharply defined; the spots on the primaries are all ochraceous; the bands on the tail are broader though of the same number. Beneath the longitudinal blotches do not appear; but the rusty rufous covers nearly the whole surface, leaving the median portion only white, and this not well defined; the rusty shows ragged minute transverse bars of blackish. The whitish collar round the nape is also better defined than in the type. Total length 5·20 inches, wing 4·20, tail 2·10.

Another specimen, from the same locality, also apparently immature, is just like the preceding in plumage. It measures—wing 4 inches, tail 1·90. (*Id. l. c.*)

Hab. The same as that of the genus.

Subfam. II. SYRNIINÆ.

Key to the Genera.

a. Feathered tufts on head very distinct; cere much longer than culmen .... 15. *Asio*, p. 225.

b. No tufts of feathers on head; cere not so long as culmen.

  a'. Bony shell of ear-conch the same on both sides of the skull; toes sometimes thickly feathered or at least with hair-like bristles, sometimes entirely bare .... 16. *Syrnium*, p. 244.

  b'. Bony shell of ear-conch entirely asymmetrical; toes thickly feathered .... 17. *Nyctala*, p. 284.

15. ASIO.

	Type.
Asio, *Briss. Orn.* i. p. 28 (1766)	A. otus.
Otus, *Cuvier, Leçons Anat. Comp.* tabl. ii. (1799)	A. otus.
Bubo, *Duméril, Zool. Anal.* p. 34 (1806)	A. otus.
Nyctalops, *Wagler, Isis*, 1832, p. 1221	A. stygius.
Brachyotus, *Gould, P. Z. S.* 1837, p. 10......	A. galapagensis.
Pseudoscops, *Kaup, Isis*, 1848, p. 769........	A. grammicus.
Phasmoptynx, *Kaup, Isis*, 1848, p. 769	A. capensis.
Rhinoptynx, *Kaup, Contr. Orn.* 1852, p. 114..	A. mexicanus.
Nisuella, *Bp. Rev. et Mag. de Zool.* 1854, p. 542	A. madagascariensis.

Cf. *Sharpe, Ibis*, 1875, p. 324.

Foot of *Asio otus*.

Range. The entire world, excepting W. Africa, Malaisia, Australia, and Oceania: present in the Sandwich Islands.

Key to the Species.

a. Light bands on quills less than ten in number.
 a'. Under surface of quills barred across with brown, and the tips entirely brown; face more or less fulvescent, the toes also more or less plumed.
  a''. Ear-tufts very long, about equal to middle toe and claw.
   a'''. Breast streaked with black, with lateral bars or vermiculations of black also .... *otus*, p. 227.
    α. Underneath orange-buff; the breast-feathers white, mesially streaked with dull brown, each feather mottled with vermiculations on each side of the brown central stripe. *α. otus*, p. 227.
    β. Underneath orange-buff, the breast-feathers white, not only mesially streaked with dark brown, but also laterally barred with the same; the whole tone of plumage darker than in *A. otus*. *β. americanus*, p. 229.
   b'''. Breast-feathers broadly streaked with black, without lateral bars or vermiculations.
    a[4]. Inner lining of quills fulvous, barred with dark brown, the first primary with five dark bars; face whitish, rufous below the eye .... *mexicanus*, p. 231.
    b[4]. Inner lining of quills dark brown, with fulvous bars, five in number, on first primary; face dusky buff, with white shaft-streaks........ *madagascariensis*, p. 232.
  b''. Ear-tufts short, less than the middle toe and claw.
   c'''. Upper surface fulvous or tawny, streaked with brown down the centre of the feather........... *accipitrinus*, p. 234.
   d'''. Upper surface uniform brown, with here and there slight rufescent vermiculations, scarcely perceptible.. *capensis*, p. 239.
 b'. Under surface of quills nearly uniform chocolate-brown, with one band of yellowish white about halfway up the first primary; toes partially bare, the inner one entirely so .... *stygius*, p. 241.
b. Light bands on quills ten in number, and about the same number on the tail-feathers; face foxy red; bill yellow; toes bare .... *grammicus*, p. 242.

1. Asio otus*.

The Horn Owl, *Albin, Birds*, ii. pl. 10.
Le moyen Duc ou Le Hibou, *Briss. Orn.* i. p. 486; *Buff. Pl. Enl.* 29.
Le Hibou d'Italie, *Briss. Orn.* i. p. 491.
Strix otus, *Linn. S. N.* i. p. 132; *Wolf & Meyer. Vog. Deutschl.* Taf. to p. 131; *Vieill. Faun. Franç.* p. 42, pl. 19. fig. 2; *Naum. Vog. Deutschl.* i. p. 451, pl. 45. fig. 1; *Roux, Orn. Prov.* i. p. 71, pl. 47; *Werner, Atlas*, Rapaces, pl. 41; *Nordm. in Demid. Voy. Russie Mérid.* iii. p. 111; *Yarr. Brit. B.* i. p. 117; *Schl. Vog. Ned.* pl. 45; *Sundev. Sv. Fogl.* pl. xxiv. fig. 1.
Otus albicollis, *Daud. Traité*, ii. p. 213.
Otus italicus, *Daud. Traité*, ii. p. 213.
Bubo otus, *Savign. Descr. Egypte*, p. 292.
Otus asio, *Leach, Syst. Cat. Mamm. &c. B. M.* p. 11.
Otus europæus, *Steph. Gen. Zool.* xiii. pt. 2, p. 57.
Otus vulgaris, *Fleming, Brit. An.* p. 56; *Gould, B. Eur.* pl. 39; *Gray, Gen. B.* i. p. 40; *Bp. Consp.* i. p. 50; *Kaup, Contr. Orn.* 1852, p. 113; *id. Tr. Z. S.* iv. p. 232; *Horsf. & Moore, Cat. B. Mus. E. I. Co.* i. p. 206; *Hewits. Eggs Br. B.* i. p. 55; *Fritsch, Vog. Eur.* Taf. 11. fig. 3; *Jaub. et Barth. Lapomm. Rich. Orn.* p. 81; *Schl. Mus. P.-B.* Oti, p. 2; *Jerd. B. Ind.* i. p. 125; *Degl. & Gerbe, Orn. Eur.* i. p. 138; *Gould, B. Gt. Br.* i. pl. xxxi.; *Loche, Exp. Sci. Algér. Ois.* p. 96; *Heugl. Orn. N.O.-Afr.* i. p. 107; *Hume, Rough Notes*, ii. p. 363; *Salvad. Faun. Ital. Ucc.* p. 32.
Asio otus, *Less. Man. d'Orn.* i. p. 116; *Macgill. Brit. B.* iii. p. 453; *Strickl. Orn. Syn.* p. 205; *Newt. Ooth. Wolley.* p. 154; *id. ed. Yarr. Brit. B.* i. p. 158; *Shelley, B. Egypt*, p. 178.
Otus otus, *Cuvier, Règne An.* i. p. 341; *Schl. Rev. Crit.* p. xiv.
Ulula otus, *Macgill. Brit. B.* Raptores, p. 403.
Otus communis, *Less. Traité*, p. 110.
Otus aurita, *Rennie, ed. Mont. Orn. Dict.* p. 262.
Otus sylvestris, *Brehm, Vog. Deutschl.* p. 121.
Otus arboreus, *Brehm, l. c.* p. 122.
Otus gracilis, *Brehm, l. c.* p. 123.
Ægolius otus, *Keys. & Blas. Wirb. Eur.* p. xxii; *Borggr. Vogelf. Norddeutschl.* p. 63.
Otus major, *Brehm, Naum.* 1855, p. 270.
Otus minor, *Brehm, l. c.*
Otus assimilis, *Brehm, l. c.*
Otus verus, *Finsch, J. f. O.* 1859, p. 381.

---

* The following is a species recognized by Von Heuglin, but at present unknown to me:—

ASIO ABYSSINICUS.

Otus abyssinicus, *Guérin-Ménev. Rev. Zool.* 1843, p. 321; *id. et Lafr. Voy. Abyss. Zool.* iii. *Ois.* p. 185, pl. 3; *Gray, Gen. B.* i. p. 40; *Finsch & Hartl. Vog. Ostafr.* p. 111, note.
Asio abyssinicus, *Strickl. Orn. Syn.* p. 211.
Ægolius montanus, *Heugl. Syst. Uebers.* p 89
Otus habessinicus, *Heugl. Orn. N.O.-Afr.* p. 107.

Similar to *A. otus*, but larger, with shorter ear-tufts, the upper surface of a more obscure smoky brown, and less waved with ash-colour; toes half bare; iris reddish yellow; bill dusky horn-colour, paler at tip; toes pale lead-colour, cere fleshy grey. Total length about 14 inches, wing 1·1–1·3, tail 5·4, tarsus 1·7–2. (*Heuglin.*)

Hab. Abyssinia.

Nestling. Covered with dusky brown down, the filaments of which are ashy white at their tips, causing a general hoary appearance; the sprouting plumes fulvous, with brown cross bars; quills ashy grey, barred and vermiculated with dark brown; the secondaries dark brown, slightly mottled with fulvous brown; tail fulvous, with whitish tip and crossed with dark brown bars; feathers of under surface yellowish buff, with narrow cross bars of dark brown.

Adult female. Above blackish brown, mottled all over with orange-buff, the bases of the feathers being generally of this colour; all the dorsal plumes likewise silvered with white, over which vermiculations of dark brown are more or less distributed; wing-coverts coloured like the back, the greater series as well as the scapulars having a large oval spot of white on the outer web; edge of the wing white; primary-coverts blackish brown, inclining to orange-buff near the base of the outermost feathers; quills greyish brown, with hoary tips, and barred with darker brown, the bands much broader on the primaries, which are more or less inclining to orange-buff on the interspaces, especially bright near the base; the secondaries more silvery grey, mottled with dark brown vermiculations, the darker brown bars more obsolete and dissolved on the innermost; upper tail-coverts orange-buff, dark brown in the centre and vermiculated with the same towards the tips; tail-feathers greyish or fulvous brown, inclining to orange-buff near the base, and crossed by seven bands of darker brown, more or less decomposed and vermiculated towards the tip, the outer feathers more dusky and having about ten narrow bands of dark brown; head pale orange-buff, with blackish brown longitudinal centres to the feathers, many of which are whitish on their margins and vermiculated with dusky brown lines, the plumes on the sides of the neck much whiter, and the cross lines almost obsolete; frontal feathers greyish white, very minutely freckled with brown bars; ear-tufts 1·65 inch long, blackish, inclining to orange-buff on the outer web and to white on the inner one; face dusky white, obscured with hair-like lines of black; feathers round the eye blackish; ear-coverts yellowish, with very faint vermiculations of dusky brown; facial ruff white at base, all the feathers tipped with black, forming a frill; chin whitish; throat-feathers long and stiff, in continuation of the ruff, spotted and centred with dark brown; rest of under surface of body orange-buff, the flanks and leg-feathers entirely of this colour; the breast-feathers for the most part white, with a broad centre of dark brown, and having a few smaller frecklings of brown on either web; under wing-coverts white, washed with orange-buff, a few of the feathers narrowly lined with brown, the greater series white at base and broadly brown at tip, forming a bar across the wing and resembling the inner lining of the quills, which are ashy brown barred with darker brown, the interspaces more or less yellowish and the bases entirely orange-buff; bill dusky horn-colour; claws horn-colour; iris orange-yellow. Total length 13·5 inches, wing 11·6, tail 6, tarsus 1·6.

Adult male. Similar to the female. Total length 14 inches, wing 11·5, tail 6·5, tarsus 1·6.

Hab. The whole of Europe, ranging across Siberia to Japan and China: Himalayas and N.W. India. Egypt and N.E. Africa in winter.

a. Pull. sk.	England.	J. Gould, Esq.
b. Ad. st.	England.	
c. Var. st.	England.	
d. Ad. st.	England.	J. Warniford, Esq. [P.].
e. Ad. st.	Oxfordshire.	Col. Montagu [P.].
f. ♂ ad. sk.	Christchurch, Hants, Jan. 19, 1874.	Baron A. von Hügel [P.].
g, h. ♀ ad. sk.	Avington, Hants, Nov. 10, 1873.	R. B. Sharpe, Esq. [P.].
i. Ad. st.	Essex.	Purchased.
k. ♀ juv. sk.	Skåne, S. Sweden, Jan. 9, 1874.	Purchased.
l. ♂ ad. sk.	Skåne, S. Sweden, Jan. 9, 1874.	Purchased.
m. ♀ ad. sk.	Skåne, S. Sweden, Feb. 7, 1874.	Purchased.
n. ♂ juv. sk.	Skåne, S. Sweden, Feb. 7, 1874.	Purchased.
o. ♂ ad. sk.	Belgrade, Turkey, March 27, 1870 (*T. Robson*).	R. B. Sharpe, Esq [P.].
p. ♀ ad. sk.	Petnahore, Turkey, Dec. 30, 1870 (*T. Robson*).	R. B. Sharpe, Esq. [P.].
q. Ad. sk.	Algeria.	M. A. Bouvier.
r. ♂ ad. sk.	Granada, April 1870.	Howard Saunders, Esq. [C.].
s. Skeleton.		Purchased.
t, u, v. ♂ ♀ skulls.	Rauhe Alp.	Dr. Gunther [C.].
w. Sternum.	England.	T. Ingall, Esq. [P.].
x. Sternum.	England.	W. Yarrell, Esq. [C.].
y. Sternum.	England.	J. Ray, Esq. [P.].

Subsp. *a.* **Asio americanus** *.

Strix otus, *Wils. Am. Orn.* vi. p. 73, pl. 51. fig. 3; *Sw. & Rich. Faun. Bor.-Am. Birds*, p. 72; *Audub. B. Amer.* pl. ccclxxxiii., et *Orn. Biogr.* iv. p. 572.

Otus americanus, *Steph. Gen. Zool.* xiii. pt. 2, p. 57; *Bp. Comp. List B. Eur. & N. Amer.* p. 7; *id. Consp.* i. p. 50; *Kaup, Contr. Orn.* 1852, p. 113, et *Tr. Z. S.* iv. p. 232; *Dekay, Zool. N. Y.* p. 27, pl. xi. fig. 24; *Gray, Hand-l. B.* i. p. 57.

Otus wilsonianus, *Less. Traité*, p. 110; *Cass. B. Calif.* p. 181; *Cass. in Baird, B. N. Am.* p. 53; *Baird, ed. Cooper, B. Calif.* p. 426.

Otus vulgaris, *Jardine, ed. Wils. Am. Orn.* ii. p. 278.

Ulula otus, *James. ed. Wils. Am. Orn.* i. p. 104.

Otus wilsonius, *Gray, Gen. B.* i. p. 40.

Otus zonurus, *Kaup, Contr. Orn.* 1852, p. 232 (*ex Gray, MS.*); *id. Tr. Z. S.* iv. p. 232; *Gray, Hand-l. B.* i. p. 50.

---

* *Strix peregrinator* of Bartram (Trav. Calif. p 285), allowed by Strickland to be the oldest title of this species, appears to me to be quite unrecognizable from the description.

Asio peregrinator, *Strickl. Orn. Syn.* p. 207.
Otus vulgaris americanus, *Schl. Mus. P.-B.* Oti, p. 2.
Otus vulgaris, var. americanus, *Ridgw. in Coues, Key N.-Amer. B.* p. 204; *id. N. Amer. B.* iii. p. 18; *Coues, B. N.W.* p. 304.

Adult. Above dusky blackish, with only a slight shade of orange-buff, where the bases show through; all the feathers of the upper surface freckled with white on either web, causing a very distinctly waved appearance, very plainly exhibited on the scapulars; crown of the head and neck dark, like the rest of the upper surface, the white cross markings rather darker than on the back; wing-coverts also uniform with the back, the white wavy markings obscure, the greater coverts with tolerably large oval spots of white on the outer web; quills ashy brown, as also the primary-coverts, barred with darker brown, the interspaces also mottled with the same, the secondaries with wavy lines of white, resembling the dorsal plumes, the primaries more broadly banded than the secondaries and inclining to orange near the base of the outer web; tail-feathers ashy brown, tipped with greyish white and crossed with nine bars of darker brown, the bars narrowing in extent towards the tips of the feathers, where they become more or less merged in the mottlings which cover the interspaces, the bases of all the feathers more or less inclining to orange-buff; forehead dark brown, mottled with small greyish white bars; ear-tufts dark brown, inclining to orange-buff on the outer web; the inner web fulvous, with white notches, as on the rest of the upper surface; lores and feathers over the eye white, with black hair-like centres; face deep orange-buff; the feathers round the eye black, the facial disk black, the bases to the feathers white, the ruff succeeded by a second line of small greyish feathers mottled with black cross lines behind the ear-coverts; chin whitish; throat-feathers long and stiff, dark brown, mottled with white on either web, in continuation of the facial ruff; rest of under surface slightly washed with orange-buff, most of the feathers being white with a broad longitudinal centre of dark brown, from which radiate on each web some broad and distinct bars of the same colour; leg-feathers orange-buff, as also the under tail-coverts, which incline to white near the tips and are barred with brown; under wing-coverts orange-buff, spotted with brown, and streaked with the same near the outer margin, which is white, the greater series yellowish at base, dark brown at tip, thus resembling the inner lining of the wing, which is ashy brown, barred with darker brown and inclining to orange-buff near the base; soft parts as in *A. otus.* Total length 14·5 inches, wing 12, tail 6·5, tarsus 1·65.

Hab. North America, as far south as Mexico.

a.	Ad. st.	Fort Simpson.	Bernard R. Ross, Esq. [P.].
b.	Ad. sk.	Sumass, British Columbia.	J. K. Lord, Esq. [C.].
c.	Ad. st.	Mexico.	Purchased.
d.	Ad. sk.	W. Mexico.	Alex. Dorman, Esq. [C.].
e.	Ad. sk.	W. Mexico.	Alex. Dorman, Esq. [C.].
f.	Ad. sk.	Valley of Mexico.	M. A. Bouvier.
g.	Ad. st.		Type of *Otus zonurus.*

2. Asio mexicanus*.

Le Hibou du Brésil, *Briss. Orn.* i. p. 499.
Mexican Eared Owl, *Lath. Gen. Syn.* i. pt. 1, p. 123.
Strix mexicana, *Gm. Syst. Nat.* i. p. 288 (*ex Lath.*); *Audub. Orn. Biogr.* iv. p. 574.
Nacurutú chorreado, *Azara, Apunt.* i. p. 202.
Bubo clamator, *Vieill. Ois. de l'Amér. Sept.* p. 52, pl. 20.
Strix maculata, *Vieill. N. Dict. d'Hist. Nat.* vii. p. 45.
Strix clamata, *Max. Beitr.* iii. p. 281; *Bonn. et Vieill. Enc. Méth.* iii. p. 1279.
Strix longirostris, *Spix, Aves Bras.* p. 20, pl. 9 *a*.
Otus mexicanus, *Cuvier, Règne Anim.* i p. 341 (1829); *Bp. Oss. Regn. An. Cuv.* p. 42; *Less. Traité,* p. 110; *Gray, Gen. B.* i. p. 40; *Hartl. Ind. Azara,* p. 3; *Kaup, Tr. Z. S.* iv. p. 235; *Cass. B. Calif. & Tex.* p. 195; *Gray, Hand-l.* i. p. 51; *Pelz. Orn. Bras.* pp. 10, 400.
Otus americanus, *Jard. Ann. N. H.* xviii. p. 118; *Burm. Th. Bras.* ii. p. 123; *Scl. & Salv. P. Z. S.* 1869, p. 634; *iid. Nomencl. Av. Neotr.* p. 116.
Scops clamator, *Gray, Cat. Accipitr.* 1844, p. 45.
Rhinoptynx mexicanus, *Kaup, Contr. Orn.* 1852, p. 114; *id. Tr. Z. S.* iv. p. 235; *Bp. Rev. et Mag. de Zool.* 1854, p. 541.
Asio mexicanus, *Strickl. Orn. Syn.* p. 208.
Bubo mexicanus, *Ridgw. B. N. Am.* iii. p. 61.

Adult. General colour above golden buff, everywhere more or less obscured with blackish, the centre of the feathers being entirely of the latter colour and producing a longitudinally striped appearance, the plumes having lateral vermiculations and wavy bars of blackish, so that the back has rather an obscure appearance; scapulars less vermiculated externally, so that they appear to have oval patches of golden buff on the outer web; wing-coverts golden buff, vermiculated with dark brown, the latter colour predominating on the outermost coverts; the primary-coverts almost entirely blackish brown, with bars of lighter brown obscured by darker vermiculations; quills ashy brown, with very distinct cross bands of darker brown, breaking up into vermiculations on the secondaries, all the interspaces being more or less obscured by frecklings of dark brown; the primaries lighter than the secondaries, and inclining to pale orange-buff towards their external base; upper tail-coverts brownish, slightly mottled with golden buff; tail dull fulvous, with seven tolerably broad bands of darker brown, narrower and more indistinct towards the apex, the inner webs banded with narrower bars of dark brown, about ten in number, the bases of the tail-feathers paler and more distinctly golden buff; crown of head and neck deeper golden than the back, the central streaks of black rather narrower and the vermiculations more scanty, being generally represented by thread-like bars branch-

* ASIO MIDAS.

Otus midas, *Licht. Nomencl. Av* p. 6; *Schl. Mus. P.-B.* Oti, p. 2, *Gray, Hand l. B.* i. p. 50.

Professor Schlegel says this is a large form of *A. mexicanus.* He does not unite it to the latter bird; and I therefore think it better to keep it distinct, at least for the present. The following are his measurements taken from the type in Berlin—total length 18 inches, wing 11·6, tail 5·9.

Hab. Montevideo

ing from the central streak; ear-tufts two inches long, dark brown, narrowly margined with golden buff on the external and broadly on the inner web; forehead rufous, with dark brown centres, eyebrow white and the lores white, the latter with black bristles; ear-coverts dull chestnut, some of the feathers whitish; round the face a disk of blackish brown feathers, with which the hinder margins of the ear-coverts coalesce, and behind the dark ruff appears another less characterized one of bright orange; chin whitish; the throat-feathers stiff, rufous with dark brown shaft-stripes; rest of under surface bright orange-buff, with broad dark brown centres, narrower on the abdomen and absent on the under tail-coverts; leg-feathers yellowish white; under wing-coverts whitish, somewhat tinged with buff and having a few brown spots near the outer margin, the greater series whitish at base, dark brown at tip, resembling the inner lining of the wing, which is ashy brown, barred with darker brown and inclining to yellowish buff near the base of the feathers. Total length 13·5 inches, wing 10·9, tail 6·5, tarsus 2·3.

Obs. The specimen described is a dark example with a chestnut face, which is also exhibited in a Bahian specimen. In other respects, however, the latter differs in being everywhere more rufous, the central stripes to the feathers of the upper surface being by no means so distinct, while the vermiculations are more generally dispersed over the upper surface; the eyebrow and lores are pure white, contrasting with the chestnut face.

Two other examples, the one from Brazil the other from Bogotá, are much more mealy in general appearance, and are much whiter underneath, with very little tinge of orange-buff or yellow; and consequently the black streaks stand out in bold relief. Above the general aspect is lighter, the scapulars having large external patches of white, as also the greater wing-coverts. The tail-bands, eight in number, are more broken up, and mingle with the vermiculations of the interspaces.

Hab. **Mexico southwards through Columbia to Brazil.**

a. Ad. st.	Brazil.	M. Claussen [P.].
b. Ad. st.	Brazil.	Purchased.
c. Ad. sk.	Bahia.	Dr. Wucherer [C.].
d. Ad. sk.	Bogotá.	R. B. Sharpe, Esq. [P.].
e. Ad. sk.	Panama.	M. E. Arcé [C.].

3. Asio madagascariensis.

Otus madagascariensis, *Smith, S.-Afr. Q. Journ.* ii. no. 4, pt. 1, p. 316; *Gray, Cat. Accipitr.* 1848, p. 106; *Cass. Cat. Strigidæ Philad. Mus.* p. 3; *Gurney, Ibis,* 1869, p. 453; *Schl. N. T. D.* iv. p. 1; *id. Revue Accipitr.* p. 2.

Bubo madagascariensis, *Bp. Consp.* i. p. 44; *Pucher. Arch. du Mus.* iv. p. 328, pl. 23; *Hartl. Faun. Madag.* p. 23; *Grand. Rev. et Mag. de Zool.* 1867, p. 353; *Gray, Hand-l. B.* i. p. 44.

Rhinoptynx madagascariensis, *Kaup, Tr. Z. S.* iv. p. 235; *id. Contr. Orn.* 1852, p. 114.

Nisuella madagascariensis, *Bp. Rev. et Mag. de Zool.* 1854, p. 542.

Asio madagascariensis, *Strickl. Orn. Syn.* p. 212; *Newt. P. Z. S.* 1865, p. 834.

Adult (type of species). Above brown, with a slight purplish gloss, the feathers of the head rather distinctly margined with dull rufous, those of the hind neck with fulvous, producing a distinctly ocellated appearance; the back and scapulars irregularly mottled with fulvous and brown vermiculations, producing an irregularly barred appearance on the inner scapulars and lower back; wing-coverts dark brown, the inner ones with irregular bars of rufous buff, the greater series with faint bars of lighter brown, and with a large ocellum of buffy white at the tips, forming an ill-defined band, the same ocelli being smaller and less developed on some of the outer scapulars; primary-coverts dark brown, with faint bars of lighter brown; quills dark brown, tipped with ashy fulvous, externally chequered with square fulvous spots, those on the secondaries smaller and somewhat shaded with brown; lower surface of quills brown, barred with light fulvous, the bars not conterminous and much more distinct on the secondaries; upper tail-coverts brown, barred with fulvous; tail brown, tipped with ashy brown, and crossed with five bands of fulvous mottled with brown, the basal one very indistinct; ear-tufts 1·8 inch long, dark brown largely margined with fulvous; loral plumes buffy white; feathers round the eye black; ear-coverts deep fulvous, with black hair-like shaft-lines; facial ruff dark brown, the feathers margined and tipped with bright buff; under surface of body bright buff, inclining to orange, some of the feathers whitish buff on their margins, their centres dark brown, giving a streaked appearance, these streaks nearly absent on the thighs and less developed on the under tail-coverts; under wing-coverts deep orange-buff, marked longitudinally with brown, especially near the margin of the wings, which have a buffy white patch along the outer edge, the lower coverts dark brown, buff-coloured at base. Total length 17·5 inches, culmen 1·6, wing 10·8, tail 6·5, tarsus 1·9.

Hab. Madagascar.

Foot of *Asio madagascariensis*.

a. Ad. st.	Madagascar.	Sir A. Smith [C.].	Type of species.
b. Ad. sk.	Madagascar.	Mr. A. Crossley [C.].	

4. Asio accipitrinus.

La grande Chouette, *Briss. Orn.* i. p. 511.
Strix accipitrina, *Pall. Reise Russ. Reichs*, i p. 455.
Noctua minor, *Gm. N. Comm. Petrop.* xv. p. 447, pl. 12.
Strix brachyotus, *Forst. Phil. Trans.* lxii. p. 384; *Gm. S. N.* i. p. 289; *Wils. Am. Orn.* iv. p. 65, pl. 33. fig. 3; *Vieill. Faune Franç.* pl. 20; *Naum. Vog. Deutschl.* i. p. 459, pl 45. fig. 1; *Roux, Orn. Provenç.* i. p. 75, pl. 49; *Werner, Atlas*, Rapaces, pl. 42; *Sw. & Rich. Faun. Bor.-Am. Birds*, p. 75; *Audub. B. N. Am.* pl. ccccxxxii.; *id. Orn. Biogr.* v. p. 273; *Schl. u. Susem. Vog. Eur.* Taf. 51; *Yarr. Brit. B.* i. p. 121; *Schl. Vog. Nederl.* pl. 46; *Sund. Sv. Fogl.* p. 183, pl. xxiv. fig. 2; *Homeyer, J. f. O.* 1868, p. 256.
Caspian Owl, *Lath. Gen. Syn.* i. p. 147.
La Chouette, *Buff. Pl Enl.* ii. pl. 438.
Strix arctica, *Sparrm. Mus. Carls.* ii. pl. 51.
Strix ulula, *Gm. S. N.* i. p. 294; *Pall. Zoogr. Rosso-Asiat.* i. p. 322.
Strix tripennis, *Schrank, Fauna Boica*, i. p. 112.
Strix palustris, *Bechst. Naturg. Deutschl.* ii. p. 906.
Strix caspia, *Shaw, Gen. Zool.* vii. p. 272.
Strix ægolius, *Pall. Zoogr. Rosso-Asiat.* i. p. 309; *Nordm. in Demid. Voy.* iii. p. 112.
? Otus microcephalus, *Leach, Syst. Cat. Mamm. &c. Brit. Mus.* p. 11.
Strix brachyura, *Nilss. Orn. Suec.* i. p. 62.
Otus brachyotus, *Steph. Gen. Zool.* xiii. pt. 2, p. 57; *D'Orb. Voy. Am. Mérid.* p. 134; *Gould, B. Eur.* pl. 40; *Gray, Gen. B.* i. p. 40; *Cass. Cat. Strigidæ*, p. 3; *Lemb. Aves Isla de Cuba*, p. 21, pl. 4. fig. 1; *Kaup, Contr. Orn.* 1852, p. 114, et *Tr. Z. S.* iv. p. 236; *Horsf. & Moore, Cat. B. Mus. E. I. Co.* i. p. 79; *Cass. B. Calif. & Tex.* p. 182; *Burm. Th. Bras.* ii. p. 125; *Hewits. Eggs Br. B.* i. p. 58, pl. xvii. fig. 2; *Jaub. & Barth. Lap. Rich. Orn.* p. 82; *Scl. P. Z. S.* 1860, p. 384; *Jerd. B. Ind.* i. p. 127; *Schl. Mus. P.-B.* Oti, p. 3; *Newt. in Baring-Gould's Iceland*, p. 408; *id. Ooth. Wolley.* p. 167; *Pelz. Reis. Novara*, p. 27; *Degl. & Gerbe, Orn. Eur.* i. p. 136; *Gurney, Ibis*, 1868, p. 150; *Heugl. Orn. N.O.-Afr.* i. p. 108; *Gray, Hand-l. B.* i. p. 51; *Hume, Rough Notes*, ii. p. 364; *Finsch, Tr. Z. S.* vii. p. 211; *Swinh. P. Z. S.* 1871, p. 344; *Pelz. Orn. Bras.* pp. 10, 400; *Salvad. Faun. Ital. Ucc.* p. 33; *Allen, Bull. Harv. Coll.* ii. p. 341; *Scl. & Salv. Nomencl. Av. Neotr.* p. 116; *Schl. Revue Accipitr.* p. 5; *Ridgway, in Baird's B. N. Am.* iii. p. 23.
Asio ulula, *Less. Man. d'Orn.* i. p. 116.
Otus agrarius, *Brehm, Vög. Deutschl.* p. 124.
Otus palustris, *Brehm, Vög. Deutschl.* p. 124; *Darwin, Voy. Beagle, Birds*, p. 33.
Ulula brachyotus, *James. ed. Wils. Am. Orn.* i. p. 106.
Brachyotus galapagoensis, *Gould, P. Z. S.* 1837, p. 10; *Bp. Consp.* i. p. 51; *Cass. B. Calif. & Texas*, p. 183.
Brachyotus palustris, *Bp. Comp. List B. Eur. & N. Am.* p. 7; *Bp. Consp.* i. p. 51; *Fritsch, Vog. Eur.* Taf. 11. fig. 6; *Gould, B. Gt. Br.* pl. xxxii.; *Gurney, Ibis*, 1868, p. 150; *Coues, Key N.-A. Birds*, p. 204; *Taczan. J. f. O.* 1872, p. 350.
Asio brachyotus, *Macgill. Brit. B.* iii. p. 461; *Strickl. Orn. Syn.* p. 209.
Ægolius brachyotus, *Keys. & Blas. Wirb. Eur.* p. 143; *Borggr. Vogelf. Norddeutschl.* p. 63.

Otus galapagoensis, *Darw. Voy. Beagle, Birds*, p. 32, pl. 3; *Gray, Gen. B.* i. p. 40; *id. Hand-l. B.* i. p. 51.
Brachyotus palustris americanus, *Bp. Consp.* i. p. 51.
Otus breviauris, *Licht. Nomencl.* p. 6; *Schl. Mus. P.-B.* Oti, p. 4.
Brachyotus ægolius, *Bp. Rev. et Mag. de Zool.* 1854, p. 541; *Loche, Expl. Sci. Algér.* Ois. i. p. 97.
Brachyotus agrarius, *Brehm, Naum.* 1855, p. 270.
Brachyotus leucopsis, *Brehm, Naum.* 1855, p. 270.
Brachyotus cassinii, *Brewer, N. Am. Ool.* i. p. 68; *Cass. U. S. Expl. Exp.* p. 108; *id. in Baird, B. N. Am.* p. 54; *Baird, ed. Cooper, Orn. Calif.* p. 428; *Dall & Bann. Tr. Chic. Acad.* i. p. 273; *Gundl. J. f. O.* 1871, p. 375.
Asio sandvicensis, *Blyth, Ibis*, 1863, p. 27.
Otus cassinii, *Gray, Hand-l.* i. p. 51.
Asio accipitrinus, *Newt. ed. Yarr. Brit. B.* i. p. 163; *Shelley, B. Egypt*, p. 179; *Sharpe, ed. Layard, B. S. Afr.* p. 80.

Nestling. Covered with down of a yellowish buff colour, the plumes of a dull brown colour, with ochraceous or whitish buff tips, the quills largely spotted and barred with ochre, all the feathers round the eye and on the sides of the face uniform dark brown; frontal plumes light buff.

Fully grown young. Above more or less uniform brown, somewhat streaked by reason of the ochraceous buff margins to most of the feathers, all of which have concealed oval spots of ochre, nearly all these, however, being invisible; head and hind neck much more distinctly streaked, the centres being dark brown, the margins broadly ochraceous; wing-coverts coloured like the back, but varied with larger spots of ochre, which occupies the base of most of the feathers, the greater series with large ovate spots of white on the outer web, the outer scapulars also similarly marked, but the ochraceous markings much indented by brown mottlings; primary-coverts uniform dark brown; quills brown, whitish at tip, and barred with buffish orange, the primaries entirely of the latter colour for two thirds of their length, and broadly barred with dark brown; the secondaries irregularly mottled with spots and barrings of light ochre, the margins to these quills being also of this colour; all the quill-feathers when spread light ochre on the inner web, narrowly barred with brown; upper tail-coverts nearly all orange-buff, with brownish margins and tips; tail dark brown, shading into whitish at tip, and crossed with about six bars of whitish ochre on the centre feathers, these bars not continuous and the interspaces mottled with brown, the outer feathers richer ochraceous, the brown bars much narrower and about six in number, disappearing towards the base of the tail; ear-tufts 0·85 inch long, brown, with narrow margins of ochre; ruff brown, speckled with ochraceous, the plumes above the eye whitish buff, with brown central streaks to the feathers; facial plumes white, the lores and most of the feathers with hair-like black shafts, especially round the eye, the feathers surrounding which are entirely black; under surface of body light buff or ochraceous straw-colour, inclining to whitish on the abdomen and under tail-coverts, the latter, as well as the leg-feathers, being quite uniform excepting a few linear brown shaft-stripes; the breast clouded with dark brown centres to the feathers, rather narrower on the lower breast, abdomen, and flanks, these having a distinctly striped appearance; under wing-coverts light ochraceous buff, with a few narrow brown shaft-streaks, the lower series dark brown, with ochraceous base, forming a bar across the inner side of the wing.

Adult male. General colour above pale ochraceous buff, deeper on the centre of the back, the centres of the feathers longitudinally dark brown, producing a broadly streaked appearance, the scapulars much paler on their outer margins, the brown markings more irregular and branching into bars, so that the interspaces represent large ovate spots of whitish buff; wing-coverts rather darker than the back and more of a reddish ochre, the central markings to the feathers dark brown as on the back, but more irregular and forming bars, on the median and greater series more confined to the inner web, the outer ones having very large and distinct ovate whitish spots, which are very conspicuous on the wing; primary-coverts dark brown, with a few rufous ochre spots, principally near the base; quills rufous ochre, with a whitish tip, inclining also to fulvous near the base of the outer primaries, all the feathers very plainly chequered with dark brown bars, which are much narrower on the inner webs; upper tail-coverts nearly uniform ochraceous, shaded slightly with brown near the tips; tail-feathers ochraceous, tipped with whitish and crossed with seven brown continuous bars on the centre feathers, only five in number on the outer ones, where the bars are much narrower and disappear near the base; frontal plumes dark brown, narrowly margined with ochraceous; ruff whitish, slightly washed with ochre, and minutely speckled with triangular spots of brown; facial aspect dull white, the lores shaded with brown, the ear-coverts with narrow blackish shaft-lines, all the feathers round the eye black; ear-tufts 0·5 inch long, resembling the feathers of the crown; chin whitish; the ruff below, extending upwards behind the ear-coverts, ochraceous, mottled with dark brown centres to the feathers; rest of under surface buffy white, washed with golden buff on the breast and sides, the breast-feathers streaked with brown down the centre, these streaks narrowing into linear shaft-lines on the lower breast and abdomen, and disappearing entirely on the legs and under tail-coverts; under wing-coverts white, very faintly tinged with ochre, the outer greater coverts blackish, forming the characteristic spot, the rest of the wing-lining whitish; the secondaries with obsolete brownish bars, which are nearly absent on the primaries, these being blackish on tips and only slightly barred with ochraceous, the outer feathers having a blackish mark about halfway up. Total length 14 inches, wing 12·4, tail 6·5, tarsus 1·75.

Adult female. Of the same general appearance and markings as the male, but always much deeper in colour, the under surface being of a rich ochre; bands on tail six on the middle and four or five on the outer feathers. Total length 15·5 inches, wing 12·5, tail 6·7, tarsus 1·85.

Obs. From the examination of a large series of Short-eared Owls,

I believe that the change from the young or darker plumage is accomplished more by the gradual dissolution of the brown markings on the feathers than by a direct moult; this change affects the tail also, the bars narrowing by reason of the dissolution of the brown colour, remains of which are always more or less apparent on the interspaces. Another tolerably good indication of an adult bird can also be found in the amount of brown on the inner side of the wing, the younger birds having the end of the wing clouded with brown without transverse bars, and showing also a distinct spot or bar of brown about halfway down on the outer primary. This bar is absent in an adult bird, and the end of the feather has buff-coloured bars on the inner, spots on the outer web. Another difference in the adult bird is the attenuation of the shaft-stripes on the breast-feathers, which become narrower with age; and these narrow breast-markings are generally contemporary with the disappearance of the wing-spot above mentioned; but this is not always the case, for, as in other birds of prey, the exact sequence in the change of the parts seems to vary in individuals.

Hab. The temperate and tropical regions of both hemispheres, extending far into the Arctic circle: absent in West Africa, Australia, and Oceania, but present in the Sandwich Islands.

a. Asio accipitrinus.

a. Ad. st.	England.	Mr. J. Baker [C.].
b. Juv. st.	England.	Mr. J. Baker [C.].
c. Ad. st.	London Market.	Mr. J. Baker [C.].
d. Ad. st.	Hills of Offed.	Mr. J. Baker [C.].
e, f, g. Pull. st.	Hills of Offed.	Mr. J. Baker [C.].
h. ♀ ad. sk.	East Grinstead, Surrey, Dec. 1861.	J. E. Harting, Esq. [P.].
i. ♀ juv. sk.	Pagham, Sussex, Sept. 3rd, 1872.	R. B. Sharpe, Esq. [P.].
k. Ad. sk.	Christchurch, Hants, Jan. 1874.	Baron A. von Hugel [P.].
l. ♂ ad. sk.	Monmouthshire, Dec. 4th, 1873.	Rev. A. Morgan [P.].
m. ♀ juv. sk.	Skåne, Sept. 1867.	Purchased.
n. ♂ ad. sk.	Skåne, Nov. 27, 1873.	Purchased.
o. ♂ juv. sk.	Wermland, June 30, 1873.	Purchased.
p. Ad. sk.	France.	M. A. Bouvier.
q, r. Ad. st.	Archangel.	R. B. Sharpe, Esq. [P.].
s. Juv. st.	Archangel.	R. B. Sharpe, Esq. [P.].
t. ♂ ad. sk.	Malaga, Jan. 3, 1872.	Howard Saunders, Esq. [C.].
u. ♀ ad. sk.	Seville, Nov. 8, 1869.	Howard Saunders, Esq. [C.]
v. Ad. st.	Athens.	C. W. L. Merlin, Esq. [P.].
w. Ad. sk.	Corfu.	J. Rocke, Esq. [P.].
x. ♀ ad. sk.	Havankoi, Asia Minor, Sept. 19, 1871 (*T. Robson*).	R. B. Sharpe, Esq. [P.].
y. Juv. sk.	Plain of Nimroud, Mesopotamia.	
z, a', b', c'. Ad. sk.	Nepal.	B. H. Hodgson, Esq. [P.].
d'. Juv. st.	China.	J. R. Reeves, Esq. [P.].

e'. Imm. sk.	Shanghai.	— Webb, Esq. [P.].
f'. Imm. sk.	Siberia.	Moscow Museum [P.].
g'. Skeleton.	England.	Purchased.
h'. Skull.	S. Germany.	Dr. Gunther [C.].
i'. Sternum.		W. Yarrell, Esq. [P.].
k'. Sternum.		T. Ingall, Esq. [P.].
l'. Sternum.		J. Ray, Esq. [P.].
m', n', o'. Sternums.		Purchased.

β. *Asio cassini.*

p'. Juv. sk.	"At sea, after passing the Aleutian Isles, 300 miles off land."	Capt. Kellett [P.].
q'. Ad. st.	North America.	Hudson's Bay Co. [P.].
r'. Juv. st.	North America.	Admiral Sir George Back [P.].
s'. ♂ ad. sk.	North America.	Lord Odo Russell [P.].
t'. Ad. sk.	Arctic America.	John Ray, Esq. [P.].
u'. Juv. sk.	Fort Simpson, Sept. 15, 1861.	B. R. Ross, Esq. [P.].
v'. Ad. st.	Canada.	Mrs. Mauger [P.].
w', x'. Ad. sk.	British Columbia.	J. K. Lord, Esq. [C.].
y', z'. Juv. sk.	British Columbia.	J. K. Lord, Esq. [C.].
α, β. Juv. sk.	Sumass Prairie, B. C.	J. K. Lord, Esq. [C.].
γ. Ad. sk.	Valley of Mexico.	M. A. Bouvier.
δ. Ad. sk.	Bogotá.	M. A. Bouvier.
ε, ζ. Ad. sk.	Chili.	Purchased *.
η. Ad. sk.	Chili.	E. C. Reed, Esq. [C.].
θ. Ad. sk.	Province of Colchagua.	E. C. Reed, Esq. [C.].
ι. Ad. st.	Falkland Islands.	Sir W. Burnett and Admiral Fitzroy [P.].
κ. Ad. sk.	Falkland Islands.	Capt. Kellett and Lieut. Wood [P.].
λ. Ad. st.	Falkland Islands.	Antarctic Expedition [C.].
μ, ν. ♀ ad. sk.	Berkeley Sound, Falklands, April 1842.	Antarctic Expedition [C.].

γ. *Asio galapagensis.*

ξ. Juv. st.	Galapagos Islands (*Darwin*).	Zoological Society. Type of *Otus galapagoensis.*
ο. Ad. st.	Galapagos Islands.	Sir W. Burnett and Admiral Fitzroy [P.].
π. ♂ ad. sk.	Indefatigable Island, August 24, 1869.	Dr. A. Habel [C.].

δ. *Asio sandwichensis.*

ρ, σ. ♂ ♀ ad. sk.	Honolulu, Sandwich Islands.	R. B. Sharpe, Esq. [P.].

Obs. The excellent series of Short-eared Owls enumerated above enables me to speak with some certainty concerning the various races which have been raised to specific distinction by various ornithologists. In the foregoing list of the specimens in the national collection I have arranged them under four headings, each form thus

* One of these specimens is labelled "*galapagoensis*" by Dr. Kaup; but it is by no means a very dark-coloured one.

classified having slight peculiarities, but none of them possessing sufficient characters to allow of their being specifically separated. All over Europe and Asia there is but one type, according to my observations; but in America the general run of the specimens is rather darker and more ochraceous: hence the Short-eared Owl of the New World has been called *A. cassini*. I have seen specimens from different parts of America which I could not separate in the least from European examples; and in comparing specimens from the Old and New World care must be taken that birds of the same sex are examined, as the females of both are always more rufous than the males; thus if a female of the American Short-eared Owl were compared with a male of the European kind, the impression would be that the former was more rufous, and *vice versâ*. I have seen, however, birds of both sexes from British Columbia and from Chili which were absolutely similar to British-killed specimens.

In the Falkland Islands the Short-eared Owls seem permanently rufescent, but cannot be distinguished from some Chilian skins; and as some of the latter agree perfectly with European examples, no line for specific separation can be drawn: I must say, however, that at present I have never seen a light-coloured bird from the Falklands.

The Owl from the Galapagos is by far the most different of any, by reason of its small size, dark coloration, golden forehead, and striped thigh-feathers. I have never seen a specimen from any other locality exhibiting the latter character; at the same time a Bogotá skin in every other respect approaches it in appearance; and therefore, although I admit that at present it would appear to be specifically distinct, I should like to be sure that the New-Granadan Short-eared Owl would not form a connecting-link.

The specimens from the Sandwich Islands form an interesting pair on account of their locality; but they cannot be separated as a species. They are rather small, and have a very dusky frontal patch: this I have found in other Asiatic specimens; and therefore the Sandwich-Island Owl can only be considered a small race of *A. accipitrinus*. The following are the measurements of the pair:—

	Total length.	Wing.	Tail.	Tarsus.
	in.	in.	in.	in.
a. ♂ ad. Honolulu	14·8	12	6·2	1·65
b. ♀ ad. Honolulu	15·5	11·9	6	1·7

5. Asio capensis.

Otus capensis, *Smith, S.-Afr. Q. Journ.* ser. 2, no. 4, part 1, p. 316; *id. Zool. S. Afr.* pl. 67; *Gray, Gen. B.* i. p. 40; *Bree, B. Eur.* 2nd ed. i. p. 148; *Schl. Mus. P.-B.* Oti, p. 5; *Gray, Hand-l. B.* i. p. 51; *Heugl. Orn. N.O.-Afr.* i. p. 109; *Finsch & Hartl. Vög. Ostafr.* p. 109.

Brachyotus capensis, *Bp. Consp.* i. p. 51; *Fritsch, Vog. Eur.* tab. 11. fig. 7.

Phasmoptynx capensis, *Kaup, Contr. Orn.* 1852, p. 114; *id. Tr. Z. S.* iv. p. 236; *Loche, Expl. Sci. Algér.* Ois. i. p. 99.

Asio capensis, *Strickl. Orn. Syn.* p. 211; *Drake, Ibis,* 1867, p. 424; *Irby, B. Gibralt.* p. 61; *Sharpe, ed. Layard, B. S. Afr.* p. 78.
Ægolius capensis, *Heuglin, Syst. Uebers.* p. 14.
Otus capensis major, *Schl. Mus. P.-B. Revue Accipitr.* p. 5.

Adult. General colour above earthy brown, narrowly vermiculated with fulvous, these markings more or less concealed, becoming a little coarser on the wing-coverts and on the hinder neck; greater wing-coverts ovally spotted on the outer and barred on the inner web with ochraceous buff; primary-coverts dark brown, excepting some spots of rufous ochre near the base; quills dark brown, inclining to whitish buff at the tips of the feathers, the secondaries with only a few indications of bars in the shape of ochraceous vermiculations; the innermost secondaries uniform with the back, the primaries deep ochre at base, and chequered with deep ochre spots on the outer web, all the quills broadly barred on the inner web with paler ochre; upper tail-coverts slightly more rufous brown than the back, with narrow wavy vermiculations of brown, and having large oval spots of fulvous white; tail brown, the middle feathers with two bands of deep ochre near the base, but merely vermiculated with the same colour towards the end of the feathers, the remainder of the feathers barred with ochraceous buff and brown, the tips of the feathers being broadly buffy white, and the brown bars six in number, the two basal ones concealed; feathers over the eye and on the forehead very light rufous brown, mottled with darker brown; feathers of the face dusky white, the plumes above and below the eye blackish, the hinder tips of the ear-coverts brown; facial ruff brown, thickly spotted with rufous ochre; under surface

Foot of *Asio capensis* *.

of the body very pale ochraceous brown, mottled with lines and vermiculations of light fulvous or whitish; the breast and flanks barred with white, the terminal bars broadest; abdomen, leg-feathers, and under tail-coverts light ochre, shading off into whitish on the two latter; under wing-coverts fulvous, slightly spotted with brown, the median series with a few bars of brown, the greater series whitish at base, dark brown at tip, forming a bar across the wing;

* In the specimen figured the toes are slightly feathered, but they are generally quite bare.

inner lining of wing yellowish ochre, with two or three dark brown bands, merging into a dark brown tip; iris light reddish brown. Total length 13·5 inches, wing 11·11, tail 7, tarsus 2·35.

Hab. Southern Africa, ranging into North-western Africa, and even penetrating into Southern Spain: said also to have occurred in the Sinaitic peninsula.

a. Ad. st.	South Africa.	South-African Museum. Type of species.
b. Ad. st.	South Africa.	Purchased.
c. Ad. st.	South Africa.	Massena collection.
d. ♂ ad. sk.	Bloemfontein, Sept. 10, 1872 (*Dr. Exton*).	F. A. Barratt, Esq. [C.].
e. Ad. sk.	Lydenburg, S.E. Africa.	F. A. Barratt, Esq. [C.].
f. Ad. sk.	Shupanga, River Shiré.	Dr. Kirk [P.].
g. Ad. sk.	Angola.	J. J. Monteiro, Esq. [C.]. *Cf.* Ibis, 1862, p. 336.
h. Ad. sk.	Tangiers.	M. Olcese [C.].

Obs. Professor Newton has kindly lent me a specimen of this bird from Madagascar (*Otus capensis major*, Schl.), where it has been supposed to constitute a large race of the African species. This example measures as follows—total length 15·5 inches, wing 12·7, tail 6·9, tarsus 2·25. The difference in size is not sufficient, however, to make one think that the Madagascar bird is really distinct.

6. **Asio stygius.**

Nyctalops stygius, *Wagler, Isis,* 1832, p. 1221; *Gray, List Gen. B.* p. 6: *Scl. & Salv. Nomencl. Av. Neotr.* p. 116.
Otus siguapa, *D'Orb. in Ramon de la Sagra's H. N. Cuba,* Ois. p. 31, pl. 2; *Gray, Gen. B.* i. p. 40; *Bp. Consp.* i. p. 50.
Otus stygius, *Pucher. Rev. Zool.* 1849, p. 29; *Gray, Gen. B.* i. p. 40; *Kaup, Contr. Orn.* 1852, p. 113; *id. Tr. Z. S.* iv. p. 232; *Schl. Mus. P.-B.* Oti, p. 6 (1862); *Gray, Hand-l. B.* i. p. 50; *Pelz. Orn. Bras.* pp. 10, 400; *Ridgw. N.-A. Birds,* iii. p. 17 (note).
Otus melanopsis, *Licht. Nomencl. Av.* p. 6.
Asio stygius, *Strickl. Orn. Syn.* p. 207.
Asio siguapa, *Strickl. Orn. Syn.* p. 212.

Adult. Above of a nearly uniform chocolate-brown, the hinder neck and wig with a few longitudinal spots of light ochre, rather more oval in shape on the latter, the rest of the upper surface having concealed spots and vermiculations of whitish buff, rather larger and more distinct on the outer margin of the scapulars; wing-coverts uniform with the back, with the same more or less concealed vermiculations, these being absent on the primary-coverts, which are uniform brown; quills chocolate-brown like the back, with obsolete bars of lighter brown, more distinct on the secondaries, where they are often replaced by ochraceous spots or vermiculations, the innermost uniform with the back, and slightly freckled with vermiculations of whitish buff or light ochre, the primaries with a few spots of deep ochre on the outer web, only distinct or of any size near the base; upper tail-coverts brown, with distinct bars of bright ochre;

tail deep chocolate-brown, with a whitish tip, and crossed with five or six additional bars of bright ochre, these bars more numerous (seven or eight) on the inner web when the tail is spread; forehead and feathers above the eye brown, streaked with silvery grey; ear-tufts 2 inches long, chocolate-brown, with outer margins of light ochre; sides of face dingy brown, the cheeks streaked with fulvous, and the ear-coverts fulvous at their bases; ruff brown, mottled with light ochre, the hinder feathers almost entirely ochraceous, with brown margins and shaft-stripes; chin dingy brown, mottled with pale ochraceous, the ruff across the throat composed of white feathers with dark brown centres; rest of under surface ochraceous, mottled with brown, this colour more prevalent on the upper breast, where it occupies the centre of the feathers; the lower breast and abdomen streaked with brown down the middle of the feathers, with dark brown lateral bars to each, the interspaces being oval spots of white; leg-feathers deep ochre, spotted with triangular brown markings; under tail-coverts deep ochre, the longest ones streaked with brown; under wing-coverts deep ochre, the outermost spotted and margined with brown, the greater series light ochraceous at base, dark brown at tip, thus resembling the inner lining of the wing, which is almost entirely dark brown, excepting a few irregular bars of ochraceous, these being almost entirely absent near the primaries. Total length 20 inches, wing 13·7, tail 7·8, tarsus 1·9.

Hab. Brazil, Cuba.

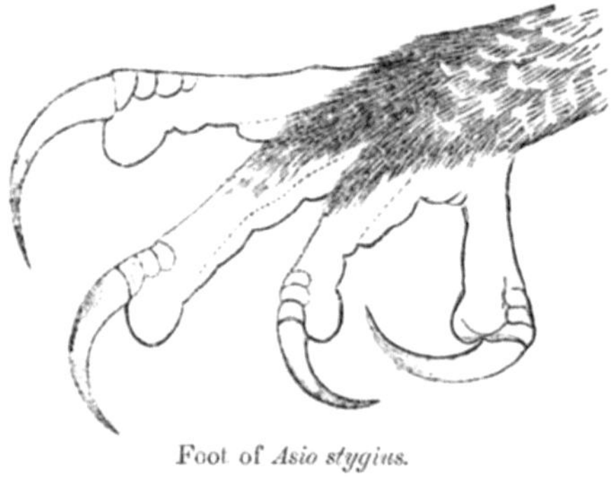

Foot of *Asio stygius*.

a. Ad. sk.	Brazil.	Purchased.

7. **Asio grammicus*.**

Ephialtes grammicus, *Gosse, B. Jamaica*, p. 19, pl. 4.

---

* I am unacquainted with the following species, and reproduce here Dr. Kaup's original description:—

Asio macrurus.

Otus macrurus, *Kaup, Tr. Z. S.* iv. p. 232; *Gray, Hand l. B.* i. p. 51.

"*Diagn.* The tail brown, 148 mm long, with five small whitish bands at the tip.
"*Descr.* The upper parts dark brown, with two to three irregular rusty-yellow

Pseudoscops grammicus, *Kaup, Isis*, 1848, p. 769; *id. Contr. Orn.* 1852, p. 113; *id. Tr. Z. S.* iv. p. 231; *Scl. & Salv. N mencl. Av. Neotr.* p. 116.
Scops grammicus, *Bp. Consp. Av.* i. p. 46; *Strickl. Orn. Syn.* p. 205.
Otus grammicus, *Gray, Hand-l. B.* i. p. 50.

Adult female. General colour above sandy buff, transversely vermiculated with dark brown, more distinctly on the back and scapular feathers, some of which are mesially streaked with dark brown, some of the outermost rather more broadly barred with sandy colour, but not exhibiting any trace of white or buff spots; head and neck rather lighter sandy colour, the transverse black lines very regular, especially on the ear-tufts, which are coloured like the rest of the head, but are nearly uniform rufous on their inner webs; entire facial aspect foxy red, the hindermost of the ear-coverts whitish, tipped with black, merging in the ruff, which is composed of sandy-buff feathers, black at tip and at base; the stiff gular feathers sandy rufous, streaked and indistinctly barred with black; chin-feathers buffy white; rest of the under surface deep sandy rufous, with faint indications of dusky vermiculations of brown, the abdominal plumes more or less verging on white and showing very little of the mesial black streaks which are so distinct on the breast-feathers; leg-feathers uniform tawny buff, as also the under tail-coverts; under wing-coverts buff, slightly washed with sandy rufous, the lower series black, with fulvous bases, forming a bar across the wing, and resembling the inner lining of the quills, which are fulvescent towards the base of the inner web, on which they are broadly barred with black; upper wing-coverts resembling the back, and very coarsely vermiculated with dark brown, the sandy-buff bars being pretty apparent here and there, and especially distinct on the outer web of the spurious quills; primary-coverts dark brown, irregularly barred across with sandy buff; quills banded alternately for their entire length with blackish brown and sandy buff, these bars less distinct on the secondaries, the light interspaces obscured with dark brown vermiculations, especially the innermost, which consequently resemble the back; tail sandy buff, barred across with dark brown, about eleven bars being traceable on the centre feathers, the interspaces more or less mottled with vermiculations of brown, the exterior rectrix paler and more fulvous, crossed with about thirteen brown bars; cere blackish grey; bill pale blue-grey; feet dull lead-

---

cross bands on each feather; the lower parts white, on each feather a broad dark brown shaft-stripe. The under tail-coverts pure white. Wings on the inner sides blackish, with from three to five whitish pronged bands and spots. Tarsi rusty yellowish white, with rusty yellow cross lines. The bristle-feathers of the *lorum* white, with black shafts. The stripe over the eye dirty white, extending to the ear-horns, which are not very well distinguished. Veil white, each feather with rusty yellow, black-margined, shaft-spots.

"*Dimens.* Head 70 mm.; bill from the root 20, from the gape 34 mm.; wing 250 mm.; tarsus 60 mm; middle toe without the nail 39 mm.; left ear-hole (taken on a stuffed specimen) 14, right ear-hole 22 mm. long.

"*Hab.* Mexico (Wurzburg Museum)."

colour; claws horny grey; iris hazel. Total length 12·2 inches, wing 8·4, tail 5·1, tarsus 1·55.

Obs. Mr. Gosse's typical specimen differs somewhat from the one described in the less rufescent tone of the plumage, especially underneath, where the bird is also plainly vermiculated with brown cross lines; the bars on the tail about nine in number.

Hab. Jamaica.

Foot of *Asio grammicus.*

a.	Ad. st.	Jamaica.	P. H. Gosse, Esq. [C.]. Type of species.
b.	Juv. st.	Jamaica.	P. H. Gosse, Esq. [C.].
c.	Ad. st.	Jamaica.	P. H. Gosse, Esq. [C.].
d.	♀ ad. sk.	Near Spanishtown, Jamaica, Oct. 1862 (*W. T. Marsh*).	Smithsonian Institution [P.].
e.	Sternum.	Jamaica.	P. H. Gosse, Esq. [C.].

16. SYRNIUM.

	Type.
Syrnium, *Savign. Descr. Egypte,* p. 298 (1809)	S. aluco.
Ulula, *Cuvier, Règn. Anim.* i. p. 329 (1817)	S. uralense.
Aluco, *Kaup, Natürl. Syst.* p. 190 (1829, *nec Flem.*) ..	S. aluco.
Ciccaba, *Wagler, Isis,* 1832, p. 1222................	S. huhulum.
Bulaca, *Hodgs. As. Research.* xix. p. 169 (1836)......	S. newarense.
Scotiaptex, *Swains. Classif. B.* ii. p. 217 (1837)	S. cinereum.
Meseidus, *Hodgs. J. A. S. B.* x. p. 28 (1841)	S. newarense.
Pulsatrix, *Kaup, Isis,* 1848, p. 771	S. torquatum.
Ptynx, *Bp. Consp.* i. p. 53 (1850)................	S. uralense.
Myrtha, *Bp. Rev. et Mag. de Zool.* 1854, p. 541	S. seloputo.
Macabra, *Bp. l. c.* p. 541........................	S. hylophilum.
Gisella, *Bp. l. c.* p. 541	S. harrisi.

Range. The whole of the New World, and the entire Old World also, excepting Madagascar, the Austro-Malayan subregion, Australia, and Oceania.

Key to the Species.

a. Crown of head mottled or barred, never entirely uniform.

a'. Toes plumed*.

a''. Belly with mesial streaks of blackish.

---

* *Syrnium fulvescens,* Scl. & Salv., seems to be the only exception.

a'''. Chest with regular longitudinal streaks of black or dark brown, and having no complete cross bands, sometimes a few mottlings on each side of the longitudinal streaks being present.

a^4. Head and neck streaked down the centre of the feathers, all of them laterally barred with dark brown.

a^5. Size small: wing 12·1–14·6 inches; facial disk grey or rufous, but with streaks of buff or a few dusky bars.

a^6. Rump and upper tail-coverts fulvous or greyish white, strongly vermiculated with brown cross lines.

a^7. General appearance of upper surface *streaked*, the cross vermiculations very insignificant; centre tail-feathers grey or rufous, with a few brown mottlings; no bars excepting the indications of a few narrow ones at the tip ...... *aluco*, p. 247.

b^7. General appearance of upper surface mottled, the central streaks very irregular and obscured in the lateral barrings and vermiculations of the feathers; central tail-feathers dark brown, with five or six broad but irregular bars of fulvous .... *nivicolum*, p. 250.

b^6. Upper tail-coverts and base of tail uniform brown; wing of male bird 14·6 inches .... *davidi*, p. 251.

b^5. Size large: wing 18·6–19 inches; facial disk grey, with very regular bars of dusky brown.................. { *cinereum*, p. 252. *lapponicum*, p. 254.

b^4. Head and neck distinctly streaked down the centres of the feathers, these streaks being perfect and having no lateral bars { *uralense*, p. 255. *fuscescens*, p. 256.

b'''. Chest with regular broad cross bands.

c^4. Toes feathered; bands on tail six, including the terminal one; general colour light brown, with white cross bands........ *nebulosum*, p. 257.

d^4. Toes bare; bands on tail five, the terminal one included; general colour rufous brown, with fulvous cross bands ...... *fulvescens*, p. 258.

b''. Belly regularly barred, with no distinct mesial streaks.

c'''. Bill yellow; dark bars on breast broad, and quite equal to intervening interspace.

e^4. Legs white, with a few dusky brown bars; spots on wing-coverts whitish; face dull greyish white, with a few obsolete brown cross markings and shaft-streaks of white.................. *occidentale*, p. 260.

f^4. Legs uniform yellowish buff; spots on

wing-coverts fulvous; face dusky, with broad cross markings of dark brown; feathers in front of the eye blackish brown, with a patch of tawny rufous below the eye.................... *rufipes*, p. 261.

d'''. Bill black or bluish black; the dark bars on the breast very narrow, much less than the interspaces.

g⁴. Upper surface uniform brown, with spots or bars of white; facial disk uniform fulvous .... *sinense*, p. 261.

h⁴. Upper surface brown, vermiculated with white; facial disk whitish, washed with orange and barred with dark brown .. *ocellatum*, p. 263.

b'. Toes bare*.

c''. Leg-feathers white or buff, sometimes with a few faint remains of brown bars.

e'''. Belly barred, but not mesially streaked.

i⁴. Chest uniform chestnut-red .... *leptogrammicum*, p. 264.

k⁴. Chest barred.

c⁵. Head and neck dark brown, spotted with white; spots on scapulars white.

c⁶. Back and wing-coverts light brown, spotted with white and vermiculated with darker brown .... *woodfordi*, p. 267.

d⁶. Back and wing-coverts uniform dark brown, with scanty white spots and no vermiculations at all.......... *nuchale*, p. 265.

d⁵. Head and neck tawny buff, barred with black; scapulars barred with tawny buff .... *hylophilum*, p. 269.

f'''. Belly mesially streaked with dark brown or black.

l⁴. Size small: wing about 8·5 inches; tail brown, mottled with lighter brown and spotted with orange-buff, but not regularly barred across; breast with no streaks.......................... *albigulare*, p. 270.

m⁴. Size larger: wing not less than 10·5 ins.; tail distinctly barred.

e⁵. Forehead and eyebrow white, with an additional superciliary band of orange; ear-coverts orange, with white shaft-streaks, the hinder margin blackish.. *superciliare*, p. 271.

f⁵. Eyebrow whitish or buffy white; forehead brown, spotted or barred with rufous.

e⁶. Tail brown, barred with lighter brown, but not showing white when spread; abdomen streaked with rufous brown............. *suinda*, p. 272.

f⁶. Tail blackish, barred with white, these white bars sometimes mottled with brown; abdomen streaked with black .... *virgatum*, p. 273.

* The amount of feathering varies much in *S. woodfordi*.

d''. Leg-feathers black, barred with white.
g'''. Crown of head and back black, with numerous white wavy lines .... *huhulum*, p. 275.
h'''. Crown of head and back uniform black { *nigrolineatum*, p. 276. *spilonotum*, p. 277.
b. Crown of head uniform brown.
c'. Ear-coverts whitish, narrowly barred with dark brown; chest brown, narrowly barred across with whitish .... { *newarense*, p. 281. *indranee*, p. 282.
d'. Ear-coverts uniform; chest also perfectly uniform.
e''. Chest chocolate-brown; forehead uniform with crown.
i'''. Breast uniform buff; toes feathered .... *perspicillatum*, p. 277.
k'''. Breast buff, with broad rufous bars; toes bare .... *melanonotum*, p. 280.
f''. Chest fulvous like the rest of the under surface, with no pectoral band; forehead buffy white; fore part of ear-coverts fulvous, the hinder part blackish .... *harrisi*, p. 283.

1. Syrnium aluco.

The Brown Owl, *Albin, Birds*, i. pl. 9.
Le Chat-huant, *Briss. Orn.* i. p. 500; *Buff. Pl. Enl.* 437.
La Hulotte, *Briss. Orn.* i. p. 507; *Buff. Pl. Enl.* 441.
Strix aluco, *Linn. S. N.* i. p. 132; *Pall. Zoogr. Rosso-Asiat.* p. 321; *Vieill. Faun. Franç.* p. 45, pl. 21. fig. 2; *Naum. Vög. Deutschl.* p. 473, pls. 46, 47. fig. 1; *Roux, Orn. Provenç.* i. p. 76, pls. 50–52; *Werner, Atlas*, Rapaces, pl. 35; *Schl. u. Susem. Vög. Eur.* Taf. 45; *Nordm. in Demid. Voy.* iii. p. 110; *Sundev. Sv. Fogl.* p. 184, pl. xxv. fig. 4; *Newton, ed. Yarr. Brit. B.* i. p. 146; *Shelley, B. Egypt*, p. 176.
Strix stridula, *Linn. S. N.* i. p. 153; *Vieill. Faun. Franç.* p. 44, pl. 21. fig. 1; *Yarr. Br. B.* i. p. 131.
Austrian Owl, *Lath. Gen. Syn. B.* p. 136.
Solognese Owl, *Lath. Gen Syn B.* p. 137.
Strix sylvestris, *Scop. Ann.* i. p. 21; *Gm. S. N.* p. 292.
Strix soloniensis, *Gm. S. N.* i. p. 292; *Lath. Ind. Orn.* i. p. 62.
Strix austriaca, *Shaw, Gen. Zool.* vii. p. 247.
Strix sylvatica, *Shaw, Gen. Zool.* vii. p. 253.
Syrnium ululans, *Savign. Descr. Egypte*, p. 298.
Syrnium stridulum, *Steph. Gen. Zool.* xiii. pt. 2, p. 62; *Brehm, Naum.* 1855, p. 27.
Syrnium aluco, *Boie, Isis*, 1828, p. 315; *Less. Man. d'Orn.* i. p. 114; *Brehm, V. D.* p. 115; *Gould, B. Eur.* pl. 47; *Gray, Gen. B.* i. p. 39, pl. 14. fig. 2; *Bp. Consp.* i. p. 51; *Kaup, Contr. Orn.* 1852, p. 121; *Schl. Vog. Nederl.* pl. 42; *Strickl. Orn. Syn.* p. 186; *Fritsch, Vog. Eur.* tab. 11. fig. 8; *Jaub. & Barth. Lap. Rich. Orn.* p. 84; *Gould, B. Gt. Br.* i. pl. xxiv.; *Newt. Ooth. Wolley.* p. 176; *Degl. & Gerbe, Orn. Eur.* i. p. 127; *Loche, Expl. Sci. Algér.* Ois. i. p. 94; *Gray, Hand-l. B.* i. p. 48; *Salvad. Faun. Ital. Ucc.* p. 27.
Aluco aluco, *Kaup, Naturl. Syst.* p. 190.
Ulula aluco, *Keys. & Blas. Wirb. Eur.* p. 33; *Macgill. Brit. B.* iii. p. 338; *Schl. Mus. P.-B.* Striges, p. 12; *Borggr. Vogelf. Norddeutschl.* p. 63.
Aluco stridulus, *Macgill. Brit. Rapt.* p. 367.

Ulula stridula, *Selby, Ill. Brit. Orn.* i. p. 102, pl. 25.
Syrnium œdium, *Brehm, Naum.* 1855, p. 270.
Syrnium rufescens, *id. t. c.* p. 270.
Syrnium macrocephalum, *id. t. c.* p. 270.

Young in down. Covered with white down, the sprouting feathers yellowish, with dark brown cross markings; bill ivory-white for the apical half of both mandibles.

Nestling. General colour above golden buff, everywhere banded with dark brown, all the feathers tipped with white, producing a hoary appearance; wings much lighter than the back, the golden interspaces and the terminal white margins both broader than on the dorsal feathers; primary-coverts dark brown, banded with lighter and more reddish brown, inclining near the base to golden buff; quill-feathers, as far as developed, dark brown, inclining to greyish at their tips, barred with ochraceous or reddish brown, inclining to golden buff on their outer webs; the secondaries tipped with whitish and less distinctly barred, the interspaces thickly mottled with brown vermiculations; tail reddish, tipped with white and barred across with brown, these bars dissolving on the inner web into brown vermiculations; head more dusky than the rest of the upper surface; feathers round the eye whitish; remainder of under surface golden buff, inclining to white on the breast, the whole narrowly banded across with dark brown; leg-feathers almost entirely buffy white, with a few bars of brown on the upper part of the tarsus.

Adult female (grey phase), General colour above ashy grey, the dorsal feathers slightly washed with rufous, and most of them streaked down the centre with dark brown, with lateral zigzag lines of the same, the interspaces between the latter being often white, this arrangement giving a generally vermiculated appearance to the upper aspect of the bird; the outer scapulars more coarsely vermiculated, and having a large oval spot of pure white externally; wing-coverts more rufescent than the back, the least ones reddish brown, centred and vermiculated with dark brown, the median and greater series rather greyer, more coarsely lined, barred with white externally, and having a large ovate spot of white on the outer webs; primary-coverts reddish brown, barred across with dark brown; primaries ochraceous brown, inclining to white at their tips and barred across with six broad bands of dark brown, not conterminous near the base of these quills, the interspaces paler on the external web and inclining to white towards the base; the secondaries regularly barred with light ochraceous brown and dark brown, the interspaces being mottled all over with the latter colour, all the feathers tipped with white; the inner secondaries greyer and more like the back, mesially streaked with dark brown and vermiculated with the same, the dark brown bars complete only on the inner web, and entirely broken up on the innermost quills; tail ashy rufous, tipped with whitish and varied with dark brown zigzag markings, the outer feathers broadly barred with dark brown, though most of the bars are more or less dissolved; crown of head dark brown, the feathers notched on both webs with white or fulvous; the hind neck and wig whiter than the rest of the upper surface, the centres of the feathers only being brown with lateral margins of the same colour, causing a somewhat streaked appearance; facial aspect whitish, the plumes in front of the eye pure white, the lores terminating in blackish bristles; ear-coverts slightly mottled with a few cross markings of brown; facial ruff blackish brown, the feathers tipped with silvery white, the plumes of the upper part also notched with the latter colour; chin rufous, with black streaks down the centre of the feathers; the ruff below the chin whitish, the feathers also having a broad mesial streak of brown; fore neck white; rest of under surface white, broadly streaked with blackish brown down the centre of the feathers, the breast-feathers having also some zigzag lateral bars of brown, the central streaks also not continuous to the base of the plumes, and much paler on the abdomen and under tail-coverts; leg-feathers white, spotted and minutely barred with brown; under wing-coverts and margin of wing white, streaked with dark brown, the lower series ashy brown, whitish at base, resembling the inner lining of wing, which is ashy brown, barred on the inner web and spotted on the outer with yellowish white, the secondaries barred right across; bill whitish horn-colour; claws horny white at base, darker towards the tip; iris blue-black. Total length 19 inches, wing 12·1, tail 8, tarsus 2·1.

Adult female (rufous phase). Similar in character and markings to the grey phase of plumage, but ferruginous instead of grey, the face being also washed with ferruginous. Total length 19·5 inches, wing 12, tail 9, tarsus 2·1.

Adult male (rufous phase). Total length 18·5 inches, wing 11·1, tail 7, tarsus 1·95.

Foot of *Syrnium aluco*, to show extent of feathering on toes.

Obs. The rufous and grey plumages of the Tawny Owl are not dependent upon age or sex; but the idea has been entertained by some ornithologists that the rufous birds were the young and the grey ones the old. As a rule, the nestlings have all red tails; but the Museum possesses a young bird still with remains of down on it, but yet entirely in the grey phase of plumage.

Hab. Europe generally, breeding as high as 64° N. lat., but becoming rarer in Russia and Eastern Europe; found as far south as Palestine and in North-eastern Africa.

a. Pull. st.	Great Britain.	Mr. Bullock [C.].
b. Pull. st.	Great Britain.	Mr. Bullock [C.].
c, c'. Pull. sk.	Great Britain.	J. Gould, Esq. [C.].
d. ♂ ad. st.	Cambridgeshire.	Mr. J. Baker [C.].
e. Juv. st.	Cambridgeshire.	Mr. J. Baker [C.].
f, g. Pull. st.	Cambridgeshire.	Mr. J. Baker [C.].
h. Ad. st.	Wiltshire.	Col. Montagu [P.].
i. Ad. st.	Carmarthenshire.	Col. Montagu [P.].
k. Ad. st.	Carmarthenshire.	Col. Montagu [P.].
l. Ad. st.	Devonshire.	Sir W. Elford, Bart. [P.].
m. Pull. sk.	Skåne, S. Sweden, May 11th, 1869.	Purchased.
n. Pull. sk.	Wermland, Sweden, June 9th, 1863.	Purchased.
o. ♂ juv. sk.	Stockholm, June 12th, 1873.	Purchased.
p. ♀ ad. sk.	Stockholm, June 11th, 1873.	Purchased.
q. ♀ ad. sk.	Stockholm, Nov. 2nd, 1873.	Purchased.
r. Imm. sk.	Germany.	R. B. Sharpe, Esq. [P.].
s, t. ♂ ♀ ad. sk.	Switzerland.	Howard Saunders, Esq. [P.].
u, v. Ad. st.	Fontainebleau, France.	R. B. Sharpe, Esq. [P.].
w. Ad. sk.	France.	M. A. Bouvier [P.].
x, y. Ad. sk.	Algeria.	M. Olcese [C.].
z. Ad. sk.	Russia.	M. A. Bouvier.
a'. Skeleton.	North Europe.	Purchased.
b'. Sternum.	England.	J. Ray, Esq. [P.].
c'. Sternum.	England.	T. Ingall, Esq. [P.].

2. Syrnium nivicolum.

Mesomorpha nivicola, *Hodgs. in Gray's Zool. Misc.* p. 82.

Syrnium nivicolum, *Blyth, J. A. S. B.* xiv. pp. 185, 550; *id. Cat. B. Mus. A. S. B.* p. 41; *Gray, Gen. B.* iii. App. p. 3; *Bp. Consp.* i. p. 51; *Kaup, Contr. Orn.* 1852, p. 121; *Horsf. & Moore, Cat. B. Mus. E. I. Co.* i. p. 84; *Strickl. Orn. Syn.* p. 187; *Jerd. B. Ind.* i. p. 124; *Gray, Hand-l. B.* i. p. 48; *Hume, Rough Notes,* ii. p. 359; *Swinh. P. Z. S.* 1871, p. 344.

Urrua nivicola, *Hodgs. J. A. S. B.* xiv. p. 185.

Adult (rufous phase). General colour above brown, with numerous zigzag bars and mottlings of orange-rufous, coarser on the upper back and scapulars, the outer webs of the latter being for the most part white or fulvous white, causing a conspicuous shoulder-patch; wing-coverts dark brown, minutely spotted with orange-rufous, the median coverts more coarsely mottled towards the tips, the greater coverts with a large oval white spot near the tips of the feathers; primary-coverts blackish brown, with faint indications of lighter brown bars; quills dull brown, narrowly margined with whitish at the tips and crossed with bars of paler brown, almost obsolete on the inner webs, but inclining to fulvous brown on the outer, giving a spotted appearance to the wing, the primaries marked on the outer web with five large notches of rufous ochre; tail-feathers dark brown, narrowly tipped with ashy white and crossed with five bands of lighter brown, inclining to whity brown on both margins; forehead blackish brown, a few of the feathers tipped with orange-rufous; facial aspect orange-rufous, with rather distinct white shaft-streaks, the loral plumes and those on the fore part of the cheeks white, with blackish shaft-lines; ruff orange-rufous, mottled with dark brown and buff, the bases of the feathers and the terminal margins being dark brown, the plumes adjoining the throat also dark brown down the centre; chin whitish, as well as the fore neck; rest of under surface of body bright yellowish, washed and banded with orange-buff, all the feathers streaked with dark brown down the centre, the breast-feathers crossed with white bars, giving a paler appearance to this portion of the body; leg-feathers fulvous, mottled with orange-buff; under tail-coverts deep fulvous, with faint brown centres and terminal margins to the feathers; under wing-coverts deep orange-buff, with a few small brown markings near the outer edge of the wing, the lower series dull brown, yellowish at base, resembling the inner lining of the wing, which is brown, with yellowish white bars on the inner web; "legs and feet densely feathered to the terminal joints of the toes, which have two or three moderate transverse scales of a dull plumbeous colour; claws brown, paler at tips and bases; iris dark brown; bill pale fleshy yellow; cere very faintly marked, dark brown" (*Hume*). Total length 18 inches, wing 12·7, tail 7·7, tarsus 2·15.

Young. Similar to the adult, but much more mottled above and on the chest, all the feathers being barred on each web with orange-buff and brown, producing a more waved appearance than in the adults; bands on the tail seven in number.

Adult (grey phase). Similar in character and markings to the rufous phase, but of a greyish brown shade generally on the upper surface; the face very white, as also the lower portion of the facial ruff adjoining the throat; the white bars on the lower surface of the body much closer together, the brown cross markings very narrow. Total length 19 inches, wing 12·4, tail 7·5, tarsus 1·9.

Hab. Himalaya Mountains from Murree eastwards, extending into Moupin in Western China.

a. Ad. st.	Nepal.	B. H. Hodgson, Esq. [P.].
b, c. Jun. sk.	Nepal.	B. H. Hodgson, Esq. [P.].
d, e, f. Ad. sk.	Nepal.	B. H. Hodgson, Esq. [P.].
g. Ad. sk.	Kulu (*Hay*).	John Gould, Esq. [P.].

3. Syrnium davidi.

Ptynx fuscescens, *David, N. Arch. Mus.* vii. *Bull.* p. 4.
Ptynx fulvescens, *Swinhoe, P. Z. S.* 1871, p. 344.
Syrnium davidi, *Sharpe, Ibis,* 1875, p. 256.

Adult male. Back brown, lighter on the margins of the dorsal feathers, which are also slightly streaked or spotted with white, more especially on the scapulars, which have large oval spots of white on the outer web, the dark centres to the feathers of the upper surface giving the bird a somewhat streaked appearance; head dark brown, the forehead nearly uniform; the feathers of the hind part of the crown mottled with buffy white, most of the feathers being spotted with this colour, much more distinctly on the hind neck, the spots being especially large on the sides of the latter; wing-coverts brown like the back, the least series with a few whity brown spots of small size, the greater coverts largely spotted with white on the outer web; primary-coverts dark brown, with two rows of fulvous spots; quills brown, barred with pale brown on both webs, the bars not conterminous and represented on the primaries by a distinct notch of buffy white, the bars on the secondaries fulvous near the base of the inner web, the bars paler and less distinct on the innermost; lower back, rump, and upper tail-coverts nearly entirely uniform brown, without any of the white spots which mark the interscapulary region; tail brown, tipped narrowly with whitish, the basal part almost entirely uniform, the exterior feathers barred with lighter or more fulvous brown towards their tips; facial aspect grey or whitish, much obscured with brown, this colour predominating on the ear-coverts; lores and feathers in front of the eye whitish, obscured by blackish hair-like bristles; the hinder part of the ear-coverts greyish, mottled with brown and a little resembling the facial ruff, which is very distinct and is composed of dark brown feathers spotted with fulvous, being white streaked with brown just above the eyes, where it forms two very distinct patches on each side of the forehead, and behind the ear-coverts it is white, also centred and tipped with brown; on the throat the feathers of the ruff are whitish with dark brown centres; under surface of body greyish white, streaked with brown, the centres to the feathers being mostly of the latter colour, broader on the chest, which is consequently more sombre; on the flanks the feathers are somewhat barred near the base as well as streaked down the centre with brown; vent tinged with fulvous; under tail-coverts white, with broad brown bars to the feathers, radiating from the central streak; leg-feathers fulvous, whiter on the lower part, with remains of narrow brown bars on the latter; under wing-coverts buffy white, streaked with brown near the outer edge of the wing; the lower series yellowish at base, greyish brown at tip, and resembling the inner lining of the quills, which are greyish brown below, barred with buffy white, more plainly on the inner web; bill yellow; claws pale horn-colour; iris brown. Total length 20 inches, wing 14·6, tail 11·4, tarsus 2·1. (*Mus. Paris.*)

Hab. Moupin.

4. Syrnium cinereum.

Cinereous Owl, *Lath. Gen. Syn.* i. p. 134.

Sooty Owl, *Penn. Arctic Zool.* ii. p. 232.

Strix cinerea, *Gm. S. N.* i. p. 291; *Vieill. Ois. Am. Sept.* p. 48; *id. N. Dict. d'Hist. Nat.* vii. p. 23; *Swains. & Rich. Faun. Bor.-Am. Birds*, p. 77, pl. xxxi.; *Audub. B. N. Am.* pl. cccli., et *Orn. Biogr.* iv. p. 364; *Tyzenh. Rev. et Mag. de Zool.* 1831, p. 577.

Strix acclamator, *Bartr. Trav. Carol.* p. 285.

Syrnium cinereum, *Audub. Syn. B. Amer.* p. 26; *De Kay, Zool. N. Y. Orn.* p. 26, pl. xiii. fig. 29; *Kaup, Contr. Orn.* 1852, p. 121; *Cass. B. Calif.* p. 184; *Strickl. Orn. Syn.* p. 188; *Kaup, Trans. Z. S.* iv.

p. 256; *Cass. B. N. Am.* p. 56; *Gray, Hand-l. B.* i. p. 48; *Dall & Bann. Tr. Chicago Acad.* i. p. 273; *Cooper, B. Calif. ed. Baird,* p. 433; *Maynard, Nat. Guide,* p. 130; *Ridgw. N.-Am. B.* iii. p. 30; *Coues, B. N. W.* p. 307.

Scotiaptex cinerea, *Swains. Classif. B.* ii. p. 217.

Ulula cinerea, *Bp. Consp.* i. p. 53.

Syrnium lapponicum, var. cinereum, *Coues, Key N.-Am. B.* p. 204.

Adult. General colour above dusky greyish brown, everywhere mottled with white in the form of zigzag bars and markings, the white on the outer web of the scapulars more distinct, but not forming a very distinct shoulder-patch; wing-coverts exactly like the back, and marked with white in the same way; primary-coverts dark brown, barred across with lighter brown; primaries dark brown, barred with lighter brown, these pale bars inclining slightly to whitish on the outer webs, the innermost secondaries deep brown, mottled with white in the same manner as the back; tail-feathers brown, darker towards the tip, crossed with about eight feebly indicated greyish white bands, all of which are plentifully mottled with brown; head and neck greyish brown, mottled with numerous white bars, not very distinct or continuous; face dusky cinereous, indistinctly banded with dark brown, the feathers immediately over the eye black; lores and an indistinct eyebrow whitish; ruff chocolate-brown, very slightly spotted with buffy white, the light markings more distinct on the lower parts of the ruff, particularly on the feathers adjoining the sides of the neck; chin whitish, the ruff-feathers immediately below being brown, with rather broad whitish margins, some of them buffy white, narrowly barred with brown; rest of under surface greyish brown, mottled with dull white, generally in the form of rather broad bars; the leg-feathers very dingy brown, mottled with whitish; sides of body and under wing-coverts rather whiter than the rest of the under surface and more distinctly banded with brown; the lower series of the greater under wing-coverts brown, barred with yellowish white, thus resembling the inner lining of the wing; bill yellow; iris yellow. Total length 30 inches, wing 19, tail 13·5, tarsus 2·35.

Obs. As in the case of the Long-eared Owl, the American and European birds are not exactly alike, but yet approach each other so closely that it is difficult to give exact specific characters. The American bird in the present instance is altogether darker than *S. lapponicum* both above and below, the central brown streaks to the feathers of the breast being much less distinct in *S. cinereum.* The best character that I have seen as yet for distinguishing the two is by lifting the wing, when the European species immediately discloses a distinct white patch at the base of the primaries, which is absent, or at least scarcely discernible, in the American birds. This is caused by the much broader character of the white bars at the base of the primaries in *S. lapponicum*—these existing, it is true, in *S. cinereum,* but much less distinctly and generally clouded over and obscured by greyish-brown mottlings.

Hab. The northern part of North America, descending rarely to the United States in winter, but found generally on the western side of the continent.

a. Ad. st.	North America.	Admiral Sir George Back [P.].
b. Ad. st.	North America.	Hudson's Bay Company [P.].
c. Ad. sk.	British Columbia.	J. K. Lord, Esq. [C.].
d. Juv. sk.	British Columbia.	J. K. Lord, Esq. [C.].
e. Skeleton.	North America.	

5. Syrnium lapponicum.

Strix lapponica, *Retz. Faun. Suec.* p. 79; *Temm. Man. d'Orn.* i. p. 81; *Werner, Atlas,* Rapaces, pl. 33; *Tyzenh. Rev. et Mag. de Zool.* 1831, p. 576; *Schl. u. Susem. Vög. Eur.* Taf. 40; *Sundev. Sv. Fogl.* p. 184, pl. xxiv. fig. 1; *Naum. Vög. Deutschl. Nachtrag,* Taf. 349.
Strix fuliginosa, *Shaw, Gen. Zool.* vii. p. 245.
Strix barbata, *Pall. Zoogr. Rosso-Asiat.* i. p. 318; *Radde, Reis. Sibir.* p. 121.
Strix microphthalmos, *Tyzenh. Orn. Pow.* i. p. 86, frontisp.
Ulula lapponica, *Less. Man. d'Orn.* i. p. 113; *Degl. & Gerbe, Orn. Eur.* i. p. 131.
Surnia lapponica, *Gould, B. Eur.* i. pl. 42.
Syrnium cinereum, *Bp. Comp. List B. Eur. and N. Am.* p. 6; *Gray, Gen. B.* i. p. 39; *Fritsch, Vög. Eur.* tab. 12. fig. 6.
Ulula barbata, *Keys. u. Blas. Wirb. Eur.* p. xxxii; *Borggr. Vogelf. Norddeutschl.* p. 62.
Syrnium microphthalmum, *Gray, Gen. B.* i. p. 39.
Syrnium lapponicum, *Strickl. Orn. Syn.* p. 188; *Newt. Ooth. Wolley.* p. 170; *Gray, Hand-l. B.* i. p. 48; *Taczan. Journ. f. Orn.* 1873, p. 303.
Ulula cinerea, *Schl. Mus. P.-B.* Striges, p. 9.
Syrnium cinereum, var. lapponicum, *Ridgw. N.-A. Birds,* iii. p. 29; *Coues, B. N.W.* p. 308.

Adult. General colour above ashy brown, the feathers varied with white on both webs in the shape of bars or vermiculations, causing an indistinctly streaked appearance to the whole of the upper surface, the centres of the dorsal feathers being longitudinally brown; head and neck rather lighter than the back, the feathers being for the most part whitish, tinged with fulvous near the base, broadly streaked down the centre and barred with brown; scapulars whiter than the rest of the upper surface, being almost entirely white on the outer web with a few bars of brown; wing-coverts darker than the back, brown, mottled with a few irregular whitish markings here and there, the greater series with large oval spots of white on the outer web; primary-coverts dark brown, barred across with ashy brown; quills dark brown, the primaries barred with whity brown, washed with fulvous on the inner web, and all the light bars slightly mottled with zigzag vermiculations of brown; the secondaries also barred across with lighter brown, inclining slightly to whitish on the outer web, the innermost secondaries exactly like the back; tail ashy brown, darker brown at tip, and crossed with five lighter bands, none of them plainly distinguishable, and for the most part consisting only of broad whitish lines of irregular shape, the interspaces also mottled with whitish, especially on the inner web; facial aspect ashy whitish, barred with broad wavy lines of blackish; lores and feathers above the eye whitish, the plumes immediately over the latter black; facial ruff dark chocolate-brown, most of the feathers behind the ear tipped and barred with tawny buff, producing a highly mottled appearance, those on the lower parts whitish, with central streaks of dark brown; chin-feathers dark brown, with a few spots of whitish on the margins of some of the feathers; rest of under surface dull whitish, with broad streaks of brown, most of the feathers having obsolete bars of light brown; the leg-feathers very thickly barred with brown; under wing coverts dull white, plentifully streaked and barred with brown, the lower series dark brown, yellowish at base, resembling the inner lining of the wing, which is dark brown, barred with yellowish near the base of the inner web. Total length 28 inches, wing 18·6, tail 13·5, tarsus 3·1.

Hab. Northern Europe, extending but rarely into the central portion.

a. Ad. st.	Europe.	
b. Ad. st.	North Europe.	Le Comte de Tyzenhauz [P.].

6. Syrnium uralense.

Strix uralensis, *Pall. Reise,* i. p. 455; *Temm. Man. d'Orn.* i. p. 84; *id. Pl. Col.* 27; *Naum. Vog. Deutschl.* i. p. 422, pl. 42. fig. 1; *Werner, Atlas,* Rapaces, pl. 32; *Schl. u. Susem. Vög. Eur.* Taf. 42; *Midd. Sibir. Reise, Zool.* p. 129; *Schrenk, Reis. Amurl.* p. 245; *Sundev. Sv. Fogl.* p. 186; *Radde, Reis. Sibir.* p. 121.

Strix liturata, *Retz. Faun. Suec.* p. 79; *Sundev. Sv. Fogl.* pl. xxiv. fig. 2.

Strix macrura, *Meyer u. Wolf, Taschenb.* p. 84 (*ex Natt. MS.*); *iid. Vog. Deutschl.* pl. to p. 184.

Strix macrocephala, *Meisner, Syst. Verz. Vög. Schweiz.* p. 34.

Surnia uralense, *Steph. Gen. Zool.* xiii. pt. 2, pl. 40; *Gould, B. Eur.* i. pl. 44.

Ulula liturata, *Steph. Gen. Zool.* xiii. pt. 2, p. 60.

Syrnium uralense, *Boie, Isis,* 1828, p. 315; *Brehm, Vög. Deutschl.* p. 115; *Bp. Comp. List B. Eur. & N. Am.* p. 6; *Gray, Gen. B.* i. p. 39; *Cass. Cat. Strigidæ Philad. Mus.* p. 5; *Kaup, Contr. Orn.* 1852, p. 121, et *Tr. Z. S.* iv. p. 257; *Strickl. Orn. Syn.* p. 189; *Fritsch, Vög. Eur.* tab. 12. figs. 1, 2; *Gray, Hand-l. B.* i. p. 48.

Noctua uralensis, *Cuv. Règne An.* i. p. 344.

Syrnium macrocephalum, *Brehm, Vög. Deutschl.* p. 115.

Scotiaptex uralensis, *Swains. Classif. B.* ii. p. 217.

Ptynx uralensis, *Gray, List Gen. B.* p. 8 (1841); *Bp. Consp.* i. p. 53; *id. Rev. et Mag. de Zool.* 1854, p. 541; *Degl. et Gerbe, Orn. Eur.* i. p. 129.

Ulula uralensis, *Keys. u. Blas. Wirb. Eur.* p. xxxii; *Schl. Mus. P.-B.* Striges, p. 10; *Borggr. Vogelf. Norddeutschl.* p. 62.

Ptynx liturata, *Brehm, Naum.* 1855, p. 270.

Adult female. General colour above dull white, streaked with dark brown, all the feathers being brown in the centre with broad lateral margins of white; the scapulars purer white externally, the central brown streaks narrower, so that these plumes stand out in relief against the dorsal feathers; wing-coverts rather browner than the back, the whitish margins not so broad, the greater series dull brown, barred with lighter brown, inclining to white at the tips; quills brown, tipped with whitish and broadly banded across with lighter brown, these bands being paler and more ashy white on the outer webs, especially of the primaries, where the external white markings are very large and give a chequered appearance to the outer aspect of the wing; upper tail-coverts pale brown, mottled all over with irregular spots and bars of ashy white; tail-feathers dull brown, and crossed with six broad bands of ashy brown, the last subterminal and preceding a whitish tip to all the feathers; frontal plumes dark brown and not distinctly streaked like the hind neck and back; facial aspect dusky white, with a few indistinct dark brown streaks on the ear-coverts, the loral plumes with black shaft-lines; facial ruff very distinct and complete, white, with brown bases and central streaks branching out into small bars near the apex of the plumes, and thus producing somewhat of a mottled appearance on the lower portion of the ruff, the brown being entirely confined to a central streak; rest of under surface whitish, some of the feathers washed with yellow, and all of them very broadly streaked down the centre with dull brown, blacker on the flanks and abdomen; leg-feathers uniformly whitish; under wing-coverts white, streaked with brown at a little distance from the outer edge of the wing, the greater series dark brown, whitish at base, resembling the inner lining of the wings, which are dark brown, broadly barred with yellowish or yellowish white on the inner web. Total length 27·5 inches, wing 15·7, tail 12·5, tarsus 2·3.

Obs. A specimen from the Ussuri River, collected by Dr. Dybowski and now in the Museum, differs from the European birds examined by me in being a much darker bird everywhere, the brown predominating and the white margins not so broad, the scapulars, secondaries, and upper tail-coverts inclining to clear ashy grey on the margins; sides of the neck and facial ruff purplish brown, the latter mottled and spotted with white; under surface of body much more broadly streaked than in the European specimens. It measures as follows—total length 24·5 inches, wing 14·3, tail 11·5, tarsus 2·2. The feathers of the foot seem to me to be thicker and longer than in the Swedish and Styrian birds in the Museum; and the Japanese specimen also seems to me to be equally thickly clothed as regards the feet.

Hab. Northern Europe and Siberia.

a.	♀ ad. st.	Lappmark, Sweden.	Prof. Wahlberg [C.].
b.	♀ ad. st.	Styria.	Purchased.
c.	♂ ad. sk.	Ussuri River, Nov. 5th, 1873 (*Dybowski*).	Warsaw Museum [E.].

Subsp. α. **Syrnium fuscescens** *.

Strix rufescens, *Temm. & Schl. Faun. Japon.* Aves, p. 30; *Midd. Sibir. Reise*, p. 130.

* This species is called *rufescens* in the text and *fuscescens* on the plate in the

Syrnium fuscescens, *Temm. & Schl. Faun. Japon.* Aves, pl. 10; *Gray, Gen. B.* i. p. 39; *Cass. Cat. Strigidæ Philad. Mus.* p. 5; *Strickl. Orn. Syn.* p. 190.
Ptynx fuscescens, *Bp. Consp.* i. p. 53.
Ulula fuscescens, *Schl. Mus. P.-B.* Striges, p. 11.
Syrnium fulvescens, *Gray, Hand-l. B.* i. p. 45 (*nec Scl. & Salv.*).
Syrnium rufescens, *Sharpe, Ibis,* 1875, p. 256, note.

This Owl is so very similar to *S. uralense* that Professor Schlegel only gives us its specific characters—a slightly smaller size and general darker coloration, inclining to rufous. I have only examined one specimen in the Museum, and this measures as follows—total length 23 inches, wing 12·7, tail 10·5, tarsus 2.

Hab. Japan.

a. Ad. sk. Hakodadi, Dec. 12th, 1865 (*H. Whitely*). J. Gould, Esq. [P.].

7. Syrnium nebulosum.

Strix nebulosa, *Forster, Phil. Trans.* xxii. p. 386; *Vieill. Ois. de l'Am. Sept.* pl. xvii.; *Wilson, Am. Orn.* iv. p. 61, pl. xxxiii. fig. 2; *Temm. Man.* i. p. 88; *Werner, Atlas,* Rapaces, pl. 34; *Audub. B. Amer.* pl. xlvi., et *Orn. Biogr.* i. p. 242; *Swains. & Rich. Faun. Bor.-Am. Birds,* p. 81; *Gould, B. Eur.* i. pl. xlvi.
Barred Owl, *Penn. Arctic Zool.* ii. p. 234, pl. 11.
Strix varius, *Barton, Frag. N. H. Penn.* p. 11.
Ulula nebulosa, *Steph. Gen. Zool.* xiii. pt. 2, p. 60; *Less. Man.* i. p. 113; *James. edit. Wils. B. N. Amer.* i. p. 107; *Gould, B. Eur.* pl. 46; *Bp. Consp.* i. p. 53; *id. Rev. et Mag. de Zool.* 1854, p. 540; *Schl. Mus. P.-B.* Striges, p. 11.
Syrnium nebulosum, *Boie, Isis,* 1828, p. 315; *Gray, Gen. B.* i. p. 39; *Cass. Cat. Strigidæ Philad. Mus.* p. 4; *Kaup, Contr. Orn.* 1852, p. 256; *id. Tr. Z. S.* iv. p. 121; *Strickl. Orn. Syn.* p. 189; *Cass. B. Calif.* p. 185; *id. B. N. Amer.* p. 56; *Gray, Hand-l. B.* i. p. 39; *Baird, in Cooper's B. Calif.* p. 431; *Coues, Key N.-A. Birds,* p. 204; *id. B. N. W.* p. 308; *Ridgw. N.-Am. B.* iii. p. 34.

Adult. General colour above brown, the head rather darker than the back, the entire upper surface barred across with broad bands of yellowish buff, rather whiter on the hind neck and wig, and decidedly white on the back, where the markings are less distinctly in the form of bars, being often confined to white notches on each side of the feather; wing-coverts brown, the least series with minute yellowish or white spots, the median and greater series more largely spotted with white, the latter being also faintly barred across with lighter brown; primary-coverts dark brown, barred across with yellowish brown; quills dark brown, barred with lighter brown, the primaries externally spotted with yellowish white, producing a distinctly chequered appearance; the secondaries regularly barred with light brown, inclining to white on the outer web, and there forming

'Fauna Japonica;' so that some *lapsus calami* has occurred on the part of the original describers. Under these circumstances it seems better to adopt the latter title, which is the only one applicable to the species and is the one which Prof. Schlegel, in his later works, has applied to it.

a spot, the innermost ones barred with white; upper tail-coverts light brown, banded with white or pale yellowish; tail brown, rather broadly tipped with white and crossed with five other bands of yellowish, inclining to white on the outer webs; frontal plumes dark brown; facial aspect dull white, obscurely barred with brown; the feathers in front and over the eye and the loral plumes purer white, the latter terminating in blackish bristles, the plumes round the eye dark brown; facial ruff dark brown, banded with white or yellowish, especially on the lower portion; chin dusky brown, with

Foot of *Syrnium nebulosum*, to show extent of feathering.

whitish shaft-lines; rest of under surface yellowish white, the breast barred with brown and the flanks broadly streaked down the centre with the same colour, the abdomen and under tail-coverts much more narrowly streaked; leg-feathers dull white, everywhere minutely spotted and barred with brown; under wing-coverts white, streaked with brown, the inner ones washed with yellowish; the lower series ashy brown, barred with yellowish white on the inner web, the base being entirely of the latter colour; and hence this series resembles the inner lining of the wing, which is dull or ashy brown, regularly and broadly barred with yellowish white; bill yellow; iris brownish black. Total length 20 inches, wing 12·8, tail 9, tarsus 2·5.

Hab. Eastern region of the United States.

a. Ad. st.	North America.	Lord Odo Russell [P.].

8. Syrnium fulvescens*.

Syrnium fulvescens, *Scl. & Salv. P. Z. S.* 1868, p. 58; *Gray, Hand-l. B.* i. p. 48; *Scl. & Salv. Nomencl.* p. 116.
Ulula fulvescens, *Schl. Mus. P.-B. Revue Accipitr.* p. 19.

---

* Syrnium sartorii.

Mr. Ridgway separates a race from Mirador, Mexico, as *Syrnium nebulosum*, var. *sartorii*, with the following diagnosis:—"Colours blackish sepia and clear white; face without any darker concentric rings." He describes the race fully as follows (N.-Am. B. iii. p. 29):—"In general appearance like *S. nebulosum*, but the brown very much darker and less reddish, that of the markings below very nearly black: the pattern, however, precisely the same; and there is about

Syrnium nebulosum, var. fulvescens, *Ridgw. N.-A. Birds*, iii. p. 29; *Coues, B. N.W.* p. 309.

Adult. Above reddish chocolate-brown, with more or less concealed bands of yellowish ochre, sometimes inclining to white, the scapulars more largely marked with ochre on the outer web; wing-coverts rather darker chocolate than the back, the least series with only a few indistinct reddish bars, the outer ones with a few white spots, the greater and median series barred across with yellowish, inclining to white on the outer web; primary-coverts dark brown, with a few reddish bars; quills dark brown, slightly margined with whitish at the tip, and barred across with dull rufous, the lighter bars inclining to white on the margins of the quills, the spots much larger on the primaries, producing a chequered appearance; upper tail-coverts chocolate-brown, barred with yellowish buff; tail chocolate-brown, narrowly tipped with whitish and crossed with six

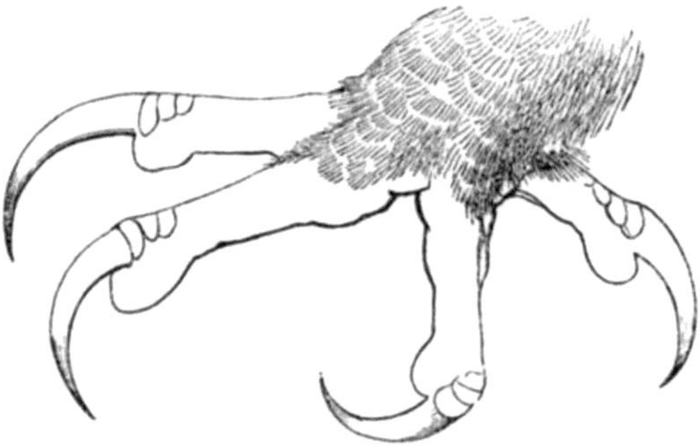

Foot of *Syrnium fulvescens*, to show the bare toes.

other rather narrow bands of ochraceous, the centre bars inclining to whitish; frontal feathers chocolate-brown; the head, hind neck, and wig bright golden ochraceous, all the feathers tipped with chocolate-brown, producing a distinctly barred appearance, the plumes on the side of the neck having a whitish shaft-stripe; facial aspect dull white, the feathers round the eye blackish, as also the loral plumes; the ear-coverts with mesial white shaft-lines and faint rufous bars; facial ruff dark chocolate-brown, the upper and lower portion more fulvous, the feathers being barred with the latter colour; chin dark brown; the ruff below the chin whitish, with broad dark central streaks; rest of under surface yellowish, the breast barred with chocolate-brown, the flanks and sides of the body streaked broadly down

the same proportion between the light and dark bars as is seen in *S. nebulosum*. The face is plain dirty white, without the brown bars or semicircles—a constant and conspicuous feature in *S. nebulosum*. There is the same number of spots on the primaries and bands on the tail as in the latter; the white beneath is without any ochraceous tinge; the primary-coverts are plain dark brown. Wing 14·80 inches, tail 9, culmen 1·05, tarsus 2·20, middle toe 1·60.

"*Hab.* Eastern Mexico."

the centre with chestnut; abdomen and under tail-coverts yellowish white; leg-feathers yellowish buff, minutely but thickly spotted with chestnut; under wing-coverts yellowish, spotted and narrowly barred with chestnut-brown, the lower series ashy brown, yellowish at base and resembling the inner lining of the wing; the outer primaries nearly uniform, the others with yellowish bands on the inner webs, the secondaries barred right across. Total length 18 inches, wing 13·1, tail 8, tarsus 2·25.

Hab. Guatemala.

a. Ad. sk.	Vicinity of Antigua, Guatemala (*Constancia*).	Osbert Salvin, Esq. [P.]. One of the typical specimens.
b, c. Ad. sk.	Guatemala.	Purchased.
d. Juv. st.	[America.]	Purchased.
e. Ad. sk.	[America.]	Earl of Derby [P.].

9. Syrnium occidentale.

Syrnium occidentale, *Xanthus, Pr. Philad. Acad.* 1859, p. 193; *Baird, B. N. Am.* App. pl. lxvi.; *id. in Cooper's, B. Calif.* p. 430; *Coues, Key N.-Am. B.* p. 204; *id. B. N. W.* p. 309 (note); *Ridgw. N.-Am. B.* iii. p. 38.

Adult. Above deep umber-brown, a little clearer on the head, all the feathers with concealed bars of fulvous, and having a broad subterminal spot or bar of pure white, imparting a distinctly spotted appearance to the upper surface of the body; the white bars very much broader on the hind neck and on the scapulars; wing-coverts coloured like the back, and spotted and notched in exactly the same manner; primary-coverts dark brown, with about four bars of fulvous on the outer webs; quills paler brown, tipped with whitish and barred with fulvous brown, about seven square spots being visible on the outer webs of both primaries and secondaries, inclining to whitish on the outer web of both series, the innermost secondaries vermiculated with white; upper tail-coverts brown like the back, with very distinct cross bars of white; tail pale brown, tipped with whitish, and crossed with about seven other bands of fulvous brown, obsolete near the base, but becoming lighter and more whitish towards the tips; face duller brown, with white shaft-streaks and concentric rings of darker brown, not many in number or strongly pronounced, the lower angle of the facial disk inclining to silvery white where it adjoins the ruff; the latter consisting of deep chocolate-brown feathers, spotted and barred with white or fulvous, the white bars on the long throat-feathers very distinct; under surface of body barred with white and brown, most of the feathers yellowish buff at base; some of the white bars on the flanks not complete, but consisting of pairs of large oval spots; the under wing-coverts white, barred with light brown; leg-feathers white; bill yellow. Total length 18·5 inches, wing 12·8, tail 9, tarsus 2·35.

Hab. Southern middle provinces of the United States, ranging into Mexico.

a. ♀ ad. st.	Mexico.

10. Syrnium rufipes.

Strix rufipes, *King, Zool. Journ.* iii. p. 426.
Ulula rufipes, *Gould, Voy. Beagle, Birds*, p. 34.
Athene rufipes, *Gray, Gen. B.* i. p. 35; *Strickl. Orn. Syn.* p. 172.
Ulula fasciata, *Des Murs, Iconogr. Orn.* pl. xxxvii.; *Gray, Gen. B.* iii. App. p. 30 *a*.
Syrnium fasciatum, *Bp. Consp.* i. p. 52; *Strickl. Orn. Syn.* p. 192; *Gray, Hand-l. B.* i. p. 48.
Syrnium rufipes, *Gray, Hand-l. B.* i. p. 48; *Scl. & Salv. Nomencl. Av. Neotr.* p. 116.

Adult. Above chocolate-brown, barred with yellowish buff and white, the yellow bars mostly concealed, the white bars conspicuous and subterminal; round the hind neck a very distinct wig, with broad white bars; the head rather darker than the back, with distinct bars of yellowish buff and white, the latter broader than on the back; least wing-coverts not quite so plainly barred with buff, but the median and greater series very distinctly barred with tawny buff, and externally notched with white, all these markings being also very distinct on the scapulars; primary-coverts nearly uniform brown; quills brown, barred across with lighter brown, shading off into buffy white on both the inner and outer webs, the primaries with large square spots of yellowish buff; tail brown, tipped with whitish, and crossed with eight other fulvous bars, becoming whiter towards the tips of the feathers; loral plumes silvery white, the shafts developing into hair-like blackish bristles; feathers in front of the eye blackish brown; rest of facial disk dusky greyish white, the feathers below the eye tawny buff, all with obsolete concentric bars of dusky brown, the ear-coverts narrowly streaked with silvery white lines along the shaft; facial ruff deep chocolate-brown, the feathers spotted and tipped with fulvous, the white bars very distinct towards the tips of the feathers; entire under surface of body washed with yellowish, the bases of the feathers showing here and there, all crossed with broad transverse bars of white and pale brown, about equal in width but less distinct on the under tail-coverts; leg-feathers yellowish buff; under wing-coverts yellowish buff, the median ones and the axillaries barred with brown and white to a greater or less extent; greater series dark brown, inclining to yellowish at base, and resembling the inner lining of the quills, which are dark brown below, barred with fulvous. Total length 18 inches, wing 11·1, tail 7·8, tarsus 2.

Hab. Patagonia and Chili.

a. Ad. sk.	Port Famine.	Capt. King [C.].	Type of species.
b. Ad. st.	Straits of Magellan.	J. Gould, Esq.	

11. Syrnium sinense.

China Owl, *Lath. Gen. Syn. Suppl.* ii. Add. p. 368.
Strix sinensis, *Lath. Ind. Orn. Suppl.* p. xvi.
Strix orientalis, *Shaw, Gen. Zool.* vii. p. 257 (*ex Lath.*).
Strix seloputo, *Horsf. Tr. Linn. Soc.* xiii. p. 140; *Less. Man.* i. p. 108.

Strix pagodarum, *Temm. Pl. Col.* i. pl. 230.
Noctua pagodarum, *Steph. Gen. Zool.* xiii. pt. 2, p. 65.
Ulula pagodarum, *Less. Man. d'Orn.* i. p. 113.
Syrnium pagodarum, *Cuvier, Règne An.* i. p. 343.
Bulaca sinensis, *Jerd. Madr. Journ.* x. p. 88; *Hume, Rough Notes,* ii. p. 357.
Syrnium seloputo, *Gray, Gen. B.* i. p. 39; *Cass. Cat. Strigidæ Philad. Mus.* p. 6; *Horsf. & Moore, Cat. B. Mus. E. I. Co.* i. p. 84; *Strickl. Orn. Syn.* p. 191; *Gray, Hand-l. B.* i. p. 49.
Bulaca seloputo, *Kaup, Contr. Orn.* 1852, p. 122.
Ciccaba pagodarum, *Bp. Consp.* i. p. 44.
Myrtha pagodarum, *Bp. Rev. et Mag. de Zool.* 1854, p. 541.
Ciccaba seloputo, *Kaup, Tr. Z. S.* iv. p. 254; *Wall. Ibis,* 1868, p. 26; *Salvad. Ucc. Born.* p. 21.
Ulula seloputo, *Schl. Mus. P.-B.* Striges, p. 22.

Adult. Above of a rufous chocolate-brown, the head deeper brown, all the upper surface barred with white, these bars being generally in the form of two spots, each of which has a narrow margin of black, the white bars on the outer scapulars much broader, thus forming a distinct shoulder-patch; the wig distinctly barred with white, but the back more spotted; least wing-coverts uniform chocolate-brown, the rest coloured like the back, and spotted with white in the same fashion, the greater series with irregular oval markings of white, two or three in number on the outer web, these feathers slightly vermiculated with brown towards their tips; quills light rufous chocolate, darker at their tips, barred across with light brown on the outer and with dark brown on the inner webs, the light bars terminating externally in a more or less distinct spot of whitish; the secondaries much paler and more uniform with the back, the white external spots very distinct, and the innermost barred with white exactly like the back; all the quills tipped with white, and the spots on their outer webs generally vermiculated slightly with brown; upper tail-coverts of the same colour as the back, and barred across with white; tail dull rufous chocolate, inclining to brown on the outer feathers, all the rectrices dull whitish at tip and crossed with seven bands of fulvous brown, inclining to whitish towards the external margins, the bars on the inner webs much broader and bright ochraceous in colour; all the sides of the face dull ochraceous; the ruff chocolate-brown, narrowly barred with white near the tip, and with ochraceous towards the base, the feathers being entirely ochraceous at the extreme base; the ruff-feathers encircling the throat chocolate-brown, narrowly barred with ochraceous; chin fulvescent; rest of under surface of body white, barred across with deep brown, these bars becoming narrower on the abdomen and under tail-coverts, and gradually disappearing into zigzag markings on the leg-feathers; on all the under surface a shade of golden buff is apparent in places, the breast being of this colour, transversely barred with white and dark brown; under wing-coverts white, very narrowly barred with brown in the form of zigzag lines, the outer feathers rather more broadly barred in this way and washed with golden buff; the greater series fulvous at base, ashy brown at tip, thus resembling the inner lining of the wing, which is ashy brown, fulvous at the base of the feathers, which are also barred with the same colour on the inner webs. Total length 22·5 inches, wing 15·6, tail 9, tarsus 2·3.

Obs. A Penang specimen from Mr. Wallace's collection has the wing 14·3 inches, and is probably the male; but a Cochin-Chinese example measured only 13·3 inches; so that perhaps the bird from the latter country runs smaller. I can see no other differences.

Hab. Java, ranging northwards through the Malayan peninsula to Burmah and Cochin China.

a. Ad. st.	Java.	Zoological Society.
b. Ad. sk.	Penang.	A. R. Wallace, Esq. [C.].

12. **Syrnium ocellatum.**

Strix sinensis, *Gray & Hardw. Ill. Ind. Zool.* i. pl. 21 (*nec Lath.*).
Syrnium ocellatum, *Less. Rev. Zool.* 1839, p. 289; *Gray, Gen. B.* i. p. 35; *Gould, B. Asia*, part xxii.; *Gray, Hand-l. B.* i. p. 49.
Syrnium sinense, *Blyth, Journ. A. S. B.* xi. p. 162; *Gray, Gen. B.* i. p. 39; *Cass. Cat. Strigidæ*, p. 6; *Bp. Consp.* i. p. 52; *Horsf. & Moore, Cat. B. Mus. E. I. Co.* i. p. 82; *Strickl. Orn. Syn.* p. 190; *Jerd. B. Ind.* i. p. 123; *Gray, Hand-l. B.* i. p. 49.
Bulaca sinensis, *Gray, Cat. Accipitr.* 1844, p. 43; *Kaup, Contr. Orn.* 1852, p. 122; *Blyth, Ibis*, 1865, p. 29.
Ciccaba sinense, *Kaup, Tr. Z. S.* iv. p. 254.
Ulula sinensis, *Schl. Mus. P.-B.* Striges, p. 21.
Bulaca ocellata, *Hume, Rough Notes*, ii. p. 383; *id. Nests & Eggs Ind. B.* p. 61.

Adult. General colour above rufous orange, the feathers of the crown and hind neck mottled with black and white, each feather having a double white spot at the tip preceded by a distinct black bar; the back greyer, each feather being orange at base, but for the terminal half crossed with distinct zigzag bars of dark brown, the interspaces being white, thickly mottled with brown vermiculations; the scapulars externally white, crossed with narrow brown bars, forming a conspicuous shoulder-patch; wing-coverts coloured like the back and mottled with grey in the same manner, the greater series rather whiter than the rest; primary-coverts dark brown, with a few zigzag markings of ashy grey on the outer web; primaries dark brown, barred across with lighter brown, these bars nearly obsolete on the inner web and represented by an ashy spot on the outer one, these spots clouded and obscured by brown vermiculations, the inner webs orange for the basal half with brown cross bars; secondaries for the most part ashy white like the back, vermiculated minutely with dark brown markings and barred across with zigzag bars of brown, the inner webs orange with dark brown bars; upper tail-coverts orange, barred near the tip with white and black; tail orange, barred with ashy white, these bars vermiculated with brown, the tips of the feathers brown, crossed in the same way with ashy whitish bars, these being altogether eight or nine in number on the tail; sides of the face greyish white, mottled all over with minute black bars, the ear-coverts orange in the centre; facial ruff dark chocolate-brown, many of the feathers spotted with white, those on the lower part inclining to orange and the plumes under the chin being white with narrow black bars, in both instances assimilating to the adjacent parts of the body; chin and a patch on the fore neck white, the plumes on the latter part with a few narrow terminal bars of dark brown; rest of the under surface white, narrowly barred with dark brown, the whole of the feathers orange at base, this colour showing conspicuously on the chest; leg-feathers, as well as under wing- and tail-coverts, white, barred with dark brown, the greater series of under wing-coverts blackish, with light orange-buff bases, thus resembling the inner lining of the wing, which is orange-buff for the greater part of the inner webs of the feathers, the ends being brown, barred across with lighter brown; "bill horny black, pale and greyish on lower mandible; cere dingy, a mixture of dirty, pinkish, brownish, and yellowish horny (varies a good deal); bare part of toes pale greenish brown, soles yellowish white; iris brown, light in some specimens, dark in others" (*Hume*). Total length 21 inches, wing 13·9, tail 8·5, tarsus 2·2.

Hab. Throughout India east of the Sutledje, and of the Indus below its junction with the former, and west of the Ganges (*Hume*).

a. Ad. sk.	Madras.	Rev. H. H. Baber [P.].
b. Ad. st.	India.	
c. Ad. st.	India.	

13. Syrnium leptogrammicum*.

Strix leptogrammica, *Temm. Pl. Col.* ii. pl. 525.
Syrnium leptogrammicum, *Gray, Gen. B.* iii. App. p. 3; *Strickl. Orn. Syn.* p. 193; *Gray, Hand-l. B.* i. p. 49.
Ciccaba leptogrammica, *Bp. Consp.* i. p. 44; *Kaup, Tr. Z. S.* iv. p. 255; *Wall. Ibis*, 1868, p. 26; *Salvad. Ucc. Born.* p. 21.
Bulaca leptogrammica, *Kaup, Contr. Orn.* 1852, p. 130.
Myrtha leptogrammica, *Bp. Rev. et Mag. de Zool.* 1854, p. 541.
Ulula leptogrammica, *Schl. Mus. P.-B.* Striges, p. 20, et *Revue Accipitr.* p. 21.

Adult male. Above chestnut-brown, regularly and distinctly banded across with dark brown; the scapulars externally white or

* Professor Schlegel recognizes a larger species (of which I have not seen examples) representing the present one in Sumatra. The following will be the synonymy if it is really distinct:—

SYRNIUM MYRTHA.

Syrnium leptogrammicum, *Cass. Cat. Strigidæ Philad. Mus.* p. 7.
Ciccaba myrtha, *Bp. Consp.* i. p. 44; *Salvad. Ucc. Born.* p. 21.
Myrtha sumatrana, *Bp. Rev. et Mag. de Zool.* 1854, p. 541.
Syrnium myrtha, *Strickl. Orn. Syn.* p. 193.
Ulula myrtha, *Schl. Mus. P.-B.* Striges, p. 19.
Syrnium sumatranum, *Gray, Hand-l. B.* i. p. 49.

Hab. Sumatra.

buffy white, more narrowly banded with brown, forming a longitudinal shoulder-band; head and wig uniform chocolate-brown; the hind neck orange-chestnut, forming a distinct collar; wing-coverts dull chestnut, lighter than the back, but banded with dark brown in the same regular manner, the bars of both colours being about equal in breadth, excepting on the primary-coverts, where the brown predominates; quills dark brown, banded with dull chestnut on the outer web and with ashy brown on the inner, the inner secondaries more equally banded and uniform with the wing-coverts; the under surface of the wing dark brown, with pale rufescent bands on the inner web, paling into fulvous near the base of the feathers, where the bands are much broader; upper tail-coverts banded with chestnut and dark brown, more narrowly than on any other part of the upper surface; tail brown, tipped and banded with dull chestnut like the back, the bars fourteen in number, the under surface much paler than the upper, the bars on the outer feathers fulvous; forehead and eyebrow, as well as the sides of the face, bright orange-chestnut, all the feathers round the eye blackish; lores whitish, the feathers somewhat tinged with rufous, the shafts long and blackish at the tips; throat and chest orange-chestnut, the chin paler and streaked with brown; fore neck and chest chocolate-brown, the breast dark brown barred with rufous like the upper surface; sides of neck and facial ruff dark brown; rest of under surface of body orange-buff, narrowly barred with brown, each bar being conterminously margined with rufous, the whole of the breast varied with silvery white bars, very broad and distinct; leg-feathers dark orange-buff or light chestnut, very indistinctly barred with dark brown; under wing-coverts orange-buff, with the same narrow bars as the breast, the greater ones fulvous barred with brown, the tips entirely brown, this series of coverts thus resembling the inner lining of the wing. Total length 14 inches, culmen 1·6, wing 11·5, tail 6·5, tarsus 1·85.

Hab. Borneo.

a. Ad. st.	Borneo.	Purchased.
b. ♂ ad. sk.	Sarawak.	A. R. Wallace, Esq. [C.].

14. Syrnium nuchale.

Syrnium woodfordi, *Hartl. Orn. Westafr.* p. 21 (1857, *nec Smith*); *Cass. Pr. Phil. Acad.* 1859, p. 32.
Syrnium nuchale, *Sharpe, Ibis*, 1870, p. 487.

Nestling. General colour light tawny, with whitish tips to most of the feathers, broader on those of the scapulars and secondaries, which are also barred with white; tail dark brown, with lighter brown cross bands and tipped with whitish; feathers in front of and over the eye pure white; plumes round the eye blackish; the ear-coverts again whitish; under surface light tawny, paler on the abdomen and under tail-coverts, some of the breast-feathers with whitish tips, before which is a narrow brownish subterminal line, a few feathers appearing on the breast fawn-colour, barred with white; quills dark brown, shaded with rufous on the outer web and crossed with several bands of paler brown, which are margined above and below by a narrow blackish line.

Adult. Above deep chocolate-brown, the head blackish and decidedly darker than the back, the upper surface showing a few diamond-shaped spots or bars of white; the crown spotted, but the wig distinctly barred with the latter colour, most of the dorsal plumes having concealed fulvous bars; least wing-coverts blackish chocolate; the rest of the wing-coverts rather rufous brown, darker brown at base and ornamented with large bars of white externally, as on the scapulars; primary-coverts blackish brown, slightly washed with reddish brown; quills dark brown, washed at the extreme tip with ashy brown, the primaries banded across with paler brown, producing a chequered appearance on the outer web, where the light bars shade off into a white spot; the secondaries paler brown, more numerously barred with light brown than the primaries, the innermost being barred with white; tail dark brown, the centre feathers having four bands of paler brown, not counting the pale rufous-brown tip, the exterior feathers with five pale bands of brown on the outer web, lighter on the inner one and inclining to white; lores and feathers above the eye pure white, the former with blackish shaft-streaks, and some of the latter narrowly tipped with brown; feathers round the eye blackish; ear-coverts dusky white, barred across with brown; feathers of the ruff dark brown, with yellowish-white bars, these white bars much predominating on the lower part of the ruff; chin-feathers dark brown, washed with rufous near the tips; rest of under surface chestnut-brown, banded across with white bars, each band having a narrow conterminous bar of dark brown, the white band broader than the rufous bands, all the cross bands narrowing a great deal on the under tail-coverts; leg-feathers yellowish buff, with minute brown cross markings; under wing-coverts ochraceous buff, the outermost barred with blackish, the greater series blackish, at base fulvescent, resembling the inner lining of the wing, which is blackish, with fulvous bars on the inner web, brighter on the secondaries. Total length 15 inches, wing 10, tail 6·5, tarsus 1·9.

Obs. This species replaces in Western Africa the *S. woodfordi* of South Africa. It is altogether darker above than that species, the head being blackish, and it further differs in the uniform character of the back and least wing-coverts, which are without vermiculations. No specimen which I have seen as yet of *S. woodfordi* had the broad white bands on the secondaries; while below the prevailing aspect of the last-named species is buff or whitish, banded with pale rufous, in *S. nuchale* it is rich rufous with distinct white cross bands. The title was originally suggested by the white bands on the nape appearing much more distinct than in *S. woodfordi*, where their small size and the general pale colour of the bird renders them not very conspicuous.

Hab. West Africa: forests of Denkera, in the interior of the Fantee country, and Gaboon.

a. Juv. sk.	Fantee (*Ussher*).	Capt. G. E. Shelley [P.].
b. Ad. sk.	Denkera (*Aubinn*).	R. B. Sharpe, Esq. [P.].
c. Ad. sk.	Denkera.	M. Aubinn [C.].
d. Ad. sk.	Cape-Coast Castle.	H. F. Blissett, Esq. [P.].
e. Ad. sk.	Ogowé River, Gaboon.	MM. Marche et De Compiègne [C.].
f. Ad. sk.	Africa.	

15. **Syrnium woodfordi.**

Noctua woodfordi, *Smith, S.-Afr. Q. Journ.* 2nd ser. pt. 1, p. 312.
Athene woodfordi, *Smith, Ill. Zool. S. Afr.* Aves, pl. 71; *Gray, Gen. B.* i. p. 34; *Strickl. Orn. Syn.* p. 170.
Syrnium woodfordi, *Bp. Consp.* i. p. 52; *Gurney, Ibis*, 1859, p. 242, et 1862, p. 37; *Kirk, Ibis*, 1864, p. 307; *Layard, B. S. Afr.* p. 42; *Gray, Hand-l. B.* i. p. 48; *Heugl. Orn. N.O.-Afr.* i. p. 122; *Finsch & Hartl. Vög. Ostafr.* p. 108; *Sharpe, ed. Layard's B. S. Afr.* p. 76.
Ciccaba woodfordi, *Kaup, Contr. Orn.* 1852, p. 120; *id. Tr. Z. S.* iv. p. 253.
Strix woodfoordi, *Grill, Zool. Anteckn.* p. 47.
Ulula woodfoordti, *Schl. Mus. P.-B.* Striges, p. 12.
Syrnium umbrinum, *Heugl. J. f. O.* 1863, p. 12; *Gray, Hand-l. B.* i. p. 48.

Nestling. Covered with down of a dull white colour, the sprouting feathers brown, banded with dingy ochraceous and broadly tipped with whitish buff; the under feathers coloured in the same manner, the breast-feathers especially brown.

Young. Of a dull tawny colour, many of the feathers barred with fulvous and tipped with white, giving the head a hoary appearance; wing-coverts brown, slightly vermiculated with lighter brown and ochraceous, and broadly tipped with white; quills and tail-feathers as in adult, but the bars rather narrower on the latter; under surface of body light ochraceous, inclining to whitish on the belly, everywhere transversely barred with light brown.

Full-grown young bird. Above dark chocolate-brown, the hind neck with a few cross lines of ochraceous buff, the dorsal feathers with a few vermiculations of ochraceous, not disturbing, however, the general brown appearance of the upper surface; head darker brown than the back, scantily marked with a tiny white streak in the centre of the feather, these markings larger and more star-shaped on the hind neck and wig, and present, though in a smaller degree, on most of the dorsal feathers; scapulars externally pale orange-buff, shading into white subterminally, and forming a distinct longitudinal shoulder-patch; wing-coverts chocolate-brown, a little darker than the back, with a few nearly obsolete cross markings of fulvous, here and there relieved with a narrow black line, the median and greater series being the palest of all, on these alone the star-shaped median spot being apparent as on the back, the greater coverts having a distinct spot of ochraceous white on the outer web; primary-coverts almost entirely brown, with obsolete bars of lighter brown near the tip; quills brown, with a terminal edging of fulvous white; the primaries darker than the secondaries, and crossed with six bands of light ochre, deeper and more obscured with brown towards the tip; the secondaries paler brown, and crossed with four or five bands of ochraceous brown; rump and upper tail-coverts coloured like the back, but barred rather plainly with ochraceous buff; tail dark brown, with a pale whitish tip, and crossed with seven pale brown bands, inclining to ochraceous near the base, these including the subterminal light brown bar, the bands on the inner web broader and paler ochraceous; lores and feathers over the fore part of eye white, all those round the eye dark brown; the ear-coverts whitish, obscured with narrow brown bars; feathers of the ruff brown, minutely banded with ochraceous, the bars much broader on the gular portion of the ruff; chin whitish; rest of under surface of body ochraceous buff, with broad bands of fawn-brown and white, most of these bands having a conterminous line of blackish brown above and below; leg-feathers fulvous, with a few narrow zigzag lines of brown; under tail-coverts fulvous, very narrowly barred with brown; under wing-coverts deep ochraceous, with a few narrow streaks of dark brown near the outer margin of the wing, the outer coverts of the lower series dark brown at tip, thus resembling the inner lining of the wing, which is dark brown, with ochraceous bands on the inner web, especially of the secondaries.

Adult. Above chocolate-brown, regularly banded across with dull fulvous, these bars dilating in the centre into a diamond-shaped spot, indistinct on the dorsal feathers, more plainly developed on the scapulars, and very distinct on the head, hind neck, and wig, where the bars are broader and whiter; crown of the head rather darker than the back, and the bars somewhat smaller; scapulars externally white, narrowly barred with brown, forming a conspicuous longitudinal shoulder-patch; wing-coverts brown like the back, but rather more uniform, the greater and median series with narrow fulvous bars and vermiculations, and scantily marked with diamond-shaped spots of white like the back; quills brown, fulvous at tip, and distinctly banded with ochraceous brown, inclining to white on the outer web, the secondaries barred with ochraceous brown and fulvous much more narrowly than the primaries; upper tail-coverts brown like the back, but distinctly barred with white; tail brown, tipped with whitish, and crossed with six bands of paler brown, not counting the pale subterminal shade of brown which precedes the whitish tip, but which can scarcely be called a band; facial aspect dull brown, with narrow fulvous shaft-streaks, and circularly barred with dusky whitish; lores and feathers over the eye more distinctly white, with blackish shaft-stripes and a few narrow bars of brown; ruff composed of brown feathers, slightly barred with whitish, the plumes on the lower portion yellowish white, with narrow brown bars; chin dull white, barred across with brown; rest of under surface white, barred narrowly with brown in regular transverse bands, and varied with rufescent tips to the feathers, the white bars always the broadest and most distinct; leg-feathers fulvous, with narrow zigzag bars of brown; under tail-coverts white, with narrow brown cross bars; under wing-coverts fulvous, thickly barred across with brown, the outermost more broadly barred, the greater series fulvous at base, dark brown at tip, thus resembling the inner lining of the wing, which is dark brown, barred on the inner web with pale ochraceous, the secondaries more distinctly barred with the latter colour; bill, cere, and feet light yellow; iris varying from light brown to yellow. Total length 15·5 inches, wing 9·5, tail 6·3, tarsus 1·95.

Foot of *Syrnium woodfordi*, to show extent of occasional feathering*.

Hab. Southern and North-eastern Africa.

a. Ad. st.	South Africa.	John Gould, Esq. [P.].
b. Pull. st.	South Africa.	South-African Museum.
c. Ad. sk.	South Africa.	South-African Museum. Type of species.
d. Juv. sk.	South Africa.	Sir A. Smith [C.].
e. Imm. st.	South Africa.	Purchased.
f. Ad. sk.	Zambesi.	Dr. Livingstone [C.].
g. Ad. sk.	Potchefstroom, Transvaal.	F. A. Barratt, Esq. [C.].

16. Syrnium hylophilum.

Strix hylophilum, *Temm. Pl. Col.* ii. pl. 373.
Syrnium hylophilum, *Gray, Gen. B.* i. p. 39; *Bp. Consp.* i. p. 52; *Strickl. Orn. Syn.* p. 192; *Pelz. Orn. Bras.* pp. 9, 400; *Gray, Hand-l.* i. p. 43; *Scl. & Salv. Nomencl.* p. 116.
Ciccaba hylophila, *Kaup, Contr. Orn.* 1852, p. 120; *id. Tr. Z. S.* iv. p. 253.
Syrnium albitarse, *Gray, Cat. Accipitr.* 1848, p. 103 (*descr. nullâ*); *Bp. Consp.* i. p. 52; *Scl. Tr. Z. S.* iv. p. 263, pl. lx. (*descr. princeps*); *Gray, Hand-l. B.* i. p. 49.
Ulula hylophila, *Gray, Cat. Accipitr.* 1844, p. 47; *Schl. Mus. P.-B.* Striges, p. 13.
Macabra hylophila, *Bp. Rev. et Mag. de Zool.* 1854, p. 541.
Ciccaba albitarse, *Scl. & Salv. Nomencl.* p. 117.

Adult. Above blackish brown, strongly varied with orange-tawny, especially on the wig and hind neck, where the bases to the feathers are orange and only the tips black; the scapulars irregularly

* The bird whose foot is figured is an exceptional specimen, the toes in every other specimen examined being quite bare.

barred with orange, sometimes the bars traversing the feathers, at others confined to two spots on either web by no means conterminous, the outermost scapulars more distinctly orange, forming a tolerably plain shoulder-patch; wing-coverts dark sepia-brown, with small spots of deep orange, forming larger indents on the interior greater coverts, the primary-coverts almost uniform dark brown; quills dark brown, inclining to paler and more ashy brown at the tips of the feathers and chequered externally with four or five square spots of orange-buff, the secondaries barred with the same on the inner web, the primaries with ashy brown; tail-feathers deep brown, tipped with light fulvous and crossed with five other bars of dull orange-buff; lores and feathers above the eye white, as also an eyebrow, the latter tinged with orange; sides of face dull orange-buff, with blackish bases and white shafts to the feathers; facial ruff bright orange, the bases to the feathers black; chin and lower cheeks whitish, below these a ring of orange-buff feathers tipped with black; under surface of body orange-buff, many of the feathers barred across with silvery white, producing a somewhat ocellated appearance, these white bars margined above and below with black, causing the breast to appear strongly barred with that colour; the abdomen paler, broadly barred with silvery white and narrowly barred and mottled with orange-brown on the margins of the white bars; tarsal feathers white; under wing-coverts light orange-buff, whitish along the edge of the wing, the greater series ashy brown like the wing-lining. Total length 16·5 inches, culmen 1·5, wing 11, tail 6·8, tarsus 2·1.

Hab. Brazil, Guiana, and Columbia.

a. Ad. st.	New Granada.	E. Wilson, Esq. [P.].
b. Juv. st.	Brazil.	Lord Stuart de Rothsay [P.].

17. Syrnium albigulare.

Syrnium albogulare, *Cass. Pr. Philad. Acad.* iv. p. 124; *id. Journ. Philad. Acad.* ii. p. 52, pl. 4; *Strickl. Orn. Syn.* p. 192; *Gray, Hand-l. B.* i. p. 49.
Ciccaba albipunctatum, *Kaup, Contr. Orn.* 1852, p. 120.
Syrnium macabrum, *Bp. Consp.* i. p. 53.
Macabra albigularis, *Bp. Rev. et Mag. de Zool.* 1854, p. 541.
Ulula albigularis, *Schl. Mus. P.-B.* Striges, p. 16.
Ciccaba albigularis, *Scl. & Salv. Nomencl. Av. Neotr.* p. 117.

Adult. Above dull brown, slightly shaded with rufous, and obscurely mottled with ochraceous vermiculations, very minute in character; many of the feathers of the upper surface blackish in the centre and spotted here and there with white or ochraceous buff, especially on the hind neck and over the ear-coverts, where the white spots are very distinct, though scarcely forming an eyebrow; wing-coverts rather more rufous than the back, which they otherwise resemble, and distinctly spotted with deep ochre, the greater and primary-coverts notched and spotted with ochraceous buff, showing very plainly; the scapulars like the back, but the outer ones also spotted with ochraceous buff or white, forming a tolerably distinct shoulder-patch; quills dark brown, much paler at the tip and numerously banded with rufous ochre, of which bands about ten can be counted on the outer web of the primaries, the inner web being barred with paler and more chocolate-brown; the secondaries also externally barred with rufous ochre, the bars about six in number, the innermost secondaries like the back, mottled with ashy grey; inner lining of wing ashy brown, spotted with fulvous on the outer web, the inner web barred with yellowish buff near the base; tail dark brown, lighter at tip and mottled on outer margin and crossed with ten or eleven very narrow and irregular bars of rufous ochre, inclining to fulvous underneath the tail, which is lighter brown below; loral plumes blackish, inclining to white near the base, and the feathers in front of the eye spotted with white; ear-coverts dark brown, blackish on their hinder margin and obsoletely barred with ochraceous buff; the facial ruff blackish brown, all the plumes white for more than the basal half; chin whitish; rest of under surface of body deep ochraceous buff, the breast clouded with brown vermiculations and spotted with ochraceous buff or white, thus resembling the back, the white markings larger on the sides of the chest; the lower breast and flanks varied or streaked with a few dark feathers, the centres of the plumes longitudinally blackish brown, with distinct spots of white or ochraceous buff on either web, these markings almost entirely disappearing on the vent and under tail-coverts; leg-feathers deep rufous ochre; under wing-coverts also deep ochraceous buff, slightly marked with white near the outer margin of wing, the greater series ashy brown, thus resembling the inner lining of the wing. Total length 13 inches, culmen 0·95, wing 8·6, tail 5, tarsus 1·2.

Hab. Columbia and Ecuador.

a. Ad. st.	Bogotá.	
b. Ad. st.	New Granada.	Purchased.

18. Syrnium superciliare.

Syrnium superciliare, *Pelz. Verh. z.-b. Wien,* 1863, p. 1125; *Gray, Hand-l. B.* i. p. 50.
Syrnium zonocercum, *Scl. & Salv. P. Z. S.* 1867, p. 590 (*nec Gray*).
Syrnium polygrammicum, *Gray, Hand-l. B.* i. p. 49 (1869)*.
Ciccaba polygrammica, *Scl. & Salv. Nomencl. Av.* p. 117.

Adult. Above chocolate-brown, distinctly but irregularly barred across with orange-buff, each bar having a conterminous one of blackish brown; the scapulars externally orange-buff, some few yellowish white forming a distinct longitudinal shoulder-patch; wing-

* Like *S. zonocercum*, the name of *S. polygrammicum* was published by the late Mr. G. R. Gray in the second catalogue of Accipitres (1848). No type specimen was indicated, no description given, and at least three species were included under this MS. name, viz. *S. virgatum* (Cass.), *S. superciliare*, Pelz., and *S. suinda* (V.). Either of the last two species would justify the specific name of *polygrammicum*; and it is only by hazard that I can refer to the mounted example of *S. superciliare* as being the *S. polygrammicum* of the 'Hand-list,' for no indication is on the stands of the specimens. Seeing, therefore, that it is impossible to fix the type of Gray's species (which was, in addition, never described), it appears to me safer to recognize Von Pelzeln's name, which was accompanied by a full description of the bird.

coverts coloured and barred in the same manner as the back, but the brown bars rather broader; primary-coverts dark brown, barred with dull ochraceous brown; quills dark brown, barred with very pale ochraceous brown, lighter on the outer web, where about nine can be counted on the primaries and seven on the secondaries, not including the light tip; tail brown, tipped with pale brown, crossed with seven bands of the same colour, the bars decidedly ochraceous near the base; loral feathers and those on fore part of forehead conspicuously white; lores white at base, black at tip; ear-coverts rufous, fringed with black, the shafts to the feathers white; facial ruff orange-buff, slightly barred with brown, the lower part of the ruff yellowish white below the ear-coverts, forming a conspicuous crescentic mark; entire under surface of body yellowish ochre, washed with deeper ochre on the sides of the upper breast and mottled with chocolate-brown in the form of irregular bars on the chest, and of longitudinal streaks on the flanks and abdomen, which are also sparsely mottled with zigzag brown markings; the under tail-coverts with scarcely any remains of bars; leg-feathers pale ochraceous buff; under wing-coverts rich ochraceous, the greater series ashy brown, barred with yellowish white near the base, and thus resembling the inner lining of the wing, which is also ashy brown, banded with lighter brown, changing to yellowish white near the base. Total length 17 inches, culmen 1·6, wing 10·8, tail 6·7, tarsus 1·85. (*Brazil.*)

Obs. Mr. Wallace collected a specimen at Pará, which, with some hesitation, I refer to this species. Unlike the one described, it has a black tail with only *five* bars (not counting the white terminal one), and the streaks on the belly are brown, without the lateral mottlings apparent on the other specimen. The barred character of the upper surface and the facial features, however, being the same in both, I cannot but believe the other differences to be the result of age.

Hab. Brazil.

a. Ad. st.	Brazil.	M. Claussen [P.].
b. Ad. sk.	Pará, May 1849.	A. R. Wallace, Esq. [C.].

19. Syrnium suinda.

Suindá, *Azara, Apunt.* ii. p. 120; *Hartl. Ind. Azara*, p. 3.
Strix suinda, *Vieill. N. Dict.* vii. p. 34 (*ex Azara*).
Noctua dominicensis, *Tsch. Arch. Nat.* 1844, p. 262; *id. F. P.* p. 115.
Syrnium suinda, *Bp. Consp.* i. p. 52; *Gray, Hand-l. B.* i. p. 49.
Macabra suinda, *Bp. Rev. et Mag. de Zool.* 1854, p. 541.
Ulula cayennensis (pt.), *Schl. Mus. P.-B.* Striges, p. 4*.
Ciccaba suinda, *Scl. & Salv. Nomencl.* p. 117.

---

* The name of *cayennensis* is founded on a plate of Buffon's, which I find it impossible to recognize. The synonymy is as follows:—

Chat-huant de Cayenne, *Buff. Pl. Enl.* 442.
Strix cayennensis, *Gm. S. N.* i. p. 296; *Strickl. Orn. Syn.* p. 183.
Noctua cayennensis, *Steph. Gen. Zool.* xiii. pt. 2, p. 67.
Athene cayennensis, *Gray, Gen. B.* i. p. 35.
Syrnium cayennense, *Bp. Consp.* i. 52.
Ciccaba cayennense, *Kaup, Contr. Orn.* 1852, p. 120.

Adult. Above light chocolate, obsoletely varied with dull blackish and ochraceous cross lines, the crown and hind neck more strongly marked with ochraceous; outer scapulars largely marked with yellowish buff, forming a distinct shoulder-patch; wing-coverts coloured like the back, but varied with larger and more rufous mottlings, the median series with large ocellated spots of yellowish white on the outer web; primary-coverts dark brown, with a few obsolete cross markings of dull reddish; quills brown, lighter at tip and barred across with paler brown, inclining to fulvous on the outer web; the bars on the primaries about seven in number, and about six on the secondaries, which are not so fulvous externally, and the innermost of which are coloured like the back; tail brown, lighter at tip, and crossed with six bands of pale ochraceous; lores whitish, washed with rufous, the shafts black at tip; feathers over the eye whitish buff, the feathers of the facial ruff deeper fulvous, rather lighter below the cheeks; sides of face rufous brown, slightly mottled with black bars, the shafts whitish; chin whitish; rest of under surface buffy white, the throat and chest more or less inclining to fulvous brown, the flanks washed with orange-buff, the leg-feathers entirely of the latter colour; fore neck and breast streaked with rufous brown; under wing-coverts rich orange-buff; the greater series ashy brown, barred with fulvous, and thus resembling the inner lining of the wing. Total length 15·5 inches, culmen 1·3, wing 10·8, tail 7, tarsus 1·9.

Hab. Brazil.

a. Ad. st.	S. America.	
b. Ad. st.	Brazil.	M. Claussen [P.].
c. Ad. st.	Brazil.	E. Wilson, Esq. [P.].
d. Ad. sk.	Bahia.	Dr. Wucherer [C.].

20. Syrnium virgatum*.

Syrnium virgatum, *Cass. Pr. Philad. Acad.* iv. p. 124; *id. Journ. Ph. Acad.* ii. p. 51, pl. 3; *Strickl. Orn. Syn.* p. 193; *Gray, Hand-l. B.* i. p. 49.

Syrnium zonocercum, *Gray, Cat. Accipitr. B. M.* 1848, p. 103 (*descr. nullâ*).

Strix squamulata, *Licht. Mus. Berol., undè*

Syrnium squamulatum, *Bp. Consp.* i. p. 53.

Ciccaba cayennensis (pt.), *Kaup, Contr. Orn.* 1852, p. 120, et *Tr. Z. S.* iv. p. 251.

Athene squamulata, *Licht. Nomencl.* p. 7.

Ulula virgata, *Schl. Mus. P.-B.* Striges, p. 15.

Syrnium lineatum, *Lawr. Ann. Lyc. N. Y.* vii. p. 462; *Salvin, Ibis,* 1874, p. 318.

Ciccaba virgata, *Scl. & Salv. Nomencl.* p. 117.

Adult male. Above sepia-brown, crossed with numerous vermiculations of ochraceous buff, rather obsolete on the interscapulary

* I believe this species to be the *Strix fasciata* of Vieillot (N. Dict. vii. p. 21). Not being able to entirely satisfy myself on this point, I prefer to retain the well-known name of *S. virgatum* in common use.

region and more distinct on the crown, rump, and upper tail-coverts, some of the latter irregularly barred with white, and thus resembling the tail itself; the scapulars externally bright ochre or white, forming a considerable shoulder-patch, the inner scapulars more coarsely vermiculated than the back and with more tawny ochre; wing-coverts coloured like the back and vermiculated with zigzag bars of ochraceous buff, changing to yellowish white on the greater coverts; primary-coverts and quills dark sepia-brown, banded distinctly with lighter brown, these bars much lighter on the outer webs of the primaries, and inclining to whitish or yellowish white, the bars on the primaries eight in number; innermost secondaries uniform with the back; tail blackish, broadly tipped with white and crossed with five other bands of lighter brown, inclining to white on the inner web, and showing conspicuously when the tail is spread; forehead and crown minutely but very plainly barred with ochraceous; lores whitish, slightly mottled with brown, the shafts elongated into black hair-like bristles; over the eye a broad white band; sides of face and ear-coverts ochraceous buff, with whitish shaft-streaks and mottled with numerous small bars of brown; at base of beak a few white feathers, forming the lower part of the facial ruff, which is whitish, tinged with ochraceous at the tips of the feathers, and mottled with small spots of dark brown; entire under surface of body ochraceous buff, the throat and chest much mottled with brown centres and vermiculations to the feathers; the breast, abdomen, and flanks distinctly streaked with white, many of the feathers being entirely white with broad longitudinal centres of dark brown; under tail-coverts whitish, some washed with ochraceous buff and having slight remains of brown bars; leg-feathers bright ochraceous buff, also with slight remains of barring; under wing-coverts pale ochraceous, slightly streaked and mottled with brown near the outer edge, the greater series ashy brown, more resembling the inner lining of the wing. Total length 14 inches, culmen 1·3, wing 9, tail 5·7, tarsus 1·7.

Younger female. Rather larger than the male. Total length 15 inches, wing 10·2, tail 6·4, tarsus 1·75. In colour this bird is darker brown than the one described, the spots and bars on the head inclining to rufous, and the vermiculations on the back very coarse and distinct; the under surface deep ochraceous, with the same white streaks as in the male.

Young. Covered with down of a yellowish buff colour, inclining to white on the crown, sides of face, and throat. The feathers, which are sprouting, exactly resemble those of the adult above described.

Obs. It is scarcely possible to conceive a more variable species than the present appears to be, varying in coloration from a light brown to nearly black, but always with a strongly vermiculated upper surface: the markings on the chest are difficult to comprehend; but the mottlings are more pronounced in young individuals than in old ones, the latter having the breast as well as the abdomen streaked. The tail is blacker in adults, with whiter cross bands, four in number besides the terminal one; in young birds the bars are five in number, and the tail is browner.

Foot of *Syrnium virgatum*.

Hab. From Mexico southwards to Venezuela.

a. ♀ vix ad. st.	Jalapa, Mexico.	R. de Oca [C.].
b. ♂ juv. sk.	Jalapa, Mexico.	R. de Oca [C.].
c. ♂ ad. sk.	Calovevora, Veragua.	E. Arcé [C.].
d. Ad. sk.	Costa Rica.	Purchased.
e, f. Juv. sk.	Guatemala.	M. A. Bouvier.
g, h. Ad. sk.	Guatemala.	M. A. Bouvier.
i. Ad. sk.	Volcan de Fuego, Guatemala.	M. A. Bouvier.
k. Ad. st.	Caracas.	Mr. Dyson [C.]. Type of *S. zonocercum.*
l. Ad. st.	[Brazil.]	M. Claussen [P.].

21. Syrnium huhulum.

Le Huhul, *Levaill. Ois. d'Afr.* i. pl. 41; *Sundev. Crit. om Levaill.* p. 28.
Strix huhula, *Daud. Traité*, ii. p. 190 (*ex Levaill.*).
Strix lineata, *Shaw, Gen. Zool.* vii. pl. 36.
Noctua lineata, *Steph. Gen. Zool.* xiii. pt. 2, p. 67.
Strix albomarginata, *Spix, Av. Bras.* pl. 10 *a.*
Surnia lineata, *Less. Traité*, p. 101.
Ciccaba huhula, *Wagler, Isis*, 1832, p. 1222; *Gray, List Gen. B.* 1841, p. 7; *Bp. Consp.* i. p. 43; *Kaup, Contr. Orn.* 1852, p. 120, et *Tr. Z. S.* iv. p. 250; *Strickl. Orn. Syn.* p. 174; *Scl. & Salv. Nomencl. Av.* p. 117.
Ciccaba lineata, *Gray, List Gen. B.* 1840, p. 5.
Athene lineata, *Gray, Gen. B.* i. p. 35.
Athene huhula, *Gray, List Accipitr. B. M.* 1848, p. 93; *Pelz. Orn. Bras.* pp. 8, 399.
Ulula huhula, *Schl. Mus. P.-B.* Striges, p. 17.
Syrnium huhula, *Gray, Hand-l. B.* i. p. 49.
Syrnium albomarginatum, *Gray, Hand-l. B.* i. p. 49.

Adult. Above black, somewhat inclining to brown on the scapulars, everywhere barred with white, these bars being rather wavy on the scapulars and wing-coverts, and much smaller and less pronounced on the head, which is rather blacker than the rest of the upper surface; wing-coverts black, more sparsely barred with white than the back; quills deep brown, barred across with light brown, changing to white on the outer webs, the secondaries nearly all barred with white; tail black, tipped with white and crossed with four bands of the same; lores, feathers round the eye, sides of the face, and facial ruff black, spotted and tipped with white; sides of neck banded with black and white like the back of the neck; chin black; rest of under surface of body black, banded with white, the leg-feathers black, very slightly banded with white; under wing-coverts black, banded with white like the breast, the lower series ashy brown, more or less varied with white near the base of the inner web, thus resembling the inner lining of the wing. Total length 17 inches, culmen 1·5, wing 10·3, tail 6·5, tarsus 1·9.

Hab. Brazil, Guiana.

a. Ad. st. Cayenne. H. C. Rothery, Esq. [P.].

22. Syrnium nigrolineatum.

Syrnium nigrolineatum, *Sclater, Tr. Z. S.* iv. p. 268, pl. 63; *Gray, Hand-l. B.* i. p. 49.
Ciccaba nigrolineata, *Scl. & Salv. Nomencl. Av. Neotr.* p. 117.

Young. Above silvery white, narrowly barred with dark brown, a few black feathers appearing here and there, more especially on the crown; wing-coverts black, excepting a few of the greater ones, which are buffy white barred with black, the outermost plume of the winglet tipped and barred with white; quills black, with a few scattered narrow bars of white or pale brown; tail black, crossed with six narrow white bars and tipped with white, the tip being much broader than the cross bars; lores and sides of the face entirely black, as also the chin; entire under surface of body white, washed more or less with yellowish, and numerously crossed with blackish, the under wing-coverts very thickly, but the under tail-coverts more narrowly, barred with black.

Adult. Above uniform black, the head much darker than the back, which has a slight shade of chocolate; hind neck and wig black, banded with white and forming a collar; lores, feathers round the eye, and sides of face black, the eyebrow spotted with white, and the ear-coverts minutely tipped with white; entire wing-coverts black; quills blackish brown, sparingly notched or half-barred with yellowish white on the outer web, changing more or less to pale brown, especially on the inner webs of the feathers; the lower surface of the quills ashy black, with broad creamy buff bars near the base of the inner webs; upper tail-coverts black, barred with white; tail black, tipped narrowly with white, and crossed with five other bars of white; chin black; rest of under surface white, narrowly barred with black, rather more broadly on the flanks and under tail-coverts, and more thickly on the leg-feathers and under wing-coverts, especially on the outermost, the greater series blackish, mottled or notched near the base with white, thus resembling the inner lining of the wing. Total length 15·5 inches, culmen 1·8, wing 11, tail 6·9, tarsus 1·95.

Hab. Central America, from Mexico to Veragua.

a. Ad. sk.	Veragua.	M. E. Arcé [C.].
b. Juv. sk.	Veragua.	M. E. Arcé [C.].

Subsp. *a.* **Syrnium spilonotum.**

Syrnium spilonotum, *Gray, Hand-l. B.* i. p. 49 (*descr. nullâ*).

Adult. Above black, banded or rather spotted with white, the bars not carried across many of the feathers, and very few in number on the wing-coverts and scapulars; the head rather darker than the back, the spots on the feathers small and transverse, the hind neck or wig distinctly barred with white, forming a collar round the hind neck; quills chocolate-brown, banded with lighter brown, inclining to white on most of the outer webs, and more distinct on the secondaries; upper tail-coverts black, banded with white; tail black, tipped with white and crossed with six bars of the same, one concealed by the coverts; lores, feathers round the eye, and ear-coverts black, minutely spotted with white, as also the facial ruff; chin black; rest of under surface white, thickly and regularly barred across with black; leg-feathers black, indistinctly barred with white; under wing-coverts white, barred with black, the greater series more indistinctly, thus resembling the inner lining of the wing. Total length 16·5 inches, culmen 1·4, wing 11·2, tail 6·5, tarsus 1·9.

Obs. This species appears to differ constantly from *S. nigrolineatum* in its darker under surface, the black bars being very much broader.

Hab. Columbia.

a. Ad. st.	Bogotá.	Purchased.
b. Ad. sk.	Bogotá.	Purchased.
c. Ad. sk.	Bogotá.	R. B. Sharpe, Esq. [P.].

23. **Syrnium perspicillatum.**

Spectacle-Owl, *Lath. Gen. Syn. Suppl.* i. p. 50, pl. cvii.
Strix perspicillata, *Lath. Ind. Orn.* i. p. 58.
Chouette à collier, *Levaill. Ois. d'Afr.* i. pl. 42; *Sundev. Crit. om Levaill.* p. 28.
Chouette à masque noir, *Levaill. Ois. d'Afr.* i. pl. 44; *Sundev. Crit.* p. 28.
Strix torquata, *Daud. Traité*, ii. p. 193 (*ex Levaill.*).
Strix personata, *Daud. Traité*, ii. p. 192 (*ex Levaill.*).
Strix larvata, *Shaw & Nodder, Nat. Misc.* xix. pl. 801.
Strix pulsatrix, *Max. Reis. Bras.* i. p. 366.
Noctua torquata, *Steph. Gen. Zool.* xiii. pt. 2, p. 67; *Less. Traité*, p. 105; *D'Orb. Voy. Am. Mérid. Ois.* p.
Syrnium pulsatrix, *Max. Beitr. Orn. Bras.* iii. p. 268; *Gray, Hand-l. B.* i. p. 50.

Athene torquata, *Gray, Gen. B.* i. p. 35; *Cab. in Schomb. Reis. Guian.* iii. p. 732; *Strickl. Orn. Syn.* p. 172; *Pelz. Orn. Bras.* pp. 8, 399.
Syrnium personatum, *Gray, Gen. B.* iii. App. p. 3.
Athene melanota, *Pucher. Rev. Zool.* 1849, p. 28 (*nec Tschudi*); *Strickl. Orn. Syn.* p. 172.
Ciccaba perspicillata, *Cass. Cat. Strigidæ Philad. Mus.* p. 17.
Ciccaba torquata, *Bp. Consp.* i. p. 43.
Ciccaba melanonota, *Bp. Consp.* i. p. 43.
Pulsatrix torquata, *Kaup, Contr. Orn.* 1852, p. 122, et *Tr. Z. S.* iv. p. 258; *Scl. & Salv. Nomencl.* p. 117.
Syrnium perspicillatum, *Strickl. Orn. Syn.* p. 193.
Ulula torquata, *Schl. Mus. P.-B.* Striges, p. 17.
Ulula pulsatrix, *Schl. Mus. P.-B.* Striges, p. 17, note.
Syrnium torquatum, *Gray, Hand-l. B.* i. p. 50.

Nestling. Covered with pure white down, all the feathers of the face and throat, as well as the loral bristles, black, the ear-coverts with somewhat of a chocolate shade; wing-coverts light rufous brown, with very broad white tips forming bars across the wing; quills pale brown, narrowly tipped with white and crossed with numerous bars of darker brown, of which five or six can be distinctly counted; tail dark brown, nearly uniform and broadly tipped with white, crossed with bars of lighter brown, of which about five can be counted on the outermost feathers, gradually decreasing in number towards the centre ones, on which two indistinct ones alone are visible; under surface of body entirely covered with white down.

Obs. Another bird, still in nestling plumage, but more advanced than the foregoing, differs in having a yellow tinge on the down of the wing-coverts and under surface of body; the upper margin of the wing dull brown; greater wing-coverts brown, barred across and broadly tipped with yellowish white; scapulars varied with broad ochraceous bars, causing a somewhat ocellated appearance, the innermost ones barred with white; quills and tail dark brown, barred with lighter brown and tipped with white, the bars on the latter seven in number.

Adult male. Above uniform chocolate-brown, blacker on the head; wing-coverts uniform with the back, only a few of the greater coverts with lighter bars; primary-coverts entirely uniform brown; quills brown, narrowly tipped with whitish and crossed with bars of lighter brown, nine in number on the primaries and seven on the secondaries; scapulars brown like the back, the outermost barred with buffy white, forming rather a conspicuous shoulder-spot; tail dark brown, tipped with white and crossed with six bars of lighter brown; lores white, streaked with black; eyebrow buffy white; sides of face entirely blackish brown; cheeks and sides of neck white; throat black; fore neck white, followed by a broad crescentic band of chocolate-brown; rest of under surface light ochraceous yellow, paler on the leg-feathers; under wing-coverts also ochraceous, except the outermost, which are white and form a conspicuous spot on the margin of the wing; lower under wing-coverts dark brown, like the wing-lining; iris brownish orange, the pupil dark blue. Total length 19·7 inches, culmen 1·9, wing 14, tail 8·3, tarsus 2·35.

Adult female. Rather larger than the male, but very similar, excepting that the shoulder-spot is uniform ochraceous, without any bars. Total length 23 inches, wing 15·1, tail 8·9, tarsus 2·35.

Obs. Dr. Pucheran (*l. c.*) separates the Brazilian bird as *Athene melanota* (not the same as Tschudi's species of the same name) on account of the bands on the tail being only three in number and more narrow. Professor Schlegel, in his first catalogue of the Owls in the Leiden Museum, was inclined to keep the Brazilian bird distinct on account of its larger size and paler coloration. I have not been able to find any grounds for separating the northern from the southern birds; the former are generally rather darker; but the number of bands on the tail will not constitute a specific character in these Owls, as it varies with age. In his recent list of Striges, Professor Schlegel reunites the Brazilian bird with the Central-American one.

Foot of *Syrnium perspicillatum*, to show the feathered toes.

Hab. Central America, from Mexico southwards, Guiana, Brazil, and Amazonia.

a, b. Pull. st.	South America.	Purchased.
c. Ad. st.	Santa Cruz, Bolivia.	Mr. Brydges [C.].
d. Ad. sk.	North side of River Amazon.	A. R. Wallace, Esq. [C.].
e. Ad. sk.	Veragua.	M. E. Arcé [C.].
f. ♂ ad. sk.	Chitra, Veragua (*Arcé*).	O. Salvin, Esq.
g. Juv. sk.	Guatemala.	M. E. Arcé [C.].
h. Ad. sk.	N. Guatemala.	R. B. Sharpe, Esq. [P.].
i. Imm. sk.	Tehuantepec, Mexico.	Prof. Sumichrast [C.].

24. Syrnium melanonotum.

Noctua melanonota, *Tschudi, Arch. Nat.* 1844, p. 267, et *Faun. Peruan.* p. 114, pl. 4.
Macabra melanonota, *Bp. Rev. et Mag. de Zool.* 1854, p. 541.
Athene melanota, *Pelz. Orn. Bras.* p. 9.
Syrnium melanonotum, *Gray, Hand-l. B.* i. p. 50.
Ciccaba melanonota, *Scl. & Salv. Nomencl.* p. 117.

Adult. Above chocolate-brown, with the remains of obsolete ochraceous margins to the feathers of the scapulars and upper wing-coverts; the outer scapulars with small bars of ochraceous concealed by the other plumes; quills chocolate-brown, crossed with bars of ochraceous brown, more or less obsolete on the primaries, but much more distinct on the lower surface, where they form broad ochre-coloured bands, wider and lighter towards the base of the inner webs; tail chocolate-brown, barred with light ochraceous brown, changing to white on the inner web, the bars not conterminous, the middle tail-feathers (imperfect) uniform, the bars on the outer tail-feathers six in number; over the eye a broad streak of deep ochre, extending from the base of the tail to above hinder margin of ear-coverts; feathers round the eye blackish, the ear-coverts deep chocolate-brown, loral feathers and throat ochraceous buff, the former with black hair-like tips to the shafts; facial ruff and a narrow collar across the lower throat chocolate-brown; fore neck buff, with rufous brown ends to the feathers, producing a slightly barred

Foot of *Syrnium melanonotum*, to show the bare toes.

appearance; chest chocolate-brown, more rufous on the sides; rest of under surface of body tawny buff, barred with sienna-rufous, less distinctly indicated on the flanks, and almost absent on the under tail-coverts, which are yellowish; under wing-coverts bright tawny buff like the breast, some of the outermost marked with brown, the greater series ashy brown like the inner wing-lining. Total length 19 inches, culmen 1·8, wing 12·8, tail 7·3, tarsus 2·1.

Hab. Brazil and Peru.

a. Ad. st.	Peru.	Zoological Society.
b. Ad. sk.	S. America.	Purchased.

25. Syrnium newarense.

Bulaca newarensis, *Hodgs. Asiat. Research.* xix. p. 168; *Swinh. P. Z. S.* 1871, p. 344; *Hume, Rough Notes*, ii. p. 348; *id. Nests & Eggs Ind. B.* p. 60.
Mesidus newarense, *Hodgs. J. A. S. B.* x. p. 28.
Syrnium newarense, *Gray, Gen. B.* i. p. 39, pl. 14; *Jerd. B. Ind.* i. p. 122; *Gray, Hand-l. B.* i. p. 49.
Syrnium indranee (*nec Sykes*), *Gray, Cat. Mamm. &c. Nepal Hodgs.* p. 52; *Bp. Consp.* i. p. 51; *Strickl. Orn. Syn.* p. 191; *Gray, Hand-l. B.* i. p. 49.
Ulula indrani, *Schl. Mus. P.-B.* Striges, p. 19.
Bubo caligatus, *Swinh. Ibis*, 1863, p. 218, et 1864, p. 429.
Syrnium caligatum, *Gray, Hand-l. B.* i. p. 49.

Adult. Above uniform chocolate-brown, a little darker and blacker on the head, the scapulars externally white, barred with chocolate-brown, these bars more or less inclining to rufous; wing-coverts chocolate-brown like the back, the least series with a few fulvescent bars, the median series greyish at tip and barred across with fulvous, more or less inclined to white on the outer web, the bars slightly vermiculated with brown, the greater coverts more distinctly barred and tipped with white; primary-coverts chocolate-brown, barred with lighter brown; quills brown, barred with paler brown, more numerously on the secondaries, which are distinctly tipped with white, the bars on the innermost secondaries also inclining to white on the inner web; upper tail-coverts rather paler than the back, and barred across with white; tail-feathers deep chocolate-brown, barred across with ten narrow bands of lighter brown, not conterminous on the centre feathers excepting towards the tips, where they are regular, the outer feathers with as many as twelve bands on the outer web, the inner one being more broadly banded, the bars on this latter web inclining to white; feathers over the eye whitish, the plumes surrounding the latter black; loral feathers dusky, whitish at base but obscured by blackish shaft-streaks; rest of facial aspect dusky white, more or less washed with brown and narrowly barred with the same in circular lines; feathers of facial ruff glossy chocolate-brown, more marked with white as they descend to the throat, many of the plumes on the latter part white tipped with brown; feathers of the side of the neck white, narrowly barred with brown; chin dark brown; on the fore neck a crescentic patch of white; rest of under surface dull white, everywhere narrowly banded with chocolate-brown, the leg-feathers similarly but more minutely barred, those on the under tail-coverts browner and broader than on the breast-feathers; under wing-coverts similar to the breast, with a white spot on the edge of the wing, the greater series blackish, barred with yellowish white near the base, thus resembling the inner lining of the wing, which is blackish, barred with fulvous towards the tips of the feathers, these bars much broader and clearer yellowish on the secondaries; "bare joints of toes pale plumbeous, the claws dusky lead-colour; bill greenish horny white, bluish towards base; cere, which is ill-defined, plumbeous; bare eyelids and eyeshelf very pale, fleshy plumbeous; iris deep brown" (*Hume*). Total length 28 inches, wing 17·4, tail 11·5, tarsus 2·8.

Younger birds appear to me to have the visage much more rufous, to have the scapulars less distinctly white externally and generally somewhat washed with fulvous, the sides of the neck also being fulvous or ochraceous, indistinctly barred with brown. Considerable variation is exhibited in the tail, the bars on which are lighter and more yellowish in some specimens than in others, some examples also having a more distinct white tip to the feathers. The tail-bands vary from nine to twelve in number on the Nepalese specimens.

Hab. Himalaya Mountains; Formosa.

a. Ad. st.	Himalaya Mountains.	Zoological Society.
b. Ad. st.	Nepal.	B. H. Hodgson, Esq. [P.].
c. Vix ad. st.	Nepal.	B. H. Hodgson, Esq. [P.].
d. Imm. st.	Nepal.	B. H. Hodgson, Esq. [P.].
e, f. Ad. sk.	Nepal.	B. H. Hodgson, Esq. [P.].
g. Imm. sk.	Nepal.	B. H. Hodgson, Esq. [P.].

26. Syrnium indranee.

Strix indranee, *Sykes, P. Z. S.* 1832, p. 82.
Bulaca monticola, *Jerd. Madr. Journ.* p. 167.
Bulaca indranee, *Kaup, Contr. Orn.* 1852, p. 121; *id. Tr. Z. S.* iv. p. 257; *Hume, Rough Notes*, i. p. 347.
Syrnium indranee, *Horsf. & Moore, Cat. B. Mus. E. I. Co.* i. p. 83; *Jerd. B. Ind.* i. p. 121; *Holdsw. P. Z. S.* 1872, p. 415; *Hume, Str. F.* 1873, p. 429; *Legge, Str. F.* 1874, p. 342.
Syrnium ochrogenys, *Hume, Str. F.* 1873, p. 431.

Dr. Jerdon keeps *Syrnium indranee* distinct from *S. newarense* on account of the smaller size of the former bird, the wing measuring 13–14 inches against 18 inches in *S. newarense*. The specimen of the latter described by me is a very large bird from the Himalaya Mountains; but we have five Nepalese birds, presented by Mr. Hodgson, which give the following dimensions:—

	Total length. in.	Wing. in.	Tail. in.	Tarsus. in.
a. Nepal (*Hodgson*)	24	16	9·5	2·7
b. Nepal (*Hodgson*)	22	14·9	9·5	2·35
c. Nepal (*Hodgson*)	22	14·2	9	2·3
d. Nepal (*Hodgson*)	23·5	14·8	9·5	2·4
e. Nepal (*Hodgson*)	23	14·5	9·5	2·4

Apparently, therefore, the last four of these birds ought to be called *S. indranee*. Want of material prevents my determining the precise relations of these two species, as I have no authentic examples of *S. indranee* for comparison. Mr. Gould has presented to the Museum a Wood-Owl which he received from Dr. Jerdon, but of which the precise locality has not been recorded. This bird is much more fulvous underneath, with rufous brown cross bars, very narrow and more numerous than in the Himalayan examples. The face also is rufous ochre; and I do not believe it to be the true *S. newarense*.

Mr. Hume has noticed the ochraceous disk in Ceylonese birds, and proposes the name of *ochrogenys** for the island bird. The great probability is that examples from Ceylon and the peninsula of India will be found to be of the same species, which will be the true *Syrnium indranee* (Sykes). Lord Walden has kindly lent me a Malaccan specimen from his collection, obtained by the late Dr. Maingay; and this is not to be separated from Dr. Jerdon's skin, like which it has the ochraceous disk.

Hab. Peninsula of India, Ceylon, Malayan peninsula.

a. Ad. sk. Southern India? (*Jerdon*). J. Gould, Esq. [P.].

27. Syrnium harrisi.

Nyctale harrisi, *Cass. Pr. Philad. Acad.* iv. p. 157; *id. Tr. Phil. Acad.* ii. p. 53, pl. 5; *Gray, Gen. B.* iii. App. p. 30 *a*; *Strickl. Orn. Syn.* p. 178; *Sclater, P. Z. S.* 1856, p. 25; *Ridgw. N. A. Birds*, iii. p. 40.
Ciccaba gisella, *Bp. Consp.* i. p. 44.
Gisella lathami, *Bp. Rev. et Mag. de Zool.* 1854, p. 541.
Nyctalatinus† albipunctatus, "*Kaup*," *Bp. Comptes Rendus*, 1855, p. 653.
Gisella harrisi, *Bp. C. R.* 1855, p. 653.
Ulula harrisii, *Schl. Mus. P.-B.* Striges, p. 9.
Syrnium harrisii, *Gray, Hand-l. B.* i. p. 50.
Nyctalitinus harrisi, *Scl. & Salv. Nomencl. Av.* p. 116.

Adult. Upper surface of a dull chocolate-brown, deeper on the head; a very broad frontal patch and an eyebrow buffy white; on the nape a double spot of buffy white, succeeded on the hind neck by an indistinct collar of yellowish buff; scapulars also inclining to the latter colour on the outer web, and shading off into white at the tips, the longer ones spotted with buff on the outer web, the white spot near the tip being very distinct; wing-coverts chocolate-brown, notched with white near the edge of the wing and near the tip of the outer web of the greater series; quills deep brown, spotted on the outer and broadly notched on the inner web with white; tail deep brown, spotted with white at the tip, and crossed with three other rows of white spots; sides of face yellowish buff, the ear-coverts broadly tipped with chocolate-brown, of which colour there is also a small tuft; ruff composed of yellowish buff plumes, narrowly tipped with brown, especially broad on the long gular feathers; rest of under surface of body, including the leg-feathers, deep yellowish buff, slightly inclining to tawny on the centre of the body; under wing-coverts inclining to tawny buff, the lower series broadly tipped with dark brown, and resembling the

* On the change of plumage in the Ceylonese bird, Mr. Vincent Legge's excellent article should be studied ('Stray Feathers,' 1874, pp 342–348).

† Both Prince Bonaparte and Dr. Sclater refer to this name as being proposed by Kaup; the year is given as 1854 in Mr. G. R. Gray's 'Hand-list;' and the generic name adopted by the authors of the 'Nomenclator.' Yet I have never been able to ascertain the work in which it has been published; and the specific name of *albipunctatum* was applied by Kaup to *Syrnium albogulare* (Cass.)

inner lining of the quills, which are dark brown below, notched with white on the inner web, these markings not visible on the first two primaries, and only feebly developed on the third; bill yellowish horn-colour, brown at base and on the lower mandible; toes apparently yellowish flesh-colour in life. Total length 8·5 inches, wing 6·3, tail 3·5, tarsus 1·1. (*Mus. Norvic.*)

Hab. Columbia.

17. NYCTALA*.

		Type.
Nyctala, *Brehm, Isis*, 1828, p. 1271	....	N. tengmalmi.
Ægolius, *Kaup, Naturl. Syst.* p. 34 (1829)	....	N. tengmalmi.
Scotophilus, *Swains. Classif. B.* ii. p. 217 (1837)	..	N. tengmalmi.

Range. The northern and central portions of Europe and North America.

Key to the Species.

a. Larger: entire upper surface mottled with white bars; forehead spotted with white; tail-feathers with five white bars .... *tengmalmi*, p. 284.

b. Smaller: upper surface for the most part uniform; forehead with narrow streaks of white; tail-feathers with three white bars................ *acadica*, p. 287.

1. Nyctala tengmalmi.

Strix tengmalmi, *Gm. S. N.* i. p. 291; *Temm. Man. d'Orn.* i. p. 94; *Naum. Vög. Deutschl.* i. p. 500, Taf. 48. figs. 2, 3; *Roux, Orn. Prov.* pl. 53 bis; *Vieill. et Oud. Gal. Ois.* i. pl. 23; *Werner, Atlas*, Rapaces, pl. 38; *Sw. & Rich. Faun. Bor.-Am. Birds*, p. 94, pl. 32; *Aud. B. Am.* pl. ccclxxx.; *id. Orn. Biogr.* iv. p. 599; *Schl. u. Susem. Vög. Eur.* Taf. 48; *Sundev. Sv. Fogl.* p. 186, pl. xxv. fig. 3.

Strix dasypus, *Bechst. Naturg. Deutschl.* ii. p. 972; *Wolf & Meyer, Vög. Deutschl.* pl. to p. 51.

Athene tengmalmi, *Boie, Isis*, 1828, p. 771.

Nyctale planiceps, *Brehm, Isis*, 1828, p. 127, et *Vög. Deutschl.* p. 113.

Nyctale pinetorum, *Brehm, ut suprà.*

Nyctale abietum, *Brehm, ut suprà.*

Ægolius tengmalmi, *Kaup, Naturl. Syst.* p. 34.

Noctua tengmalmi, *Cuvier, Règn. An.* 1829, i. p. 345; *Gould, B. Eur.* i. pl. 49; *Yarr. Brit. B.* i. p. 146; *Hewits. Eggs Br. B.* i. p. 66, pl. xix.

Ulula tengmalmi, *Bp. Oss. Regn. An. Cuv.* p. 53; *Macgill. Hist. Br. B.* iii. p. 445.

Syrnium tengmalmi, *Eyton, Hist. Rarer Br. B.* p. 90.

Scotophilus tengmalmi, *Swains. Classif. B.* ii. p. 217.

Nyctala tengmalmi, *Bp. Comp. List B. Eur. & N. Am.* p. 7; *Gray, Gen. B.* i. p. 40, pl. 14. fig. 7; *Strickl. Orn. Syn.* p. 175; *Newt. Ooth. Wolley.* p. 165; *Degl. & Gerbe, Orn. Eur.* i. p. 125; *Gould, B. Gt. Br.* i. pl. xxxvi.; *Dall & Bann. Tr. Chic. Acad.* i. p. 273; *Borggr. Vogelf. Norddeutschl.* p. 63; *Newt. ed. Yarr. Br. B.* i.

---

* On the curious asymmetry of the ear-bones in this genus, Mr. Collett's paper should be consulted (P. Z. S. 1871, p. 739).

p. 154; *Salvad. Faun. Ital. Ucc.* p. 28; *Shelley, B. Egypt,* p. 177; *Sharpe & Dresser, B. Eur.* pt. xiv.

Nyctale richardsoni, *Bp. Comp. List B. Eur. & N. Am.* p. 7; *Gray, Gen. B.* i. p. 40; *Bp. Consp.* i. p. 54; *Strickl. Orn. Syn.* p. 176; *Cass. B. Calif. and Texas,* p. 185; *id. B. N. Am.* p. 57; *Gray, Hand-l. B.* i. p. 51.

Nyctale dasypus, *Gray, List Gen. B.* 1840, p. 6.

Nyctale funerea, *Bp. Cat. Metod. Ucc. Eur.* p. 24; *id. Consp.* i. p. 54; *Kaup, Contr. Orn.* 1852, p. 104; *id. Tr. Z. S.* iv. p. 207; *Fritsch, Vog. Eur.* tab. 11. fig. 1, tab. 13. fig. 8; *Jaub. & Barth. Lap. Rich. Orn.* p. 83; *Gray, Hand-l.* i. p. 51.

Nyctale minor, *Brehm, Naum.* 1855, p. 370.

Nyctale bædekeri, *Brehm, Naum.* 1855, p. 370.

Ulula funerea, *Schl. Mus. P.-B.* Striges, p. 8.

Syrnium funereum, *Heugl. Orn. N.O.-Afr.* i. p. 123.

Nyctale tengmalmi, var. richardsoni, *Ridgw. Am. Nat.* 1871, p. 285; *Coues, Key N.A. Birds,* p. 206; *Ridgw. N.A. Birds,* iii. p. 40; *Coues, B. N.W.* p. 313.

Nestling. All over of a dull sooty brown, lighter on under surface of body, and inclining to chocolate on the upper; the remains of nestling-down of a white colour; on the wing-coverts some very large oval spots of white; lores, forehead, eyebrow, and chin more or less whitish, under surface of body slightly varied with whitish near the base of the feathers.

Young. Everywhere chocolate-brown, much deeper on the sides of the face and ear-coverts; the lores, anterior part of cheeks, and feathers over the eye white, the former obscured with blackish bristles; wing-coverts with distinct oval markings of white on the outer web, the quills also marked with spots of white on both webs, but more distinctly on the outer one; scapulars and upper tail-coverts with concealed white bases; tail brown, crossed with three rows of white spots, not continuous across the feather; under surface of body chocolate-brown, the breast and abdomen mottled with white, the bases of the feathers being of this colour; leg-feathers and under tail-coverts yellowish white, slightly spotted with brown.

Adult female. Above rather light brown, thickly mottled with white all over the upper surface, the back varied with broad concealed bars of white, the scapulars with a large oval spot of the same near their outer webs; fore part and sides of crown rather darker than the back, thickly mottled with triangular spots of white, the hinder part of the crown lighter brown, the white markings larger and the plumage generally more lax; hind neck and wig light brown, with very broad bars of white; wing-coverts brown, the least ones nearly uniform, the median and greater series marked on the outer web with large oval spots of white; quills brown, tipped with greyish, spotted on the outer web and broadly notched on the inner web with white; upper tail-coverts light brown, broadly barred with white; tail brown, crossed with five rows of white bars; lores, a broad eyebrow, and sides of face generally white, the feathers in front of the eye blackish, the lores also obscured by blackish bristles, many of the plumes above the eye with a black tip to the shaft-line, the ear-coverts slightly inclining to blackish on their hinder margin; a very distinct ruff, formed of dark brown feathers thickly mottled with white spots; chin white, as also the fore neck, which is slightly varied with brown cross lines, and is separated from the chin by a circlet of brown feathers in continuation of the ruff; rest of under surface of body white, mottled with brown, taking the form of streaks on the flanks; the breast rather more spotted, the sides of the latter inclining to brown; leg-feathers white, spotted with brown; the under tail-coverts almost entirely white, excepting a few brown streaks on the outermost; under wing-coverts white, with small spots of brown, the greater series ashy brown, spotted with white, resembling the inner lining of the wing, which is ashy brown barred with white, these bars much larger and more oval on the secondaries; bill dirty yellow; iris bright yellow. Total length 10·5 inches, wing 7·3, tail 4·8, tarsus 0·85.

Adult male. Similar to the female, but slightly smaller. Total length 9·5 inches, wing 6·7, tail 4·4, tarsus 0·75.

Obs. My observations confirm those of Mr. D. G. Elliot and others as to the impossibility of distinguishing the North-American bird as a distinct species. Specimens in the Museum collection measure as follows:—

	Total length. in.	Wing. in.	Tail. in.
a. Arctic America	9·5	6·9	4·2
b. Repulse Bay (*J. Rae*)	9·5	7	4·4
c. ♂. Fort Simpson (*B. R. Ross*)	10·5	7·2	4·3

Mr. Ridgway considers that American birds are separable from the European as a geographical race. He writes:—"*Nyctale richardsoni*, though without doubt specifically the same as *N. tengmalmi* of Europe, is nevertheless to be distinguished from it. The colours of the European bird are very much paler; the legs are white, scarcely variegated, instead of ochraceous, thickly spotted; the lower tail-coverts have merely shaft-streaks of brown instead of broad stripes. Very perfect specimens from Europe enable me to make a satisfactory comparison."

Hab. The whole of Northern Europe and Siberia; not found in Southern Europe, but occurring all over the central portions of the continent. In North America found in the high north, rarely occurring within the United States.

a. Ad. st.	Northern Europe.	Dr. Lidth de Jeude.
b. Ad. sk.	Northern Europe.	Purchased.
c. Pull. st.	Wermland, May 18, 1872.	Purchased.
d. Juv. sk.	Trondhjem, July 17, 1866.	F. Bond, Esq. [P.].
e. Juv. sk.	Archangel.	R. B. Sharpe, Esq. [P.].
f. Ad. st.	Archangel.	R. B. Sharpe, Esq. [P.].
g. Ad. sk.	R. Ussuri, 48° N. lat. (*Dr. Dybowski*).	Warsaw Museum [E.].
h. Ad. st.	Arctic America (*Dr. Richardson*).	Zoological Society.
i. Ad. st.	Repulse Bay.	Dr. J. Rae [P.].
k. ♂ ad. sk.	Fort Simpson.	B. R. Ross, Esq. [P.].
l. Skeleton.	Sweden.	R. B. Sharpe, Esq.

2. Nyctala acadica.

Acadian Owl, *Lath. Gen. Syn.* i. p. 149, pl. 5. fig. 2.
Strix acadica, *Gm. S. N.* i. p. 296 (*ex Lath.*); *Audub. B. Am.* pl. cxcix.; *id. Orn. Biogr.* iv. 397; *Swains. & Rich. Faun. Bor.-Am. Birds*, p. 97.
Strix passerina, *Penn. Arctic Zool.* p. 236; *Wils. Am. Orn.* iv. p. 66, pl. 34. fig. 1.
Strix acadiensis, *Lath. Ind. Orn.* i. p. 65.
White-fronted Owl, *Lath. Gen. Syn. Suppl.* ii. p. 58.
Strix albifrons, *Shaw & Nodd. Nat. Misc.* xix. pl. 171.
Bubo albifrons, *Vieill. Ois. Am. Sept.* i. p. 54.
Scops albifrons, *Steph. Gen. Zool.* xiii. pt. 2, p. 51.
Athene wilsoni, *Boie, Isis*, 1828, p. 315.
Surnia acadica, *Bp. Oss. Regn. An. Cuv.* p. 52; *Schl. Mus. P.-B.* Striges, p. 8.
Ulula passerina, *James. ed. Wils. Am. Orn.* i. p. 109.
Scotophilus acadicus, *Swains. Classif. B.* ii. p. 217.
Strix frontalis, *Licht. Abhandl. k. Akad. Berlin*, 1838, p. 430.
Nyctale acadica, *Bp. Comp. List B. Eur. & N. Am.* p. 7; *id. Consp.* i. p. 54; *Gray. Gen. B.* iii. App. p. 50; *Kaup, Contr. Orn.* 1852, p. 104; *id. Tr. Z. S.* iv. p. 206; *Strickl. Orn. Syn.* p. 176; *Cass. B. Calif. & Texas*, p. 285; *id. B. N. Am.* p. 58; *Baird, in Cooper's B. Calif.* p. 436; *Ridgw. Am. Nat.* 1871, p. 285; *Coues, Key N.A. Birds*, p. 206; *id. B. N. W.* p. 315; *Ridgw. N.A. Birds*, iii. p. 43; *Scl. & Salv. Nomencl.* p. 116.
Nyctale kirtlandi, *Hoy, Pr. Philad. Acad.* vi. p. 210; *Elliot, Ibis*, 1872, p. 48*.
Athene albifrons, *Licht. Nomencl. Av.* p. 8.
Nyctale albifrons, *Cass. B. Calif. & Texas*, p. 187; *id. B. N. Am.* p. 57; *Baird, in Cooper's B. Calif.* p. 435.

Adult male. Above nearly uniform chocolate-brown, the scapulars with a large ovate spot of white on the outer web, the dorsal feathers with small concealed spots of white, which are larger and broader on the upper tail-coverts; head streaked with white, broader on the forehead, more obsoletely on the hinder crown; the hind neck mottled with white streaks, and the sides of the neck distinctly striped with central markings of white; wing-coverts dark brown, with distinct white oval spots on the outer web; quills brown, greyer towards the tips, the outer webs spotted with white, these spots present only on the primaries, and nearly absent on all the others, all the quills, however, being notched with yellowish white on the inner webs; tail dark brown, tipped with white and crossed with three rows of white bars, not continuous across the feather; lores, eyebrow, and feathers of the face generally dull white, the ear-coverts brownish, especially towards the hinder margin; feathers in front of the eye and loral bristles blackish; behind the ear-

* Mr. Elliot, in an interesting article on this bird, conclusively showed that it was the young of some species of *Nyctala*; but he considered it to be the immature plumage of *N. tengmalmi*. Mr. Ridgway, on the other hand, refers it to *N. acadica*; and, on consideration of the facts, I believe the decision of the last-named ornithologist to be correct. *N. albifrons* (Licht.), considered to belong to *N. tengmalmi* by some European authors, has been examined by Professor Schlegel, and declared by him to be *N. acadica*.

coverts a ruff of brown feathers, thickly mottled with median longitudinal streaks of white, this ruff crossing the throat, and separating the whitish chin from the fore neck, the latter having a few reddish markings; rest of under surface of body white, broadly streaked with rufous, the flanks with a slight tinge of yellow; leg-feathers yellowish white; under tail-coverts white, with a few streaks of rufous; under wing-coverts light orange, the outermost inclining to rufous, the greater series yellowish at base, ashy brown at tips, resembling the inner lining of the wing, which is ashy brown, with a few bars of yellowish white near the base of the quills; bill black. Total length 8 inches, wing 6·1, tail 3·7, tarsus 0·85.

Adult female. Similar to the male. Total length 6·5 inches, wing 5·75, tail 3·5, tarsus 0·85.

Hab. North America, ranging into Mexico.

a. Ad. st.	North America.	Zoological Society.
b. Ad. sk.	North America.	Purchased.
c. Ad. sk.	North America.	John Gould, Esq. [P.].
d. Ad. sk.	Komux Island.	Dr. Lyall [C.].
e. ♀ ad. sk.	West side of Rocky Mountains.	J. K. Lord, Esq. [P.].
f. ♂ ad. st.	West side of Rocky Mountains.	J. K. Lord, Esq. [P.].
g. ♀ ad. st.	West side of Rocky Mountains.	J. K. Lord, Esq. [P.].
h. Ad. sk.	Western Mexico.	Purchased.

Family STRIGIDÆ.

Glaucinæ, *Sundev. Av. Tent.* p. 104.

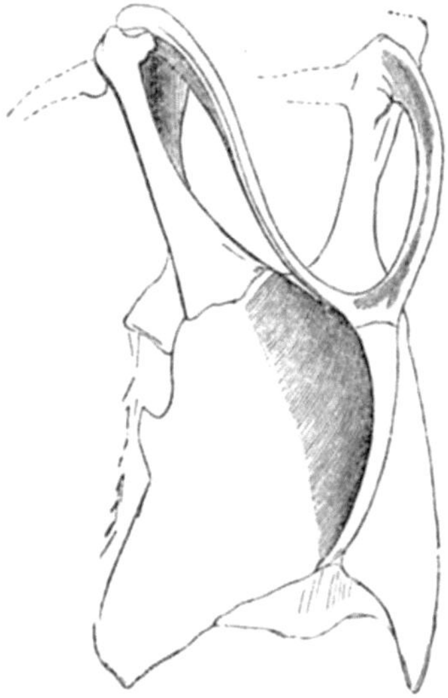

Sternum of *Strix flammea.*

Hinder margin of sternum entire, with no distinct clefts; furcula joined to keel of sternum; inner margin of middle claw serrated; inner and middle toes equal in length; between the anterior portion of the facial area a frontal patch of small stiff feathers always present and very broad.

Key to the Genera.

a. Tarsus about equal to twice length of middle toe (without claw); facial disk very perfect, the distance above the eye about equal to the same area below it; wings reaching far beyond tail .... 1. STRIX, p. 290.

b. Tarsus about equal to twice length of middle toe (without claw); facial disk irregular, the area above the eye being not nearly equal to the area below the latter; wings about reaching to tip of tail .... 2. PHODILUS, p. 390.

1. STRIX*.

	Type.
Strix, *Linn. S. N.* i. p. 133 (1766)	S. flammea.
Aluco, *Fleming, Phil. Zool.* ii. p. 236 (1828)....	S. flammea.
Hybris, *Nitzsch, Pteryl.* p. 100 (1840)	S. flammea.
Stridula, *Selys-Longch. Faun. Belge,* p. 60 (1842)	S. flammea.
Eustrinx, *Webb & Berth. Orn. Canar.* p. 8 (1844)	S. flammea.
Megastrix, *Kaup, Isis,* 1848, p. 769	S. tenebricosa.
Glaux, *Blyth, J. A. S. B.* xix. p. 513 (1850)....	S. candida.
Scelostrix, *Kaup, Contr. Orn.* 1852, p. 119	S. candida.
Dactylostrix, *Kaup, Contr. Orn.* 1852, p. 119 ..	S. novæ hollandiæ.
Glyphidiura, *Reich. Syst. Nat.* pl. xcii. fig. 2 ..	S. capensis.

Cf. *Sharpe, Ibis,* 1875, p. 324.

Middle toe of *Strix flammea*, to show the serration on the middle claw.

Range. The entire globe, excepting the extreme north of both hemispheres, New Zealand †, and certain islands of the Malayan subregions and Oceania.

Key to the Species.

a. Upper surface always vermiculated.
  a'. General colour below white or orange-buff, spotted or narrowly vermiculated with

---

* I am unacquainted personally with the following species:—

1. Strix thomensis.

Strix thomensis, *Hartl. Rev. et Mag. de Zool.* 1852, p. 2; *Müll. Beitr. Orn. Afr.* pl. 15; *Hartl. Orn. Westafr.* p. 21; *Gray, Hand-l. B.* i. p. 52; *Pelz. Journ. f. Orn.* 1872, p. 23.
Strix sancti-thomæ, *Kaup, Contr. Orn.* 1852, p. 118, et *Tr. Z. S.* iv. p. 247.

Above blackish grey, marked with minute spots of paler grey and white spots margined with black; below bright and intense orange-rufous, with rounded blackish spots; under tail-coverts and thighs unspotted; tail-feathers paler ferruginous, obscurely varied, with three broad dusky blackish bands, the tips of all more broadly dusky, marked with a minute irregular spot of white before the tip itself; quills marked in the same manner; axillaries buff; bill brown; claws blackish. Total length 14¾", wing 8" 4"', tail 4" 3"', tarsus 2" 2¼"'.
Hab. Island of St. Thomas, W. Africa.

† *Strix haasti,* Buller, is *Sceloglaux albifacies.*

brown; upper surface generally orange-buff, spotted with white and vermiculated with grey or brown.
a''. Size small; wing not exceeding 13·5 inches; bars on wing and tail four, more rarely five .... *flammea* and allies, pp. 291–298.
b''. Size large; wing 13·8–15·2 inches; bars on wing seven to eight, on tail six to eight .... *novæ hollandiæ*, p. 303.
b'. General colour above and below greyish brown, the under surface barred and spotted with white .... *tenebricosa*, p. 306.
b. Upper surface uniform brown, with spots of white.
c'. Middle tail-feathers uniform brown; primaries not barred below .... *capensis*, p. 307.
d'. Middle tail-feathers orange or white, barred across with brown; primaries barred with brown below .... *candida*, p. 308.

1. Strix flammea.

The Common Barn-Owl or White Owl, *Albin, N. H. Birds*, ii. pl. xi.
Le petit Chat-huant, *Briss. Orn.* i. p. 503.
Strix flammea, *Linn. S. N.* i. p. 133; *Meyer u. Wolf, Vög. Deutschl.* pl. to p. 185; *Wils. Am. Orn.* vi. p. 51, pl. 50. fig. 2; *Vieill. Faun. Franç.* pl. 22. fig. 1; *Temm. Man. d'Orn.* i. p. 91; *Max. Reis. Bras.* ii. 1821, p. 265; *Naum. Vög. Deutschl.* i. p. 483, pl. 47. fig. 2; *Roux, Orn. Prov.* pls. 54, 55; *Werner, Atlas*, Rapaces, pl. 36; *Brehm, Vög. Deutschl.* p. 106; *Less. Traité*, p. 112; *Audub. B. Amer.* pl. clxxi.; *id. Orn. Biogr.* ii. p. 403; *Gould, B. Eur.* i. pl. 36; *Schl. u. Susem. Vög. Eur.* Taf. 46; *Macgill. Brit. B.* iii. p. 473; *Nordm. in Démid. Voy. Russie Mérid.* iii. p. 110; *Pears. J. A. S. B.* x. p. 630; *Darw. Voy. Beagle*, Birds, p. 34; *Yarr. Brit. B.* i. p. 127; *Gray, Gen. B.* i. p. 41, pl. 15. fig. 1; *Thomps. B. Irel.* i. p. 92; *Blyth, Cat. B. Mus. A. S. B.* p. 41; *Bp. Consp.* i. p. 55; *Kaup, Contr. Orn.* 1852, p. 118; *Horsf. & Moore, Cat. B. Mus. E. I. Co.* i. p. 81; *Hartl. Orn. Westafr.* p. 21; *Schl. Vog. Nederl.* pl. 41; *Strickl. Orn. Syn.* p. 178; *Hewits. Eggs Brit. B.* i. p. 61, pl. xviii. fig. 1; *Sundev. Sv. Fogl.* p. 181, pl. lxxi. fig. 1; *Fritsch, Vög. Eur.* tab. ii. fig. 4; *Jaub. & Barth.-Lap. Rich. Orn.* p. 86; *Gould, B. Gt. Br.* i. pl. xxviii.; *Schl. Mus. P.-B.* Striges, p. 1; *Newt. Ooth. Wolley.* p. 152; *Scl. Ibis*, 1864, p. 307; *Pelz. Reis. Nov.* Vög. p. 29; *Degl. & Gerbe, Orn. Eur.* i. p. 133; *Loche, Expl. Sci. Algér.* Ois. i. p. 91; *Schl. & Poll. Faun. Madag.* Ois. p. 48; *Gray, Hand-l. B.* i. p. 52; *Heugl. Orn. N.O.-Afr.* i. p. 124; *Allen, Bull. Harv. Coll.* ii. p. 341; *Godman, Azores*, p. 22; *Finsch & Hartl. Vög. Ostafr.* p. 111; *Baird, ed. Cooper's B. Calif.* p. 415; *Salvad. Faun. Ital. Ucc.* p. 27; *Pelz. Orn. Bras.* pp. 10, 400; *Scl. & Salv. Nomencl. Av. Neotr.* p. 116; *Sharpe, ed. Layard, B. S. Afr.* p. 82.
Strix alba, *Scop. Ann.* i. p. 2.
L'Effraye ou La Fresaye, *Buff. Pl. Enl.* i. pl. 440.
Strix javanica, *Gm. S. N.* i. p. 295; *Jerd. Madr. Journ.* x. p. 85; *Gray, Gen. B.* i. p. 41, pl. 15; *Blyth, J. A. S. B.* xix. p. 513; *Bp. Consp.* i. p. 55; *Horsf. & Moore, Cat. B. Mus. E. I. Co.* i. p. 82; *Strickl. Orn. Syn.* p. 180; *Jerd. B. Ind.* i. p. 117; *Wall. Ibis*, 1868,

p. 26; *Gray, Hand-l. B.* i. p. 52; *Salvad. Ucc. Born.* p. 22; *Hume, Nests & Eggs Ind. B.* p. 59.
Lechusa, *Azara, Apunt.* ii. p. 210; *Hartl. Ind. Azara,* p. 3.
Strix perlata, *Licht. Verz. Doubl.* p. 59; *Max. Beitr.* iii. p. 263; *D'Orb. Voy. Amér. Mér.* Ois. p. 135; *Gray, Gen. B.* i. p. 41; *Schomb. Reis. Guian.* iii. p. 732; *Gray, Cat. Accipitr.* 1848, p. 109; *Cass. Cat. Strigidæ Philad. Mus.* p. 1; *Bp. Consp.* i. p. 55; *Kaup, Contr. Orn.* 1852, p. 119; *id. Tr. Z. S.* iv. p. 247; *Strickl. Orn. Syn.* p. 180; *Burm. Th. Bras.* ii. p. 137; *id. Reis. La Plata-St.* ii. p. 440; *Gray, Hand-l. B.* i. p. 52.
Aluco flammeus, *Fleming, Brit. An.* i. p. 157; *Newt. ed. Yarr. Brit. B.* i. p. 194; *Shelley, B. Egypt,* p. 176.
Strix furcata, *Temm. Pl. Col.* i. pl. 432; *D'Orb. in De la Sagra's H. N. Cuba,* Ois. p. 34; *Gundl. J. f. O.* 1871, p. 377.
Strix guttata, *Brehm, Vög. Deutschl.* p. 106.
Ulula flammea, *James. ed. Wils. Am. Orn.* ii. p. 264.
Strix delicatula, *Gould, P. Z. S.* 1836, p. 140; *id. Synops. B. Austral.* pt. 3, pl. 50. fig. 1; *id. B. Austr.* i. pl. 31; *Sturt, Exp. C. Austr.* App. p. 17; *Bp. Consp.* i. p. 54; *Kaup, Contr. Orn.* 1852, p. 119; *id. Tr. Z. S.* iv. p. 247; *Strickl. Orn. Syn.* p. 180; *Cass. U.S. Expl. Exp.* 1858, p. 105; *Gray, B. Trop. Isl.* p. 3; *Verr. Rev. et Mag. de Zool.* 1860, p. 421; *Gould, Handb. B. Austr.* i. p. 66; *Gray, Hand-l. B.* i. p. 52; *Finsch & Hartl. Vög. Cent.-Polyn.* p. 11.
Strix pratincola, *Bp. Comp. List B. Eur. & N. Amer.* p. 7; *De Kay, Zool. N. Y.* Birds, p. 31, pl. 13. fig. 28; *Gray, Gen. B.* i. p. 41; *Bp. Consp.* i. p. 55; *Strickl. Orn. Syn.* p. 179; *Cass. B. Calif.* p. 176; *id. B. N. Amer.* p. 47; *Gray, Hand-l. B.* i. p. 52.
Strix punctatissima, *Gray, in Darw. Zool. Beagle,* Birds, p. 34, pl. 4; *id. Gen. B.* i. p. 41; *Bp. Consp.* i. p. 55; *Kaup, Contr. Orn.* 1852, p. 118; *id. Tr. Z. S.* iv. p. 246; *Strickl. Orn. Syn.* p. 182; *Gray, Hand-l. B.* i. p. 52; *Scl. & Salv. Nomencl.* p. 116.
Strix poensis, *Fraser, P. Z. S.* 1842, p. 189; *Gray, Gen. B.* i. p. 41; *Strickl. Orn. Syn.* p. 182; *Hartl. Orn. Westafr.* p. 22; *Gray, Hand-l. B.* i. p. 52; *Gurney, in Anderss. B. Dam. Ld.* p. 36.
Stridula flammea, *Selys-Longch. Faun. Belge,* p. 60.
Strix bakkamma, *Forst. Descr. An.* p. 157.
Eustrinx flammea, *Webb & Berth. Orn. Canar.* p. 8.
Strix lulu, *Peale, U.S. Expl. Exp.* Birds, 1848, p. 74, pl. 21; *Gray, Hand-l. B.* i. p. 52.
Strix pusilla, *Blyth, J. A. S. B.* xviii. p. 101.
Athene forsteri, *Bp. Consp.* i. p. 42; *Gray, B. Trop. Isl. Pacif.* p. 2.
Strix glaucops, *Kaup, Contr. Orn.* 1852, p. 118; *id. Tr. Z. S.* iv. p. 246; *Gray, Hand-l. B.* i. p. 52; *Pelz. J. f. O.* 1872, p. 23.
Strix forsteri, *Hartl. Arch. Naturg.* 1852, p. 30.
Strix americana, *Brewer, ed. Wils. Am. Orn.* p. 687.
Strix africana, *Bp. Rev. et Mag. de Zool.* 1854, p. 540; *Gray, Hand-l. B.* i. p. 52.
Strix adspersa, *Brehm, Naum.* 1855, p. 270, et 1858, p. 215.
Strix margaritata, *Brehm, Naum.* 1855, p. 270.
Strix maculata, *Brehm, t. c.* p. 270.
Strix splendens, *Brehm, t. c.* p. 270.
Strix vulgaris, *Brehm, Naum.* 1858, p. 215.
Strix obscura, *Brehm, t. c.* p. 215.
Strix paradoxa, *Brehm, t. c.* p. 217.
Strix kirschoffi, *Brehm, t. c.* p. 219.
Strix indica, *Blyth, Ibis,* 1860, p. 251; *Hume, Rough Notes,* ii. p. 342; *Gould, B. Asia,* pt. xxiv.; *Holdsw. P. Z. S.* 1872, p. 415; *Hume, Str. F.* 1873, p. 163, et 1875, p. 37.

Strix flammea americana, *Schl. Mus. P.-B.* Striges, p. 4, et *Revue Accipitr.* p. 16.
Strix rosenbergi, *Schl. N. T. D.* iii. p. 181; *Wall. Ibis*, 1868, p. 26; *Gray, Hand-l. B.* i. p. 52; *Walden, Tr. Z. S.* viii. p. 41; *Schl. Mus. P.-B. Revue Accipitr.* p. 16.
Strix affinis, *Layard, B. S. Afr.* p. 43.
Strix insularis, *Pelz. J. f. O.* 1872, p. 23.
Strix flammea, var. americana, *Coues, Key N.-Am. B.* p. 202; *id. B. N. W.* p. 298.
Strix flammea, var. guatemalæ, *Ridgw. N.-Am. B.* iii. p. 11.
Strix flammea, var. furcata, *Ridgw. t. c.* p. 12.
Strix flammea, var. pratincola, *Ridgw. t. c.* p. 13.
Strix flammea, var. javanica, *Ridgw. t. c.* p. 13.
Strix flammea, var. delicatula, *Ridgw. t. c.* p. 13.
Aluco delicatula, *Sharpe, ed. Voy. Ereb & Terr.* Birds, p. 22.

Nestling. Covered with pure white down, the feathers which are appearing on the face being rufescent, those of the ruff silvery white with a very faint orange tip to a few of them; primaries orange, with faint vermiculations of grey at the tip, the secondaries more conspicuously mottled with grey as far as they are developed. (*Cambridgeshire.*)

Young bird on leaving the nest. General colour above orange, but profusely obscured with light grey, all the latter plumes vermiculated with ashy brown and having a distinct subterminal white spot margined both above and below with brown; the head and hind neck coloured like the back, but more decidedly orange, especially the sides of the neck, which are bright orange with a few brown spots; wing-coverts coloured like the back; primary-coverts orange, mottled at the tips like the rest of the coverts, but much paler externally and inclining to whitish; quills orange, mottled at the tips with greyish, and having distinct cross bars of grey mottled with white; tail pale orange, barred with greyish and mottled with the same; facial disk silvery white, the feathers rufous round the eye and especially in front of the latter; ruff glistening white, the upper plumes washed with orange, and the lower ones also tipped with clear orange; feathers of the under surface pure white, tinged with orange on the chest, but not spotted. This specimen has still plentiful remains of the nestling-down. (*Cambridgeshire.*)

Adult male. (Shropshire, Feb. 16th, 1869: *J. Gould.*) Above orange-buff, most of the feathers with white terminal and subterminal spots, each relieved by a corresponding spot of blackish; the feathers of the back and scapulars a good deal mottled with light silvery grey towards their tips, these mottlings disappearing again on the upper tail-coverts; sides of neck and least wing-coverts orange, with tiny brown spots, the rest of the wing-coverts mottled with grey at the tips, like the dorsal feathers; primary-coverts nearly uniform orange, with a few greyish mottlings at the tips of them; quills orange-buff, shading off into whitish near the base and on the inner webs, the secondaries tipped with whitish, the primaries with a few irregular spots of brown along the shaft or on the inner web, the secondaries rather deeper orange, speckled here and there with brown, the innermost mottled with grey at the tip, like the dorsal feathers; tail whitish, slightly washed with pale orange on the centre feathers, which show a few specklings of brown as well as three spots of the same along the shaft, these markings disappearing on the outermost, which are almost entirely white; facial disk pure white, with a rufous patch in front of the eye; feathers of the ruff glistening white, those on the upper parts washed with orange, the lower feathers subterminally orange with a tiny apical margin of blackish, rather more distinct on the gular feathers; rest of under surface of body pure unspotted white, as also the whole of the leg-feathers and under tail- and wing-coverts; the lower series of the latter slightly shaded with greyish and somewhat resembling the inner lining of the quills, which are greyish white below; bill nearly white; claws brown; iris black. Total length 13 inches, wing 11·9, tail 5, tarsus 2·2.

DARK RACE. *Adult male.* (Schleswig.) Above uniform bluish grey, the feathers everywhere most finely vermiculated with greyish white, and having two spots of white towards the tip of the feather, one terminal and one subterminal, each of these spots relieved by a black spot preceding it along the centre of the feather, and imparting a streaked appearance to the upper surface; on the head the markings are smaller than on the back; but in scarcely any part of the upper surface do the orange-buff bases to the feathers show through, so that the general aspect is very uniform; wing-coverts coloured like the back and similarly marked; primary-coverts resembling the quills, which are deep orange, mottled at the tips with greyish vermiculations like the back, and barred across with blackish brown, these bars slightly mottled with greyish on the outer margin, the orange interspaces also a little mottled with brown; the secondaries marked in the same way as the primaries, but more mottled on the outer webs, so that their general aspect is more obscure, the innermost being greyish, thickly vermiculated with dull white, and thus resembling the back; tail orange, with a broad mottled band of greyish at the tip, and crossed by five other bands of blackish, of which the two nearest the tip have sundry whitish mottlings; face vinous, inclining somewhat to silvery white on the hinder and lower margins of the ear-coverts; the feathers round the eye blackish, forming a black patch in front of it; ruff composed of deep orange feathers, some of which have concealed basal streaks of dark brown, the gular feathers tipped with the latter colour; rest of under surface of body, as well as the sides of the neck, deep orange, dotted with minute arrow-head spots of brown, ending in a white tip on some of the abdominal plumes and under tail-coverts, the spots becoming obsolete on the leg-feathers; under wing-coverts orange like the breast. and spotted in the same manner, the lower series whitish, tipped with greyish black, and thus resembling the inner lining of the quills, which are whitish, barred on the inner web with greyish black, and inclining entirely to greyish black at the tip. Total length 13 inches, wing 11·8, tail 5·5, tarsus 2·3.

Obs. It will be seen from the above synonymy that I have united nearly all the Barn-Owls of the world under the one heading of *Strix flammea.* It is seldom that an opportunity is afforded to the ornithologist of examining such a fine series of birds as has been permitted to me in the case of the Barn-Owls; and it would be difficult to find a more comprehensive collection than at present exists in the British Museum. Every single variation of the ordinary type has been examined by me; and I have had in all cases one or two specimens of the rarer or more obscure forms on which to found my conclusions. Mr. G. R. Gray, in the 'Hand-list,' has recognized no less than twelve species; but a very short study will suffice to show that not one half of these species can be maintained. Mr. Ridgway also separates the Barn-Owls into different races, and he considers that seven can be differentiated. The first of these is *Strix flammea*, inhabiting Europe and the Mediterranean region of Africa. In America it is represented by *S. pratincola*, the habitat of which is Mexico and southern North America, the only character given by Mr. Ridgway being, however, its larger size. A difference of an inch and a half in the wing is the extreme variation that I can find, this measuring in European examples from 11 to 12·4 inches, in American ones from 12·9 to 13·7 inches. The latter have the tarsus measuring 2·65–3·01 inches, against 2·2 to 2·5 inches in the European. The colour of the American bird is darker; and the dark phase, which is rarer in the Old-World birds, seems to be the permanent form in the New World. Indian Barn-Owls are rather darker than European and clearer grey above, in the latter respect becoming more like the Australian *S. delicatula*, whose pearly grey colour has been its chief specific character; sometimes, however, Australian examples are not to be distinguished from European; wing, in the latter, 10·8–11·3 inches, tarsus 2·4–2·6. Mr. Ridgway then separates *S. javanica* (from India and Eastern Africa) as a bird of the same colours as *S. delicatula.* The African Barn-Owl, according to my experience, is always darker than the European, especially the specimens from Southern Africa; but they are again scarcely distinguished from the dark phase of *S. flammea.* The Indian examples, on the contrary, run paler in colour, and are often nearly uniform grey on the back. In Java and the Malayan subregion a slight modification takes place, and the trace of zigzag markings on the breast, sometimes discernible in Indian birds, becomes tolerably well pronounced, especially in *S. rosenbergi*, which is a large form of *S. javanica*, and is not allied to *S. novæ hollandiæ.* The same character divides Mr. Ridgway's *S. guatemalæ* from *S. pratincola* or *S. flammea*; and this zigzag marking on the breast is extremely developed in the St.-Domingo Barn-Owl (*S. glaucops* of Kaup). The Cuban and Jamaican bird is at first sight remarkable, chiefly for its white tail; but in this respect it resembles British specimens, which, although matched by French birds, are quite different from any I have as yet seen from Germany. *Strix insularis* is a small dark form, supposed to be peculiar to the Cape-Verd Islands, but approached by some Senegambia specimens. It is closely allied to the Galapagoan *S. punctatissima.* *S. perlata* from South America is smaller than the North-American bird, and comes, therefore, very close to the European *S. flammea.*

My conclusion with regard to the Barn-Owls is, that there is one dominant type which prevails generally over the continents of the Old and New Worlds, being darker or lighter according to different localities, but possessing no distinctive specific characters. Insular birds vary, but cannot be specifically distinguished, as they can always be approached by continental specimens in a large series. Thus *S. furcata*, of Jamaica, appears different, but is scarcely to be distinguished from Bogotá specimens, which possess a white tail also. *S. insularis*, of the Cape-Verd Islands, is scarcely to be separated from some Senegambian skins. *S. glaucops* is not very different from Veraguan Barn-Owls; and these are again approached by Malayan specimens. Of these two extreme forms, *S. insularis* and *S. glaucops*, an illustration is given (Pl. XIV.). I give detailed descriptions and measurements of a few of the most striking forms of Barn-Owl which have been specifically separated by different authors; and in the list of specimens I have divided them under the headings by which they have become known to systematists.

a. Strix furcata.

Adult male. Above pale orange-buff, the feathers coarsely vermiculated at the tips with greyish brown, the markings more spot-like just before the terminal white longitudinal spot; crown of head richer and more of a golden orange, and more spotted than the back, especially on the sides of the neck, which shade off into pure white; wing-coverts orange-buff like the back, the least series minutely spotted, the median and greater coverts vermiculated near the tip with grey, disappearing on the primary-coverts, which have only a few scattered mottlings, some of the lesser coverts externally inclining to white like the edge of the wing; quills pure white, the outer primaries slightly shaded with orange-buff, and coarsely vermiculated with ashy brown at the tips, as also the secondaries, some of which have a few scanty vermiculations on the outer web, the innermost orange-buff, vermiculated with grey like the back; tail pure white; facial disk white, with a patch of rufous feathers in front of the eye; ruff composed of orange-buff plumes, inclining to white on the lower part, a few of the upper feathers with tiny spots of brown; entire under surface pure white, with the remains of a few tiny spots on the flanks. Total length 14·5 inches, wing 13·1, tail 5·8, tarsus 2·95, middle toe 1·6.

Adult female. Darker than the male, and more profusely mottled with grey on the upper surface; wings externally orange-buff, with scattered dots of black on the outer web, and indications of three blackish bars on the inner one, the ends of the feathers very coarsely vermiculated, the outermost of the secondaries inclining to white; tail white, with remains of three imperfect bands and tiny dots on most of the feathers; underneath white, tinged with orange-buff on the chest and flanks, both of which, as well as the under wing-coverts, are spotted with ashy brown. Total length 14·5 inches, wing 13·4, tail 5 2, tarsus 2·95, middle toe 1·65.

Obs. It is evident that the unspotted wing and pure white tail are gradually acquired by the disappearance of the bars and vermiculations.

β. Strix delicatula.

Adult. Above pearl-grey, with faint zigzag grey mottlings on all the feathers, which have mesial white tips preceded by a spot of dusky blackish; wing-coverts coloured like the back, the mottlings a little coarser, edge of the wing and outer margin of the bastard wing pure white, the rest of the latter resembling the other coverts; primary-coverts resembling the other series, with distinct white apical spots, and washed near the base with orange-buff; quills externally mottled with grey, with a subterminal white drop; the primaries externally washed with pale orange, inclining to white near the base, all the quills white on the inner web, showing conspicuously when the wing is spread, primaries with remains of four dark brown cross bands; tail-feathers whitish, mottled with coarse vermiculations of brown, and crossed with four bars of the latter colour, the markings disappearing on the outermost tail-feather, which is almost entirely white, with the remains of three bands only; facial disk pure white, with a dusky rufous patch in front of the eye; ruff composed also of pure white feathers, the uppermost with central streaks or zigzag cross markings of black, becoming narrower on the lower ones and disappearing on the gular plumes of the ruff, some few of which have a narrow brown tip; rest of under surface of body pure silky white, with no dusky spots on the chest, but a few on the sides of the body and abdomen, and indications of some tiny ones on the under wing-coverts, the lower series of which are dusky at tip, forming a faint wing-bar; the quills underneath white, mottled at the tip with greyish, and having three faintly indicated dusky spots or bars on the inner web; bill horn-colour; feet yellowish. Total length 13 inches, wing 11, tail 4·7, tarsus 2·35.

γ. Strix punctatissima.

Adult (type). Above dull brown, with a few scattered spots of dusky white, and minute flecks of the same on the dorsal feathers; sides of neck light golden buff, indistinctly marked with transverse lines of brown, becoming gradually obscure and disappearing on the hind neck; the wing-coverts more or less washed externally with deep golden buff, many of the white spots also tinged with the same colour, the buff predominating along the edge of the wing and at the base of the primary-coverts; quills dull brown like the back, slightly varied with minute flecks of white, and vermiculated with golden buff on the outer webs, much more distinct on the primaries, where it forms four irregularly defined bands; the secondaries with a distinct white terminal spot; the inner lining of wing ashy brown, the base of the feathers buffy white, the inner webs with two or three indistinct bands of whitish; tail golden buff, barred across with four bands of dark brown, the interspaces vermiculated with the same colour, especially towards the tips of the feathers, which are there almost entirely brown, with a white terminal spot; general colour of face vinous rufous, inclining to black round the eye; the cheeks silvery white, washed with rufous; facial ruff rich golden buff, white at base of feathers, which are coarsely vermiculated with brown, sometimes in a horseshoe pattern; crown of head deep golden buff, vermiculated with brownish; under surface of body rich golden buff, transversely vermiculated with brownish bars, irregular in shape and more distinct on the abdomen, the under wing- and tail-coverts coloured like the breast, but the latter rather paler; bill yellowish, horn-coloured at base; feet blackish. Total length 13 inches, culmen 1·7, wing 9·2, tail 4·6, tarsus 2·3.

δ. Strix rosenbergi.

Adult male. Above greyish brown, all the feathers finely speckled with whitish and tipped with white, the feathers being subterminally brown in the centre; so that the upper surface has a distinctly spotted appearance, here and there relieved with a wash of orange-buff where the bases show; wing-coverts rather greyer than the back, but spotted in the same manner, the median and greater series rich orange-buff, only varied with the greyish vermiculations towards the tips of the feathers, which are largely mottled with brown on the outer webs; quills rich orange-buff on the outer, mottled with brown and vermiculated with greyish near the tips of the feathers, all of which are barred across with dark brown, the inner webs duskier, and the light interspaces obscured with brown mottlings; the secondaries externally varied with greyish vermiculations exactly like the back, which the innermost secondaries precisely resemble; tail orange-buff, tipped with greyish white and mottled with dark brown vermiculations on the interspaces, the dark brown bars five in number; facial disk white; the lores and plumes in front of the eye rufous, those immediately in front of the latter black; feathers of the ruff rufous orange, barred and tipped with brown vermiculations; the gular plumes white, tipped with brown and subterminally washed with orange rufous; throat white; the rest of the under surface rich orange-buff, slightly washed with silvery white in the centre of the body; the lower abdomen also whitish, all the lower parts marked with triangular spots of brown, the subterminal spots being the largest, the chest at the same time slightly varied with indistinct zigzag bars of brown, all the spots vanishing on the under tail-coverts; the under wing-coverts rich orange-rufous, spotted with brown like the breast, some of the feathers washed with silvery white, the lower series yellowish white at base, ashy brown at tip, forming a wing-bar and resembling the inner lining of the quills, which are yellowish at base of inner web and barred with ashy brown, these bars disappearing towards the tips, which are entirely ashy brown. Total length 17 inches, wing 12·9, tail 6·3, tarsus 2·85.

Obs. Another specimen, collected by Dr. Meyer, measures as follows—total length 18 inches, wing 13, tail 6·5, tarsus 2·9. It is very much the same in colour as the one described, but is darker, so that the white spots show everywhere more distinctly; on the underparts there are remains of circular bars on many of the feathers, particularly those of the chest.

Hab. The entire world, ranging more rarely into the extreme north, and absent in certain localities, such as New Zealand and many groups of islands in Oceania and the Austro-Malayan sub-region, the Falkland Islands, &c.

PALÆARCTIC REGION.

A. Great Britain.

a. Light phase.

a, *b*. Pull. st.	England.	
c. Ad. sk.	England.	
d, *e*. Pull. st.	Cambridgeshire.	Mr. J. Baker [C.].
f. ♂ ad. st.	Cambridgeshire.	Mr. J. Baker [C.].
g. ♀ ad. st.	Cambridgeshire.	Mr. J. Baker [C.].
h. Ad. st.	Cambridgeshire.	Mr. J. Clarke [C.].
i. ♂ ad sk.	Shropshire, Feb. 16, 1869.	J. Gould, Esq. [P.].
k. ♀ ad. sk.	Shropshire, Oct. 16, 1868.	J. Gould, Esq. [P.].
l. ♀ ad. sk.	Sarsden, Oxon, Nov. 18, 1869.	J. Gould, Esq. [P.].
m. Ad. st.	Essex.	H. Doubleday, Esq. [P.].
n. Ad. st.	Combe Wood, Surrey.	Mr. Leadbeater [P.].
o. Pull. st.	Devonshire.	
p. Ad. st.	Devonshire.	Colonel Montagu [P.].

β. Dark phase.

q. ♀ ad. st.	Esher, Surrey, winter of 1869.	F. Bond, Esq. [P.].
r. ♀ ad. sk.	Sarsden, Oxon, Nov. 18, 1868.	J. Gould, Esq. [P.].

B. France.

a. Light phase.

s. Ad. st.	France.	R. B. Sharpe, Esq. [P.].
t. Ad. sk.	France.	M. A. Bouvier [C.].
u. Ad. sk.	Near Paris.	M. A. Bouvier [C.].
v. Ad. sk.	Nice.	M. A. Bouvier [C.].

β. Dark phase.

w. Ad. sk.	France.	M. A. Bouvier [C.].
x. Ad. st.	France.	R. B. Sharpe, Esq. [P.].
y. Ad. st.	Near Paris.	R. B. Sharpe, Esq. [P.].

C. Germany.

a. Light phase.

z. ♂ ad. st.	Westphalia.	Howard Saunders, Esq. [C.].

β. Dark phase.

a'. ♂ ad sk.	Schleswig.	Howard Saunders, Esq. [C.].
b'. ♀ ad. sk.	Schleswig, July 26, 1870.	Howard Saunders, Esq. [C.].

D. Spain.

c'. Ad.	Granada, S. Spain.	Howard Saunders, Esq. [C.].

E. Marocco.

a. Light phase.

d'. Ad. sk.	Tangier.	Signor Olcese [C.].

β. Dark phase.

e′, f′, g′. Ad. sk.	Tangier.	Signor Olcese [C.].

F. Egypt.

a. Light phase.

h′. Ad. st.	Egypt.	W. B. D. Turnbull, Esq. [P.].

G. Arabia.

a. Light phase.

i′. Ad. sk.	Red Sea, off Aden.	Purchased.

ETHIOPIAN REGION.

H. West Africa.

a. Light phase.

k′. Ad. sk.	River Gambia.	Purchased.

β. Dark phase.

l′. Ad. sk.	Cape-Verd peninsula.	M. A. Marche [C.].

γ. Strix insularis.

m′. Ad. sk.	"West Africa"*.	Rev. D. F. Morgan [P.].

I. Cape-Verd Islands. (*Strix insularis*, Pelz.)

n′. Ad. sk.	Cape-Verd Islands.	M. A. Bouvier [C.].
o′. Ad. sk.	Cape-Verd Islands.	M. A. Bouvier [C.].
p′. ♂ ad. sk.	Porto Praya, Santiago, Cape-Verd Islands, Jan. 25, 1822.	Charles Darwin, Esq. [C.].

K. South Africa. (*S. poensis.*)

q′. Ad. sk.	River Quanza, Angola.	J. J. Monteiro, Esq. [C.].
r′. Ad. sk.	Tette, Zambesi.	Dr. Kirk [C.].
s′. ♀ ad. sk.	Transvaal Republic.	A. Foresman, Esq. [P.].
t′. Ad. sk.	South Africa.	R. B. Sharpe, Esq. [P.].
u′. Ad. sk.	South Africa.	E. L. Layard, Esq. [C.].

L. Madagascar.

x′. ♂ ad. sk.	N.E. Madagascar.	Heer Van Dam [C.].
y′. Ad. sk.	Madagascar.	Purchased.
z′. Ad. sk.	Madagascar.	Purchased.

INDIAN REGION.

M. India. (*Strix indica.*)

a. Ad. sk.	India.	Colonel Cobbe [C.].

---

* Most of the Rev. Mr. Morgan's birds are from the river Gambia, and are so registered; but to the present specimen the locality is not added; and although most of the other birds received at the same time are from the Gambia, I do not think it right to affirm the Senegambian origin of this specimen (a true *S. insularis*), as Mr. Morgan may, after all, have visited the Cape-Verd Islands.

β. Ad. st.	India.	Dr. Burns [P.].
γ. Ad. st.	India.	Capt. Johnstone [P.].
δ. Ad. sk.	India.	J. Gould, Esq. [C.].
ε. Ad. sk.	India.	J. Gould, Esq. [C.].
ζ. Ad. sk.	Seetapore.	J. Gould, Esq. [C.].
η. Ad. sk.	Deccan.	Purchased.
θ. Ad. st.	Nepal.	B. H. Hodgson, Esq. [P.].
ι, κ, λ. Ad. sk.	Nepal.	B. H. Hodgson, Esq. [P.].

N. SIAM.

μ. ♂ ad.	Bangkok, March 28, 1872 (*P. Conrad*).	R. B. Sharpe, Esq. [P.].

O. MALAISIA. (*Strix javanica.*)

ν. Ad. st.	Java.	Hon. E. I. Co. [P.].
ξ. Ad. sk.	Lombock.	A. R. Wallace, Esq. [C.].

AUSTRALIAN REGION.

P. CELEBES. (*Strix rosenbergi.*)

ο. ♂ ad. sk.	Makassar.	A. R. Wallace, Esq. [C.].
π. Ad. sk.	Celebes.	Dr. A. B. Meyer [C.].

Q. AUSTRALIA. (*Strix delicatula.*)

a. Light phase.

ρ. Ad. sk.	Australia.	Purchased.
σ. Ad. st.	Australia.	Lady Carington [P.].
τ. Ad. sk.	Moreton Island.	Capt. Stanley [P.].
υ. ♀ juv. sk.	Brisbane River.	M. A. Bouvier.
φ. Ad. sk.	Queensland.	M. A. Bouvier.
χ. Ad. sk.	N.W. Australia.	J. R. Elsey, Esq. [P.].
ψ. Ad. sk.	Albert River, Aug. 30, 1856.	J. R. Elsey, Esq. [P.].
ω. Ad. sk.	N.W. Australia, Jan. 1856.	J. R. Elsey, Esq. [P.].
α'. ♂ ad. sk.	N.W. Australia, Feb. 5, 1856.	J. R. Elsey, Esq. [P.].
β'. ♂ ad. sk.	Victoria River, Feb. 1856.	J. R. Elsey, Esq. [P.].
γ'. Ad. sk.	W. Australia.	[Collector unknown.]

β. Dark phase.

δ'. Ad. sk.	Australia.	M. A. Bouvier.
ε'. Ad. sk.	Interior of S. Australia.	Capt. Sturt [C.].

R. OCEANIA. (*Strix lulu.*)

ζ'. Ad. sk.	Samoan Islands.	J. Gould, Esq. [P.].
η'. Ad. st.	Ovalau Island.	Rev. J. B. Stair [P.].
θ'. ♀ ad. st.	Ovalau Island.	Rev. J. B. Stair [P.].
ι'. Pull. sk.	Fiji Islands.	J. Brenchley, Esq. [P.].
κ'. Pull. sk.	M'bau Island, Fiji group.	J. Brenchley, Esq. [P.].

NEARCTIC REGION.

S. NORTH AMERICA.

a. Light phase. (S. pratincola.)

λ'. Ad. st.	North America.	Sir E. Parry [P.].

μ'. Ad. st.	North America.	E. Doubleday, Esq. [P.]
ν'. Ad. sk.	Canada.	Purchased.
ξ'. Ad. st.	Rio Sacramento, California.	Admiral Sir E. Belcher [P.].
ο'. ♀ ad. sk.	California.	Purchased.
π'. Ad. sk.	California.	Purchased.
ρ', σ'. Ad. sk.	San Angel, California.	Purchased.
τ'. Ad. sk.	Valley of Mexico.	M. A. Bouvier.
υ'. ♂ ad. sk.	Mexico.	R. B. Sharpe, Esq. [P.].
φ'. Ad. sk.	Mexico.	Purchased.
χ'. Ad. sk.	Guatemala.	M. A. Bouvier.
ψ'. Ad. sk.	Costa Rica.	Purchased.
ω'. Ad. sk.	Costa Rica.	Purchased.

β. *Dark race.* (Strix guatemalæ.)

a''. Ad. sk.	Mexico.	Purchased.
b''. Ad. sk.	Veragua.	M. E. Arcé [C.].
c''. Ad. sk.	Veragua.	M. E. Arcé [C.].
d''. Ad. sk.	Veragua.	M. E. Arcé [C.].

NEOTROPICAL REGION.

T. Guiana.

e''. Ad. sk.	Cayenne.	H. C. Rothery, Esq. [P.].

U. Columbia.

f''. Ad. sk.	Bogotá.	Purchased.
g''. Ad. sk.	Bogotá.	Purchased.

V. Brazil.

h''. Ad. sk.	Brazil.	M. Claussen [P.].
i''. Ad. sk.	Bahia.	Dr. Wucherer [C.].
k''. Juv. sk.	Bahia.	Dr. Wucherer [C.].
l''. Ad. st.	Bahia Blanca.	C. Darwin, Esq. [C.].

W. Chili.

m''. Ad. sk.	Chili.	Purchased.
n''. Ad. sk.	Santiago, Chili.	E. C. Reed, Esq. [P.].
o''. ♀ ad. sk.	Santiago, Chili.	E. C. Reed, Esq. [P.].

WEST INDIES.

X. Jamaica. (*Strix furcata.*)

p''. Ad. st.	Jamaica.	P. H. Gosse, Esq. [C.].
q''. Ad. sk.	Jamaica.	P. H. Gosse, Esq. [C.].
r''. ♂ ad. sk	Jamaica.	Purchased.
s''. ♀ ad. sk.	Jamaica.	Purchased.

Y. St. Domingo. (*Strix glaucops.*)

t''. Ad. st.	S. Domingo.	M. A. Sallé [C.].
u''. Ad. st.	S. Domingo.	

Z. Galapagos Islands. (*S. punctatissima*).

x''. Ad. st.	James Island, Galapagos.	Sir W. Burnett and Admiral Fitzroy [P.].

Osteological Specimens.

y''. Skeleton.	England.	Mr. E. Gerrard [P.].
z''. Sternum.	England.	T. Ingall, Esq. [P.].
a''. Sternum.	England.	J. Ray, Esq. [P.].
β''. Sternum.	England.	Purchased.
γ''. Sternum.	England.	W. Yarrell, Esq. [P.].
δ''. ♀ skeleton.	S. Germany.	Dr. Gunther [C.].
ε'', ζ. ♂ ♀ skulls.	S. Germany.	Dr. Günther [C.].
η''. Skeleton.	Nepal.	B. H. Hodgson, Esq. [P.].
θ''. Skull.	Egypt.	Dr. Christie [P.].

2. Strix novæ hollandiæ.

Mouse Owl, *Lath. Gen. Hist.* i. p. 358.

Strix novæ hollandiæ, *Steph. Gen. Zool.* xiii. pt. 2, p. 61 (*ex Lath.*); *Strickl. Orn. Syn.* p. 181; *Gould, Handb. B. Austr.* i. p. 64; *Gray, Hand-l. B.* i. p. 53.

Strix personata, *Vigors, P. Z. S.* 1831, p. 60; *Gould, Syn. B. Austr.* pt. 3, pl. 49. fig. 1; *id. B. Austr.* i. pl. 29; *Gray, Gen. B.* i. p. 41; *Sturt, Exp. Centr. Austr.* iii. App. p. 16; *Bp. Consp.* i. p. 54; *Schl. Mus. P.-B.* Striges, p. 6.

Strix cyclops, *Gould, P. Z. S.* 1836, p. 140; *id. Syn. B. Austr.* pt. 3, pl. 50. fig. 2; *Gray, Gen. B.* i. p. 41.

Dactylostrix personata, *Kaup, Isis,* 1848, p. 770; *id. Contr. Orn.* 1852, p. 119; *id. Tr. Z. S.* 1852, p. 249.

Adult (light phase). General colour golden buff, mottled with black and white, the tips of the feathers being black, thickly spotted in an irregular manner with white, the scapulars barred externally with blackish; the hind neck rather paler and more yellow; crown of head greyish black, thickly spotted with white; wing-coverts coloured exactly like the back, and mottled in the same manner; primary-coverts deep orange, tipped with white, barred with blackish brown, and mottled with the latter colour at the tip; quills orange-buff, barred with dark brown, the buff interspaces mottled with brown markings on the outer margin, and especially so on the inner web, which is much more dusky; the secondaries externally mottled with greyish white, and tipped with white, distinctly barred across with blackish brown; tail-feathers golden buff, tipped slightly with white, crossed with six bars of dark brown, the last one obscured by the blackish mottling at the tip of the feathers, which is here very distinct, all the interspaces also being mottled with black and white vermiculations; facial aspect white, slightly tinged with rufous round the eye, the feathers in front of the latter black; feathers of the ruff orange, with minute little tips of black to the upper ones, the lower ones paler buff; the gular plumes whitish, all distinctly tipped with blackish brown, forming a very distinct fringe to the face; under surface of body pure white, slightly tinged with buff on the chest, the latter, as well as the throat, minutely spotted with brown, these spots becoming larger on the lower breast and abdomen, and forming horseshoe bars on the flanks; under tail-coverts white, with a few dusky blackish subterminal spots; leg-feathers white, indistinctly barred with dull brown; under wing-coverts white, slightly washed with orange and spotted with black, as also are the axillaries, the lower series white, tipped with ashy brown, forming a small wing-bar, and resembling the inner lining of the quills, which are white below, barred with pale ashy brown. Total length 15·5 inches, wing 13·4, tail 6, tarsus 2·55.

Adult (dark phase). General colour above greyish black, the feathers distinctly spotted with white, generally along the shaft and at the tip of the feathers, which are also speckled all over with whitish; the upper surface very slightly varied with orange-buff, a few of the bases of the feathers which are of this colour showing through here and there; the scapulars externally orange-buff barred with blackish, and forming an indistinct shoulder-patch; wing-coverts coloured like the back, the spots being distinct on the least and median series, but less visible on the greater coverts, which are like the quills; the latter are dull brown, everywhere thickly and coarsely mottled with greyish white; all the quills barred across with dark brown, the secondaries tipped with white; tail greyish, thickly mottled with coarse vermiculations of dark brown, and crossed with six bands of the latter colour; facial aspect chestnut, inclining to whitish on the hinder margins of the disk, and especially on the lower parts, the feathers around and in front of the eye black; feathers of the ruff deep orange-buff, tipped with brown; the gular plumes white, tipped with the same colour; chin whitish; rest of under surface orange-buff, gradually shading off into white on the abdomen and under tail-coverts, the latter marked with brown near the tips; all the breast-feathers spotted with brown, the tips being white; on the sides of the body the markings more irregular and larger, being in the form of bars on the flanks; leg-feathers whitish, washed with orange on the tarsus, with indistinct bars of ashy brown; under wing-coverts rich orange, spotted with black and tipped with white; axillaries white, washed with orange and barred across with blackish, the lower series whitish, tipped with ashy brown, thus resembling the inner lining of the wing, which is whitish, barred with ashy brown, and much mottled with the same towards the tip; bill pale horn-colour; toes yellow. Total length 18 inches, wing 13·8, tail 7, tarsus 3.

Hab. Australia.

a. Ad. sk.	South Australia.	Sir G. Grey [P.].
b. Ad. sk.	Australia.	Rev. A. Strong [P.].
c. Ad. sk.	Australia.	J. Gould, Esq. [P.].
d. Ad. sk.	Queensland.	J. Gould, Esq. [P.].
e. Ad. sk.	Queensland.	M. A. Bouvier.
f. Ad. sk.	Interior of Queensland.	J. B. White, Esq. [C.].
g. Juv. sk.	Interior of Queensland.	J. B. White, Esq. [C.].
h. Juv. sk.	W. Australia.	
i, k. Ad. st.	Port Essington.	Capt. Chambers, R.N. [P.].
l. Ad. st.	Australia.	Lady Carington [P.].

Subsp. *a.* **Strix castanops.**

Strix castanops, *Gould, P. Z. S.* 1836, p. 140; *id. Syn. B. Austr.*

pt. 3, pl. 49. fig. 2; *id. B. Austr.* i. pl. 28; *Gray, Gen. B.* i. p. 41 *Bp. Consp.* i. p. 54; *Strickl. Orn. Syn.* p. 181; *Schl. Mus. P.-B.* Striges, p. 7; *Gould, Handb. B. Austr.* i. p. 62; *Diggles, Orn. Austr.* pl. —; *Gray, Hand-l. B.* i. p. 53.
Dactylostrix castanops, *Kaup, Contr. Orn.* 1852, p. 119; *id. Tr. Z. S.* iv. p. 248.

Adult. General colour above chocolate-brown, with a very few zigzag lines of white on the feathers of the upper surface, most of which have also a tiny arrow-head spot of white near the tip, these spots extremely small on the head and hind neck; all the upper surface varied here and there with rich golden buff, wherever the bases of the feathers show through, many of them being barred with this colour, especially on the outer aspect of the scapulars, where they form a kind of shoulder-patch; wing-coverts a little darker than the back, but marked in the same manner, the golden buff colour not showing very plainly, except on the outermost coverts and bastard wing; primary-coverts deep chocolate-brown, irregularly barred with orange-buff, and slightly vermiculated with whitish lines near the tips; quills dark brown, indistinctly barred with golden buff, inclining to fulvous on the inner web, all these buff-coloured bars more or less obscured with brown mottlings; tail dark brown, crossed with about six distinguishable bars of golden buff, plainer than on the wings, but yet more or less obscured with brown mottlings, especially towards the tips; facial aspect light chestnut, the feathers in front of the eye black, the loral plumes also slightly tipped with black; facial ruff deep golden orange, with a mesial spot of dark brown near the tip, becoming broadly tipped with brown on the lower part behind the ear-coverts, the gular plumes more rufescent but not so broadly tipped with brown; rest of under surface of body golden buff, the chest-feathers with a double spot of blackish near the tips, the flanks with larger horseshoe bars of blackish, broader but less distinct at the tips of the under tail-coverts; under wing-coverts and axillaries very deep golden buff, spotted at the tips with brown like the chest, the lower series ashy brown, whitish at base, resembling the inner lining of the quills, which are ashy brown below, barred with yellowish white near the base of the inner webs; bill yellowish brown; feet light yellow. Total length 21 inches, wing 15·2, tail 7, tarsus 3·2.

Obs. S. castanops is a large, permanently dark form of *S. novæ hollandiæ*, to the dark race of which it is very closely allied. The only differences are, in fact, the large size and the number of tail-bands, which sometimes amount to as many as eight. A comparison of a series of specimens of *S. novæ hollandiæ* and *S. castanops* gives the following result:—

	Total length. in.	Wing. in.	Tail. in.	Tarsus. in.	Mid. toe in.
S. novæ hollandiæ . .	13·5–17	12·0–13·9	5·8–6·5	2·55–3·1	1·5 –1·9
S. castanops	15·5–20	14·4–14·8	6·9–8·0	2·95–3·25	1·95–2.15

At the same time, light-coloured individuals of the Tasmanian bird are found, which in plumage are not to be told from true *Strix novæ hollandiæ*; and therefore I cannot grant full specific value to *S. castanops*.

Hab. Van Diemen's Land.

a. Ad. sk.	Van Diemen's Land.	
b. Ad. sk.	Van Diemen's Land.	Lord Lilford.
c. Ad. st.	Van Diemen's Land.	
d. Ad. st.	Van Diemen's Land.	Purchased.
e. Ad. st.	Van Diemen's Land.	Antarctic Expedition [C.].

3. Strix tenebricosa.

Strix tenebricosus, *Gould, P. Z. S.* 1845, p. 80; *id. B. Austr.* pl. 30; *Bp. Consp.* i. p. 54; *Strickl. Orn. Syn.* p. 182; *Schl. Mus. P.-B.* Striges, p. 7; *Gould, Handb. B. Austr.* i. p. 65, *Gray, Hand-l. B.* i. p. 53.

Adult. General colour smoky brown, distinctly spotted with white, more largely on the back and scapulars; on the head and hind neck remain traces of ashy white vermiculations, nearly obsolete or entirely disappearing on the back, scapulars, and wing-coverts, the latter being almost uniform brown, with spots of white here and there; primary-coverts uniform brown, excepting for a few white mottlings at the tip; quills dark brown, faintly mottled with whitish vermiculations on the outer web; the secondaries darker, especially the innermost, which have distinct spots of white like the back; tail dark brown, the two centre feathers with a distinct spot of white at the tip, and mottled with whitish vermiculations, disappearing on the outer rectrices, which have traces of darker brown bars on the inner web; face pale smoky brown, the bases to the feathers white, all the plumes round the eye dark brown; feathers of the ruff creamy white, tipped with chocolate-brown and having concealed markings of the same, the latter disappeacing on the gular feathers; general colour of under surface smoky brown, the feathers barred across with whitish, and having a distinct diamond-shaped spot of white at the tip; the abdomen and leg-feathers whitish, barred with dusky brown, disappearing on the under tail-coverts, which are almost entirely white; under wing-coverts and axillaries whitish, barred with dull brown, those near the edge of the wing nearly entirely brown, the lower series pale greyish brown, like the inner lining of the primaries; bill horn-colour; feet yellowish; iris dark brown. Total length 17·5 inches, wing 12·9, tail 7, tarsus 3.

A similar specimen sent from the neighbourhood of Brisbane is probably a male. It differs from the one described principally in having more distinct white vermiculations on the upper surface, forming distinct zigzag bars on the tail; all the white terminal spots are at the same time very distinct. Total length 14·5 inches, wing 11·7, tail 5·7, tarsus 2·6.

Hab. Australia.

a. Ad. st.	Australia.	Lady Carington [P.].
b. Ad. sk.	Interior of Queensland.	J. B. White, Esq. [C.].
c. Ad. sk.	Near Brisbane, Queensland.	M. A. Bouvier.

4. Strix capensis.

Strix capensis, *Smith, S.-Afr. Q. Journ.* 2nd ser. no. 4, pt. 1, p. 317; *id. Ill. Ind. Zool.* pl. xlv.; *Gray, Gen. B.* i. p. 41; *Bp. Consp.* i. p. 55; *Strickl. Orn. Syn.* p. 182; *Schl. Mus. P.-B.* Striges, p. 6; *Layard, B. S. Afr.* p. 44; *Sharpe, ed. Layard, B. S. Afr.* p. 81.
Strix punctata, *Licht. Nomencl. Av.* p 6; *Gray, Hand-l. B.* i. p. 53.
Aluco punctata, *Shelley, Ibis,* 1875, p. 66.

Adult. Above pale brown, darker on the wings, and much lighter on the head and neck, all the upper surface distinctly spotted with white, the scapulars and some of the wing-coverts externally washed with orange-buff, the least wing-coverts and those along the edge of the wing almost entirely bright orange-buff; primary-coverts uniform dark brown, only varied with buff at base; quills brown, barred across with darker brown, more numerously on the secondaries, which have a slight spot of white at the tips, and are much varied with orange-buff, especially near the base; the primaries less varied on the inner webs, but the external interspaces are largely marked with golden buff, gradually clearer and less vermiculated with brown towards the base; lower surface of wing buffy white, inclining to brown towards the tips, the secondaries also banded with brown below; tail brown, the two centre feathers uniform, the others barred with dull brown, the interspaces more or less washed with orange-buff, increasing in extent towards the outer tail-feathers, which are light orange-buff, inclining to whitish at tip, and barred with brown; forehead dull brown like crown of head; face fulvous, washed with rufous, deeper and inclining to black round the eye; facial disk light orange-buff, the upper part minutely spotted and the lower part broadly tipped with dark brown; sides of face and of neck, as well as entire under surface of body, bright orange-buff, spotted with brown, those on the sides of the neck and flanks rather larger and somewhat of a horseshoe shape; the under wing-coverts spotted like the breast, the tibial plumes and under tail-coverts less spotted; eye black. Total length 18 inches, culmen 1·7, wing 13·6, tail 5·8, tarsus 3·3.

Young. Very similar to the adult, but darker, especially on the face, which is dull rufous, the facial disk also dark brown on the upper part; the breast very distinctly spotted; the down (remains of which adhere) is of a greyish fulvous.

Hab. Southern Africa.

a. Ad. st.	South Africa.	South-African Museum [P.]. Type of species.
b. Ad. st.	South Africa.	South-African Museum [P.].
c. Pull. st.	South Africa.	South-African Museum [P.].
d, e. Skeletons.	South Africa.	Zoological Society.

5. Strix candida.

Strix candida, *Tickell, J. A. S. B.* ii. p. 572; *Jerd. Ill. Ind. Orn.* pl. 30; *Bp. Consp.* i. p. 55; *Kaup, Contr. Orn.* 1852, p. 119; *Strickl. Orn. Syn.* p. 181; *Jerd. B. Ind.* i. p. 118; *Schl. Mus. P.-B.* Striges, p. 6; *Gray, Hand-l. B.* i. p. 53; *Swinh. P. Z. S.* 1871, p. 344; *Gould, B. Asia*, part xxiv.
Strix longimembris, *Jerdon, Madr. Journ.* x. p. 86; *Gray, Gen. B.* i. p. 41.
Glaux candida, *Blyth, J. A. S. B.* xix. p. 513; *Ball, Str. F.* 1874, p. 381; *Hume, Nests and Eggs Ind. B.* p. 60.
Glaux javanica, *Blyth, Cat. B. Mus. A. S. B.* p. 42.
Scelostrix candida, *Kaup, Tr. Z. S.* iv. p. 248; *Blyth, Ibis,* 1866, p. 251; *Hume, Rough Notes,* ii. p. 345; *Walden, Tr. Z. S.* ix. p. 145.
Strix amauronota, *Cab. J. f. O.* 1866, p. 9, et 1872, p. 316; *Gray, Hand-l. B.* i. p. 52.
Strix pithecops, *Swinh. Ibis,* 1866, pp. 396, 397; *Gray, Hand-l. B.* i. p. 53.
Strix walleri, *Diggles, Orn. Austr.* pl. —; *Gray, Hand-l. B.* i. p. 52.

Nestling. Covered with long filamentous down of a dull orange-buff colour; the feathers of the upper part, as far as developed, of a dark brown colour, spotted near the end of the shaft of the feathers with white, the basal part of the dorsal plumes bright orange.

Adult. Above dark brown, the feathers with black shafts and minutely spotted with a subterminal mark of white, very small on the dorsal feathers and rather larger on the scapulars, the bases to the plumes orange-buff, extending for some distance upwards, and sometimes forming indistinct bars, the tips of the scapulars profusely mottled with greyish; wing-coverts dark brown, with very small white spots, the least coverts orange-buff, minutely spotted with brown, the greater series orange-buff at base on outer web, white on the inner, mottled with brown towards the tip, the outer edge of the wing white, with minute brown spots, the primary-coverts dark brown, orange-buff at base; quills orange-buff, brown at tip and on outer web, inclining to whitish on the inner one, all the quills barred with darker brown on outer web, extending nearly across the inner web on the primaries; tail whitish, slightly tinged with buff, and mottled at tip with greyish brown, the middle feathers crossed with four brown bars, the other ones with three, the latter not completely traversing the feather, and almost obsolete on the outermost one; entire face and sides of neck white, the latter with tiny spots of brown; feathers in front of the eye blackish, forming a patch; facial ruff entirely white, with narrow and almost obsolete brown shaft-stripes; under surface of body pure white, the chest, abdomen thighs, and under tail-coverts unspotted, the breast minutely spotted with brown, the spots larger and more arrow-shaped on the under wing-coverts and axillaries; inner lining of wing white, inclining to brown at tips of feathers, which are gradually notched with dark brown towards their extremities; bill horny; legs livid; iris very dark brown. Total length 14·5 inches, culmen 1·8, wing 13·3, tail 5·5, tarsus 3·1.

Obs. S. pithecops of Formosa does not seem to be a distinct species; and Mr. Swinhoe in his last list of Chinese birds unites it to *S. candida.* The specimen in the Museum differs from Indian examples only in its rather larger size, and in having five bands on the centre tail-feathers; the greater series of under wing-coverts also are barred across with greyish brown. None of these characters, however, are of specific importance in the genus *Strix.*

Hab. India, Indo-Chinese subregion and Formosa; Philippine Islands; N. Australia.

a. Ad. sk.	India.	J. Gould, Esq. [P.].
b. Ad. st.	Madras.	Dr. Jerdon [P.].
c, d. Ad. sk.	Upper Assam (*Dr. Reid*).	R. B. Sharpe, Esq. [P.].
e. Pull. sk.	Tenasserim.	J. Gould, Esq. [P.].
f. Pull. sk.	Philippine Islands.	R. B. Sharpe, Esq. [P.].
g. Ad. sk.	Formosa.	R. Swinhoe, Esq. [C.].

2. PHODILUS.

Type.

Phodilus, *Isid. Geoff. St.-Hilaire, Ann. Sc. Nat.* xxi. p. 200 (1830) .... P. badius.

Range. Eastern Himalayas, ranging through Burmah and Pegu; Ceylon; Java; Borneo.

1. Phodilus badius.

Strix badia, *Horsf. Zool. Research. Java,* pl. 37; *Temm. Pl. Col.* i. pl. 318; *Less. Traité,* p. 112.
Phodilus badius, *Isid. Geoff. St.-Hil. Ann. Sc.* xxi. p. 201; *Gray, Gen. B.* i. p. 42, pl. 15. fig. 1; *Blyth, Cat. B. Mus. A. S. B.* p. 41; *Bp. Consp.* i. p. 55; *Kaup, Contr. Orn.* 1852, p. 118; *id. Tr. Z. S.* iv. p. 245; *Mottl. & Dillw. N. H. Labuan,* p. 8; *Horsf. & Moore, Cat. B. Mus. E. I. Co.* i. p. 80; *Strickl. Orn. Syn.* p. 183; *Jerd. B. Ind.* i. p. 119; *Scl. P. Z. S.* 1863, p. 207; *Blyth, Ibis,* 1866, p. 251; *Wall. Ibis,* 1868, p 26; *Gray, Hand-l. B.* i. p. 52; *Gould, B. Asia,* pl. xxii.; *Walden, Ibis,* 1872, p. 365; *Hume, Rough Notes,* ii. p. 346; *id. Str. F.* 1873, p. 429; *Salvad. Ucc. Born.* p. 21.
Ulula badia, *Schl. Mus. P.-B.* Striges, p. 23.
? Phodilus nipalensis, *Hume, Str. F.* 1875, p. 37 *.

Adult male. Above rufous bay, the crown of the head rather darker, and separated from the back by the interscapulary region, which is deep orange-buff, sparsely spotted with dark brown, on the occiput a few feathers of orange-buff, minutely spotted with dark brown; the remainder of the upper surface dotted here and there with brown specks, and relieved on some of the scapulars and wing-coverts with lines of white running along the shafts of the feathers; the median and greater coverts with concealed blackish bars near the base of inner web, the exterior greater and primary-coverts mottled with blackish on their inner webs, the exterior feather of each series chequered alternately with white spots and black bars; quills bright bay, barred with blackish on the inner

* *Phodilus nipalensis* of Gray is *Scops spilocephalus,* Blyth! (*Vide suprà* p. 63.)

web, these bars almost obsolete on the secondaries, the outermost chequered with white on outer web and barred with deep black very distinctly, the under surface of the wing dull rufous, the cross bars very distinct; lower back, rump, and upper tail-coverts uniform bay, with very indistinct brown spots and very slight indications of white streaks; tail deep chestnut-bay, crossed with five or six blackish bars, gradually disappearing towards the tips; frontal feathers and face dull whitish, with a certain pinkish shade, especially on the former; feathers round and in front of the eye deep chestnut; facial ruff deep chestnut, with buffy white bases to the feathers and narrow blackish subterminal bars, the lower part of the ruff more varied, being whitish tipped with chestnut, and subterminally barred with black, all in strong contrast: above the eye, over the summit of the ruff, a conspicuous tuft of whitish feathers; under surface of body golden buff on breast, slightly washed with pinkish, the abdomen and flanks more decidedly glossed with a pale vinous pink shade, inclining to buffy white on the vent and under tail-coverts; the breast sparsely spotted with dull brown, becoming rather larger and darker on the abdomen and flanks; under wing-coverts orange-buff, shaded with chestnut, forming a large patch near the outer edge of the wing, the greater series buffy white, excepting the outermost, which are spotted with dusky blackish in the form of bars; tarsus buffy white, unspotted; bill yellowish; toes brown; claws yellowish; iris black. Total length 12 inches, culmen 1·4, wing 7·3, tail 3·1, tarsus 1·9.

Adult female. Decidedly larger than the male. Total length 13·5 inches, wing 8, tail 3·4, tarsus 2.

Obs. Mr. Hume is not sure of the specific identity of the *Phodili* from all the localities mentioned. Writing on a Ceylonese example he observes:—"The Ceylon bird differs from the Nepal race in its somewhat smaller size, in the much closer and darker banding of the lower surface of the quills, in the dark brown patch on the wing-lining at the base of the first two or three primaries (this patch being bright chestnut in the Nepal bird), in the dark brownish tint of the lesser wing-coverts along the ulna and of the whole crown, in the conspicuous black banding (almost obsolete in the Nepal bird) of the outer webs of the quills, in the much closer banding of the tail, and in the darker tint of the back, and especially of the lower part of it; in fact the bird seems referable rather to the Malayan than the Himalayan race."

Hab. The same as that of the genus.

a. ♂ ad. st.	Java.	Hon. East-India Co. [P.].
b. ♀ ad. st.	Java.	Zoological Society.
c. ♀ ad. st.	Java.	Earl of Ellenborough [P.].
d. ♂ ad. sk.	Sarawak, Borneo.	A. R. Wallace, Esq. [C.].
e. ♀ ad. sk.	India (*Jerdon*).	J. Gould, Esq. [P.].

ADDENDA.

Page 14. **Bubo ignavus.**

o. ♂ ad. sk.	Granada, Spain.	Howard Saunders, Esq. [C.].

Page 30. **Bubo maculosus.**

l. Ad. sk.	Kroonstadt, Orange-Tree State.	F. A. Barratt, Esq. [C.].

Page 47. **Scops giu.**

t. Pull. sk.	Granada, S. Spain.	Howard Saunders, Esq. [C.].
u. ♀ ad. sk.	Malaga, S. Spain.	Howard Saunders, Esq. [C.].

Page 53. **Scops pennatus.**

a. Bones, various.	Nepal.	B. H. Hodgson, Esq. [P.].

Page 63. Insert the following species, the type of which has been kindly lent by Dr. Finsch:—

1*. Scops leucopsis.

Athene leucopsis, *Hartl. Rev. Zool.* 1849, p. 496; *id. Beitr. Orn. W.-Afr.* p. 16, pl. 1; *Bp. Consp.* i. p. 43; *Strickl. Orn. Syn.* p. 171; *Gray, Hand-l. B.* i. p. 40.
Scops leucopsis, *Kaup, Contr. Orn.* 1852, p. 111; *id. Tr. Z. S.* iv. p. 224; *Hartl. Orn. W.-Afr.* p. 20.
Glaucidium leucopse, *Sharpe, Ibis,* 1875, p. 259.

Adult (type of species). General colour above dark brown, mottled with sandy rufous, with which all the feathers of the upper surface are very coarsely vermiculated; on the head and neck a triple set of bands, one across the occiput, very faintly indicated by a few whitish bands to some of the feathers; on the nape a second band of tawny feathers relieved by a few white spots; on the hind neck a clearer band caused by the buff or buffy-white bands, which are on most of the feathers; white scapular spots very distinct and large, the outer web of many of the feathers being entirely white with a black tip; least wing-coverts blackish, with vermiculations of sandy rufous, the markings on the median and greater series much coarser, with broad white tips like the scapulars, preceded by sandy mottlings; primary-coverts uniform blackish brown; quills brown, externally notched and mottled with sandy rufous or white; the secondaries blackish, everywhere mottled and vermiculated with sandy rufous like the back; tail brown, copiously mottled and vermiculated with pale sandy rufous, clearer at the tip, the outer feathers vermiculated on the outer web and narrowly banded on the inner one with sandy rufous; lores and a broad band over the fore part of the eye white, he feathers of the former very full and having the shafts produced into black bristles, most of these feathers edged with rufous brown; plumes round the eye chestnut, with narrow whitish shaft-streaks; ear-coverts whitish, with concentric bars of rufous and black; ear-tufts coloured like the head; facial ruff white, many of the upper feathers behind the ear-coverts barred with tawny and more narrowly with brown cross lines, the lower part of the ruff under the chin white, with a few narrow blackish cross lines; chin-feathers pure white; general aspect of under surface sandy rufous, many of the feathers with a distinct central streak of black, all the feathers vermiculated with dark brown bars and broadly barred with white, the latter prevailing on the flank-feathers, which are almost entirely white with sandy-rufous bars; the centre of the abdomen pure white; the under tail-coverts white, washed with tawny, and mesially streaked with black; leg-feathers orange; under wing-coverts buffy white, some of the outermost washed with tawny and mottled with brown markings, the lower series tipped with ashy brown, and thus resembling the inner lining of the quills, which are uniform ashy below, inclining to fulvous near the base of the inner web, the secondaries internally barred with fulvous; bill horn-brown; toes yellowish. Total length 7·5 inches, wing 5·25, tail 2·8, tarsus 1·2. (*Mus. Brem.*)

This Scops Owl is perfectly intermediate between the *S. giu* group and *S. spilocephalus*. It has the mottled upper surface of the latter bird, but differs from this species in having stripes on the under surface. The flanks in *S. giu*, *capensis*, and allies are white streaked with black and crossed with narrow zigzag cross lines of black; but in *S. leucopsis* the flanks are white with broad sandy rufous cross bars. The absence of streaks on the upper surface is also a character; and the ruff is white, with narrow black bars, but there is no black ring marking the ruff as in *S. giu*.

The following modification in the synoptic table of *Scops* becomes necessary:—

a^7. Quills barred below with lighter vermiculations, especially towards the tips; leg-feathers minutely spotted with black; upper surface of body always with distinct remains of streaks; flanks always mottled with minute wavy vermiculations of black, generally with distinct black shaft-streaks........ *giu* and allies, pp. 47–62.

b^7. Quills uniform below; leg-feathers uniform bright orange; upper surface of body without streaks; flanks white; broadly barred across with orange, and without central streaks .. *leucopsis*, p. 311.

Page 69. **Scops megalotis.**

The affinities of this species are probably with *Scops podarginus*.

Page 85. **Scops lettia.**

a. Bones, various.	Nepal.	B. H. Hodgson, Esq. [P.].

Page 89. **Scops elegans.**

Erase Spec. *a*, which is referred to *S. erythrocampa* (p. 90).

Page 102. Insert the following species:—

Scops podarginus.

Noctua podargina, *Hartl. & Finsch, P. Z. S.* 1872, p. 90; *Finsch, Journ. Mus. Godeffr.* p. ii. pl. i.
Ninox podargina, *Sharpe, anteà*, p. 151, note.

This species was referred by me to the genus *Ninox* in the absence of a personal examination of the specimen. Since the first sheets of this volume were printed, I have been favoured by the Museum Godeffroy with the loan of the typical example, which is not in the best condition. There can, however, be no doubt that the bird is a *Scops* of the yellow-billed group, and that its place will be found to be next to *Scops balli* and its allies. The following modification in the synopsis (p. 47) is therefore necessary:—

c''. Under surface very strongly vermiculated with fine zigzag lines of blackish; upper surface of a rufous chocolate tint, spotted with rufous ochre; frontal band not very broad.

e'''. Size large: wing 6·35 inches, tarsus 1·3, middle toe 1; claws deep brown ...... *podarginus*, p. 313.

f'''. Size much smaller: wing 5·45–5·6 inches, tarsus 1·15, middle toe 0·8; claws yellow *balli*, p. 100.

The following is a description of *Scops podarginus*:—

Adult (type of species). General colour above chocolate-brown, rather rufous in tint, everywhere obscurely mottled with blackish vermiculations, and relieved by bars or diamond-shaped spots of buffy white; scapulars with distinct spots of white on the outer web; the greater wing-coverts also with some distinct diamond-shaped spots of white; quills brown, externally sandy rufous, lighter than the back, the primaries with a few spots of buffy white on the outer web; the secondaries entirely sandy rufous at the tip, with a few obscure rufous bands throughout the length of the feathers; upper tail-coverts coloured like the back, many of them barred across with white or buffy white; tail dark brown, sandy rufous at tip, with more or less distinct rufous bands on the rest of the feathers; lores very full, buffy white, washed with rufous near the tip, the black shaft-streaks produced; frontal plumes and a few feathers over the fore part of the eye whitish, tipped and narrowly barred with brown; plumes round the eye rufous brown; ear-coverts sandy rufous, mesially streaked with buffy white; ruff apparently sandy rufous, narrowly barred with black, and inclining to buff near the base; some of the lower plumes tipped with white and subterminally barred with black; under surface of body sandy rufous, with narrow zigzag cross lines of black, most of the feathers subterminally barred with white or tipped with a large diamond-shaped spot of white, the flank-feathers distinctly barred with whitish; tarsi very scantily plumed, bare for more than the lower half, the plumes on the upper portion sandy rufous, with tiny bars of dark brown; under wing-coverts white, broadly barred with black or brown, especially distinct on the edge of the wing; bill yellow; feet dusky, claws dark brown. Total length 8·8 inches, wing 6·35, tail 3·6, tarsus 1·3.

Hab. Pelew Islands.

This species is extremely like *Scops balli* from the Andamans, a species which, through Mr. Hume's kindness, I am able to compare with it. It is, in fact, a large edition of it, but is apparently to be distinguished by its dark brown claws and big feet. Like *S. balli* it has a very bare leg; but the exact amount of feathering is rendered a little difficult to determine by reason of the condition of the specimen, which has been mounted after having been preserved in spirits. This fact accounts for the absence of ear-tufts (which, however, I can distinguish a little). Mr. Keulemans's plate in the Journal of the Museum Godeffroy does not represent the bird well, as, like the original describers, he has mistaken it for a *Ninox* or *Noctua*; and the peculiar feathers which he has drawn on the chest are, I believe, only two ordinary plumes of the ruff, rendered conspicuous by the falling off of all the other overlying feathers in the specimen.

Page 108. **Scops brasilianus.**

z. Ad. sk. Huiro, valley of Sta. Ana, W. Peru. H. Whitely, Esq. [P.].

Page 129. **Surnia ulula.**

a. ♀ skeleton.	Sweden.	R. B. Sharpe, Esq.
b. Skeleton.	Sweden.	Zoological Society.

Page 133. **Carine noctua.**

o, p. ♂ ♀ ad. sk.	Malaga, S. Spain.	Howard Saunders, Esq. [C.].
q. ♂ juv. sk.	Cordova, S. Spain.	Howard Saunders, Esq. [C.].
r. ♀ juv. sk.	Sierra Huetor, Granada.	Howard Saunders, Esq. [C.].
s. Skeleton.	Europe.	Purchased.
t. Skeleton.	Europe.	Zoological Society.

Page 138. **Carine brama.**

w. Ad. sk.	Nazeerah, Upper Assam (*Dr. J. Reid*).	R. B. Sharpe, Esq. [P.].
x. Bones, various.	Nepal.	B. H. Hodgson, Esq. [P.].

Page 142. **Speotyto cunicularia.**

a. Skeleton.	Ecuador (*Fraser*).	R. B. Sharpe, Esq.

Page 154. **Ninox lugubris.**

a. Bones, various.	Nepal.	B. H. Hodgson, Esq.

Page 156. **Ninox scutulata.**

r, s. Ad. sk.	Nazeerah, Upper Assam (*Dr. Reid*).	R. B. Sharpe, Esq. [P.].
t. Skeleton.	Timor.	Purchased.
u. Sternum.	Formosa.	R. Swinhoe, Esq. [C.].

Page 168. **Ninox boobook.**

a. Skeleton.	Australia.	Zoological Society.

Page 173. **Ninox novæ zealandiæ.**

a. Skeleton.	New Zealand.	Zoological Society.

Page 216. **Glaucidium castanopterum.**

a. Skeleton.	Java.	Purchased.

Page 217. **Glaucidium radiatum.**

a. Bones, various.	Nepal.	B. H. Hodgson, Esq. [P.].

Page 219. **Glaucidium cuculoides.**

a. Bones, various.	Nepal.	B. H. Hodgson, Esq. [P.].

ALPHABETICAL INDEX.

Printed by TAYLOR and FRANCIS, Red Lion Court, Fleet Street

LIST OF PLATES.

Pl. I.

$\frac{1}{3}$

J. G. Keulemans del. Mintern Bros. imp.

Scotopelia bouvieri.

J G Keulemans del

Mintern Bros. imp

Bubo shelleyi.

J G Keulemans del. Mintern Bros imp.

1. *Scops capensis.*
2. „ *stictonotus.*

J. G. Keulemans del. Mintern Bros imp.

1, *Scops malayanus.*
2, „ *gymnopodus.*

Pl. V.

J. G. Keulemans del.

Mintern Bros imp.

Scops magicus.

2/5

J. G. Keulemans del.

Mintern Bros. lith.

Scops leucospilus.

Pl. VII.

J. G. Keulemans del.

Mintern Bros lith.

1. *Scops morotensis.*
2. ....„.... *bouruensis.*

Pl. VIII.

J. G. Keulemans del.   Mintern Bros. imp.

1, *Scops albiventris.*
2, *Scops menadensis.*

Pl. IX.

J. G. Keulemans del.

Mintern Bros. imp.

Scops guatemalæ.

Pl. X.

J. G. Keulemans del.

Mintern Bros. imp.

Ninox hypogramma.

Pl. XI.

J. G. Keulemans del. Mintern Bros imp.

1, *Ninox nantu.*
2, „ *ochracea.*

J. G. Keulemans del. Mintern Bros imp.

1. *Ninox fusca.*
2. " *squamipila.*

J. G. Keulemans del.

Mintern Bros. imp.

1. *Glaucidium cobanense.*
2. „ *tephronotum.*

Pl. XIV.

J. G. Keulemans del.

Mintern Bros. imp.

Strix flammea.
a, Cape Verde Islands. b. S. Domingo.

Made in the USA
Las Vegas, NV
03 November 2026

57424066R00221